ELECTRICAL ENGINEERING MATERIALS REFERENCE GUIDE

The McGraw-Hill
Engineering Reference Guide Series

This series makes available to professionals and students a wide variety of engineering information and data available in McGraw-Hill's library of highly acclaimed books and publications. The books in the series are drawn directly from this vast resource of titles. Each one is either a condensation of a single title or a collection of selections culled from several titles. The Project Editors responsible for the books in the series are highly respected professionals in the engineering areas covered. Each Editor selected only the most relevant and current information available in the McGraw-Hill library, adding further details and commentary where necessary.

Church • EXCAVATION PLANNING REFERENCE GUIDE

Gaylord & Gaylord • CONCRETE STRUCTURES REFERENCE GUIDE

Hicks • BUILDING SYSTEMS REFERENCE GUIDE

Hicks • CIVIL ENGINEERING CALCULATIONS REFERENCE GUIDE

Hicks • MACHINE DESIGN CALCULATIONS REFERENCE GUIDE

Hicks • PLUMBING DESIGN AND INSTALLATION REFERENCE GUIDE

Hicks • POWER GENERATION CALCULATIONS REFERENCE GUIDE

Hicks • POWER PLANT EVALUATION AND DESIGN REFERENCE GUIDE

Higgins • PRACTICAL CONSTRUCTION EQUIPMENT MAINTENANCE REFERENCE GUIDE

Johnson and Jasik • ANTENNA APPLICATIONS REFERENCE GUIDE

Markus and Weston • CLASSIC CIRCUITS REFERENCE GUIDE

Merrit • CIVIL ENGINEERING REFERENCE GUIDE

Perry • BUILDING SYSTEMS REFERENCE GUIDE

Rosaler and Rice • INDUSTRIAL MAINTENANCE REFERENCE GUIDE

Rosaler and Rice • PLANT EQUIPMENT REFERENCE GUIDE

Ross • HIGHWAY DESIGN REFERENCE GUIDE

Rothbart • MECHANICAL ENGINEERING ESSENTIALS REFERENCE GUIDE

Woodson • HUMAN FACTORS REFERENCE GUIDE FOR ELECTRONICS AND COMPUTER PROFESSIONALS

Woodson • HUMAN FACTORS REFERENCE GUIDE FOR PROCESS PLANTS

ELECTRICAL ENGINEERING MATERIALS REFERENCE GUIDE

H. Wayne Beaty *Editor*

Consultant
Former Vice President, Loadmaster Systems, Inc.
Former Senior Editor, Electrical World;
Senior Member of the Institute of
Electrical and Electronics Engineers

McGRAW-HILL PUBLISHING COMPANY

New York St. Louis San Francisco Auckland Bogotá
Caracas Hamburg Lisbon London Madrid Mexico
Milan Montreal New Delhi Oklahoma City
Paris San Juan São Paulo Singapore
Sydney Tokyo Toronto

Library of Congress Cataloging in Publication Data

The McGraw-Hill electrical engineering materials reference
guide.
(The McGraw-Hill engineering reference guide series)
(McGraw-Hill handbook series)
"Data from the 12th edition of the McGraw-Hill standard
handbook for electrical engineers and from quarterly McGraw-
Hill electrical engineering reports (volume 1, number 1 and
volume 2, number 3)"--
Pref.
Includes index.
1. Electrical engineering--Materials--Handbooks,
manuals, etc. I. Beaty, H. Wayne. II. Series.
TK453.M42   1990   621.3   89-12554
ISBN 0-07-004196-2
Copyright © 1990 by McGraw-Hill, Inc. All rights reserved.
Printed in the United States of America. Except as permitted
under the United States Copyright Act of 1976, no part of this
publication may be reproduced or distributed in any form or
by any means, or stored in a data base or retrieval system,
without the prior written permission of the publisher.

1234567890   DOC/DOC   895432109

ISBN 0-07-004196-2

The editor for this book was Harold B. Crawford and the production
supervisor was Dianne L. Walber. It was set in Times Roman by
University Graphics, Inc.

Printed and bound by R. R. Donnelley and Sons Company.

*For more information about other McGraw-Hill materials,
call 1-800-2-MCGRAW in the United States. In other
countries, call your nearest McGraw-Hill office.*

CONTENTS

Contributors vi
Preface vii

1. Materials Research 1-1
2. Conductor Materials 2-1
3. Carbon and Graphite 3-1
4. Magnetic Materials 4-1
5. General Properties of Insulating Materials 5-1
6. Thermal Conductivity of Electrical Insulating Materials 6-1
7. Insulated Conductors 7-1
8. Insulating Gases 8-1
9. Insulating Oils and Liquids 9-1
10. Mica and Mica Products 10-1
11. Plastics 11-1
12. Paper, Fiber, Wood, and Insulating Fabrics 12-1
13. Inorganic Insulating Materials 13-1
14. Insulating Varnishes 14-1
15. Coating Powders 15-1
16. Impregnating and Filling Compounds 16-1
17. Structural Materials 17-1
18. Wood Products 18-1

Index follows Chapter 18.

CONTRIBUTORS

Baker, Mark A.   *Alcoa Conductor Products Company*
Bohl, Robert W.   *University of Illinois*
Croop, E. J.   *Westinghouse Research Laboratories*
Dakin, T. W.   *Consultant*
Dixon, R. R.   *Westinghouse Research Laboratories*
Harper, Charles A.   *Westinghouse Electric Corporation*
Latimer, K. L.   *The Arnold Engineering Company*
Lyon, Duane E.   *Mississippi State University*
Mattox, Douglas M.   *Westinghouse Research Laboratories*
Smith, J. D. B.   *Westinghouse Research Laboratories*
Sprengling, G. R.   *Westinghouse Research Laboratories*
Von Holle, Anthony L.   *Armoco, Inc.*

PREFACE

There has been a recent surge of interest in materials science and engineering brought on by the discovery of new material properties of ceramic oxides used in superconductivity.

This discovery has reminded us all of the importance of the study of materials in meeting the technological challenges of scientific and engineering breakthroughs.

Almost all the scientific advances ultimately involve finding new limits to existing materials or to the discovery of new materials and alloys, and the application of these new materials to engineering.

This book is designed to help meet the needs of the new interdisciplinary field of materials science and engineering by compiling data from the 12th edition of the *McGraw-Hill Standard Handbook for Electrical Engineers* and from quarterly *McGraw-Hill Electrical Engineering Reports* (Volume 1, Number 1 and Volume 2, Number 3).

Wayne Beaty

ELECTRICAL
ENGINEERING
MATERIALS
REFERENCE GUIDE

Chapter 1

MATERIALS RESEARCH

By the ELECTRIC POWER RESEARCH INSTITUTE*

1-1 Introduction 1-1
1-2 Technology Transfer and Feedback 1-3
1-3 Polymers 1-4
1-4 Electronics 1-4
1-5 Electric Utility Research 1-5
1-6 Future Trends 1-7

1-7 Strength of Materials 1-7
1-8 Corrosion 1-7
1-9 Effect of Heat 1-7
1-10 Materials for Light and Electronics 1-8
1-11 Materials on a National Scale 1-8
1-12 Suggestions for Further Reading 1-9

1-1. Introduction. Around 7000 years ago, Phoenician sailors are believed to have produced the first man-made glass, quite by accident, when it flowed from the coals of a riverside cook fire that was supported by blocks of soda from a cargo hold. When the greenish rivulet cooled, the stuff resembled the shiny, black chunks of obsidian that more-primitive peoples had prized for over 60,000 years and had shaped into arrowheads, cutting tools, and jewelry. It would be some 4700 years later before an artisan learned that similar glassy material could be made and worked into a useful form with the help of an iron blowpipe.

Today, amorphous or glassy materials remain one of the hot topics in the interdisciplinary field that has come to be called materials science and engineering, which includes amorphous silicon-based semiconductors and metal alloys that totally lack crystalline structure.

These days, fortunately, the time scales connecting discovery to invention to application are not nearly so protracted as in ancient history. Innovative new products made of amorphous metals are today a commercial reality, following on fundamental insights in the 1930s that were not conclusively and practically demonstrated until 1960.

In fact, most of modern civilization's edifices, tools, and consumer goods are based on materials discovered in recent decades (Table 1-1). Quantum physics in the 1930s illuminated the potential for semiconductors, paving the theoretical road for the invention of the transistor in 1947 and the microelectronic integrated circuit in 1959. Metallic superalloys that today offer exceptional strength, as well as superior resistance to stress and chemical attack at high temperature, first began to emerge as a family of novel nickel-based alloys in the 1950s. The coherent light of a ruby-crystal laser first appeared around 1958, laying the foundation for today's fiber-optic communications and the search for improved photonic materials.

Apart from glass, ceramic materials have probably been used by man for over 10,000 years since an anonymous soul found that a pot fashioned from clay and baked in a covered fire would later hold its shape and contents even when reheated. Purdue University's Gerald Liedl pegs this dubious date as the beginning of materials technology, the first "intentional transformation of an inorganic natural material into a new material displaying novel properties."

Today, among other uses, ceramics form the insulation packaging around microelectronic components, give a sharper edge to "new stone age" scissors and knives, and provide

---

*Adapted from *EPRI JOURNAL*, Dec. 1987, Palo Alto, Calif. Used by permission.

ELECTRICAL ENGINEERING MATERIALS REFERENCE GUIDE

TABLE 1.1 Materials Footnotes Through History

Humans have been making productive use of the materials around them since before the Stone Age, but materials science didn't really take off until the early nineteenth century, when it was spurred by the demands of the Industrial Revolution.

8000	Hammered copper		1896	Discovery of radioactivity
7000	Clay pottery		1906	Triode vacuum tube
6000	Silk production		1910	Electric furnace steelmaking
5000	Glass making			
4000	Smelted copper		1913	Hydrogenation to liquefy coal
4000–3000	Bronze Age			
3200	Linen cloth		1914	X-ray diffraction introduced
2500	Wall plaster			
2500	Papyrus		1914	Chromium stainless steels
1000	Iron Age		1923	Tungsten carbide cutting materials
300	Glass blowing			
20	Brass alloy		1930	Beginnings of semiconductor theory
105	Paper			
600–900	Porcelain		1930	Fiberglass
1540	Foundry operation		1934	Discovery of amorphous metallic alloys
late 1500s	Magnetization of iron demonstrated			
			1937	Nylon
1729	Electrical conductivity of metals demonstrated		1940s	Synthetic polymers
			1947	Germanium transistor
1774	Crude steel		1950	Commercial production of titanium
1789	Discovery of titanium			
1789	Identification of uranium		1952	Oxygen furnace for steelmaking
1800	Volta's electric pile (battery)			
			1950s	Silicon photovoltaic cells
1824	Portland cement		1950s	Transmission electron microscope
1839	Vulcanization of rubber			
1850	Porcelain insulators		mid 1950s	Silicon transistor
1850s	Reinforced concrete		1957	First supercritical U.S. coal plant
1856	Bessemer steelmaking			
1866	Microstructure of steel discovered		1958	Ruby-crystal laser
			1959	Integrated circuit
1866	Discovery of polymeric compounds		1960	Production of amorphous metal alloy
1868	Commercial steel alloy		1960	Artificial diamond production
1870	Celluloid			
1871	Periodic table of the elements		1960s	Microalloyed steels
			1960s	Scanning electron microscope
1875	Open-hearth steelmaking			
1880	Selenium photovoltaic cells		1966	Fiber optics
1884	Nitrocellulose (first man-made fiber)		late 1970s	Discovery of amorphous silicon
1886	Electrolytic process for aluminum		1984	Discovery of quasi-periodic crystals
1889	Nickel-steel alloy		1986	Discovery of high-temperature superconductors
1891	Silicon carbide (first artificial abrasive)			

the extra kick of acceleration in advanced automobile engine turbochargers, already available to car buyers in Japan. Auto makers on both sides of the Pacific, meanwhile, seem in an off-and-on race to develop almost totally ceramic engines that will run much hotter, possibly with no cooling, and be more fuel-efficient than today's models. Ceramics is also the category of materials from which an amazing new class of rare earth–based superconducting compounds are emerging that can conduct electricity with zero resistance loss at relatively high temperatures.

Advances in materials science and engineering are at the heart of nearly all technologies that pace economic productivity, from industrial technologies to biomedical science to electronics to weapons systems to transportation to energy conversion. They include the quest for new materials that bring unique or previously unattainable properties of interest to practical reality. But increasingly, materials research is focused on advanced methods of processing and fabricating existing materials so as to achieve characteristics by design.

"A fundamental reversal in the relationship between humans and materials is taking place. Its economic consequences are likely to be profound," note Joel Clark and Merton Flemings, both professors of materials science and engineering at the Massachusetts Institute of Technology. "Historically, humans have adapted such natural materials as stone, wood, clay, vegetable fiber, and animal tissue to economic uses. The smelting of metals and the production of glass represented a refinement of this relationship. Yet it is only recently that advances in the theoretical understanding of structural and biologic material, in experimental technique, and in processing technology have made it possible to start with a need and then develop a material to meet it, atom by atom."

1-2. Technology Transfer and Feedback. Although the diversity of materials in key economic sectors—metals, ceramics, semiconductors, polymers, and now composites—makes generalizations difficult, a few are worth venturing despite the exceptions to any rule.

One generalization is that discovery or development of new materials often results from intense scientific and industrial pursuit of great leaps forward in performance. This helps explain why many materials innovations have grown from weapons research, defense systems development, and the space program, where performance, not economy, is the guiding goal. Eventually, advances in materials in one realm cascade into others as industrial engineers adapt them to new markets and commercial realities.

Development of the jet engine during the Second World War provides one example. "When the first aircraft turbine was produced around 1940, it just barely ran and didn't last very long," explains John Stringer, until recently head of a materials R&D support group at EPRI and now technical director for exploratory research. "To make it into a real propulsion system it was necessary to undertake an intensive materials development program for quite some time. The development of higher-temperature materials paced the development of the aircraft turbine, which, in turn, has paced the technology of land-based turbines for power generation."

Similarly, much of the genesis of microelectronic materials can be attributed to defense requirements for more compact, miniaturized, and highly reliable command, control, and communications systems. And today, the Strategic Defense Initiative is spurring further leaps in semiconductors, superconductors, and lasers.

Another broad generalization is that there is a trend toward specialization in materials development—that is, new materials with specific properties are sought as an answer to specific application problems. For example, alloying iron with small amounts of carbon produces steel, which is harder and stronger than the base metal alone; development of improved steelmaking processes revolutionized construction in the nineteenth century. It also kicked off a string of alloying improvements that has led to an almost limitless range of choices for modern industry; nickel was added for even greater strength and shock resistance. The addition of chromium opened up a whole family of stainless steels, with greater toughness, corrosion resistance, and heat resistance. Today iron is alloyed with over a dozen elements in different proportions to produce other characteristics important for particular applications—improved hardness, stability, machinability, ductility, resiliency, and weldability, among others.

But just finding the right chemical composition is not the end-all of materials development. The question of what is now being augmented with the question of how, as researchers inquire more deeply into the potential benefits of advanced materials processing and fabrication technologies. Again, steel provides a wealth of examples. Higher purity can be achieved through such processes as secondary refining in electric ladle furnaces; because impurities and nonmetallic inclusions are greatly reduced, steels with nickel, chromium, molybdenum, and vanadium achieve exceptional strength and greater resistance to heat and corrosion. For other applications, a different approach to corrosion resistance might be used: hot isostatic pressing and powder pack diffusion make it possible to achieve good corrosion resistance by applying a thin coat of a superalloy onto a conventional steel that has otherwise desirable mechanical properties.

Directional solidification, an innovative technique now being used to make aircraft turbine blades, produces long crystals in the metal that are oriented parallel to the stress direc-

ELECTRICAL ENGINEERING MATERIALS REFERENCE GUIDE

tion for greater resistance to cracking, creep, and elongation from high centrifugal forces. Research on rapid-solidification processes has led to the manufacture of amorphous metals that lack crystal structure entirely and thus exhibit valuable magnetic properties.

These latter examples highlight the real frontier of materials research implied earlier by Clark and Flemings—complete manipulation of a material's characteristics. This means not only adding properties that are desirable but also getting around other natural properties that were long thought to be unavoidable limitations to a material's use. As mentioned, crystalline structure is one of the problems being ingeniously sidestepped in the metals area. For ceramics—porcelain, glass, tile, and a wide range of industrial refractory materials— most research is aimed at overcoming a more popularly recognized generic shortcoming, brittleness. Ceramics simply crack easily, often catastrophically and unpredictably in industrial environments.

Interest in overcoming this problem is heightened by the unique suitability of ceramics for many high-temperature applications; some forms remain stable at temperatures of 2500°F or more, where most structural metals would oxidize or melt. Research is moving toward ceramic matrix composites laminated with various types of fibers to arrest the growth of microcracks. Such ceramics as silicon carbide, silicon nitride, and oxides of zirconium are among the chief targets of interest. New chemical routes to synthesis, such as sol-gel methods, are being pursued for greater control over the microstructure of the starting material.

Brittleness is also considered one of the principal obstacles in developing useful forms of the complex rare earth barium copper oxides that are showing tantalizing signs of superconductivity at near room temperatures. The superconducting properties of these materials are also highly sensitive to oxygen content, which can be affected by high-temperature processing. "Probably most of these difficulties can be overcome by the skills of modern materials science and engineering, but this will take a great deal of time and effort," according to John Hulm, director of corporate research at Westinghouse Electric. At the moment, however, experiments with the new compounds are outrunning the theory to explain how and why.

1-3. Polymers. Polymers are perhaps the most versatile of all modern materials. Their variety of forms is impressive, including such staples of modern civilization as rubber, styrofoam, all types of plastic, glues, paints, photographic film, nylon, Kevlar, and other synthetic fibers. Expanding knowledge of the interaction between structure, properties, and processing methods is leading to polymers with unprecedented qualities.

The seemingly infinite versatility of polymers derives from the custom-tailored nature of both bulk and molecular material. These are essentially large molecules, built up from smaller molecular chains. Varying the chains and how they are linked permits the formation of a microstructure that may be amorphous or semicrystalline, uniform or inhomogeneous. Chemically distinct and separate chains can be formed to make thermoplastics, which soften when heated for easy molding and extrusion. Alternatively, the molecular bonds can be made to cross-link into extremely tight chains so the material does not deform at high temperature, a problematic characteristic of early plastics.

As research continues, new properties are being developed that promise to open up entirely new areas of application for polymers. Incorporation of double-bonded aromatic chains as fibers spun from the liquid-crystalline phase can give some polymers a tensile strength comparable to steel. Other, transparent polymers that approach optical perfection are a focus of photonic materials research for fiber-optic communications. Although polymers have traditionally been used as electrical insulators, scientists have found some forms that when doped with such impurities as arsenic pentafluoride, exhibit conductivities approaching those of metals.

1-4. Electronics. Perhaps the advanced materials area of greatest importance to the economy is that of electronics, where the steady rise of functional power in information and communications systems will continue to depend directly on materials advances. Advanced semiconductors, such as gallium arsenide, are the subject of intensive research because they offer processing speeds 10 to 100 times faster than achievable with today's silicon microcircuits. Soon, high-speed logic chips may have more than 10,000 transistors imprinted on them, while memory chips storing up to 16 million bits of information are nearing reality. Yet obtaining near-perfect crystals of gallium arsenide is proving even more difficult than obtaining crystals of silicon.

MATERIALS RESEARCH

Quite apart from their use today in microprocessors and memory devices, semiconductors are now also being pursued as solid-state lasers; unlike silicon, gallium arsenide can emit light with the right doping and thin-film architecture. Sophisticated optical sensors, meanwhile, are proliferating, leading to new streams and sources of data that some experts envision will eventually require optical computers to process.

Even the veteran transistor, which blazed the trail to microelectronics before it was upstaged by the integrated circuit, is the subject of advanced materials research. Scientists are now tentatively growing transistor materials in virtually one step from molten silicon and metal, opening the way to cheaper, more powerful transistors.

1-5. Electric Utility Research. The barriers to performance that materials science and engineering seeks to break through are all central to utility operations: temperature, pressure, corrosion, fatigue strength, conductivity—even high-speed data transmission. These frontiers of materials research cut across all lines of technology used by electric utilities, and advanced materials will most likely find important niches in many areas. Some are already entering utility use, others are under active engineering development, while plenty more are still gleams in the eyes of scientists and researchers. Together, they suggest a glimpse of how the business of generating, transmitting, and using electric power may be substantially transformed in the decades ahead.

In the 1970s utilities generally retreated from supercritical operation at coal-fired steam units because of reliability problems tied to the limits of metallic materials that were then available. Advanced metals will permit a return to supercritical steam conditions and higher operating efficiencies in these plants. In one significant advance, a class of modified 9% chromium martensitic steels containing trace amounts of niobium, nitrogen, and vanadium for high-temperature strength have been qualified for use in the advanced pulverized coal plant EPRI is helping design under an international collaboration with utilities and boiler and turbine makers. They will replace conventional ferritic steels in areas ranging from superheater/reheater tubes to pipes to heavy-section castings. The steels were originally developed by Oak Ridge National Laboratory and Combustion Engineering in the mid 1970s for piping liquid sodium in breeder reactors.

Farther downstream from the boiler in the advanced coal plant, steam turbine blades are a major area of materials focus. Titanium alloys could virtually eliminate corrosion-induced fatigue in low-pressure steam turbine blades. EPRI and Westinghouse recently qualified the titanium alloy Ti-6Al-4V (containing 6% aluminum and 4% vanadium) in next-to-last-row blades for low-pressure turbines, completing a 10-year materials application effort. The Japanese and the French are developing last-row blades of titanium, which offer increased turbine power because they can be used in longer lengths than conventional 12–13% Cr stainless steels.

In another approach, blades made of directionally solidified single crystals of superalloys may put an end to grain boundary cracking from creep and thermal stress. For gas turbines operating at temperatures up to 2300°F, EPRI is funding a pioneering effort to develop single-crystal superalloy blades by adapting the Czochralski crystal growth process now used in making silicon semiconductors. Robert Jaffee, a metallurgist and senior EPRI technical adviser, is managing experimental work at Unisil, a Silicon Valley firm, to apply the Czochralski process, in which a single crystal is pulled from the melt in a crucible.

"If we can eliminate the grain boundaries in a blade machined from a single crystal, we will eliminate cracking," says Jaffee. He adds that this cross-fertilization between metallurgy and semiconductor materials production represents "the kind of opportunities we look for. Most new developments are really transfers from one technology to another." Jaffee notes that electron-beam heating, now used extensively for scrap recovery in metals production, is being eyed by silicon crystal producers in an EPRI project for improving the Czochralski process, completing a technology transfer loop from silicon to metallic single crystals and back to silicon.

Superclean steels made by secondary refining techniques could make temper embrittlement of turbine rotors or neutron embrittlement of nuclear reactor pressure vessels problems of the past. Nearly a decade of laboratory-scale work funded by EPRI has identified a superclean low-alloy specification for low-pressure turbine rotors. Embrittlement caused by manganese and silicon in conventional low-alloy steels currently limits LP turbines from operating above about 650°F. The new superclean steel is completely resistant to temper embrittlement over the range of 700–1000°F. With manganese and silicon held to 0.02–

ELECTRICAL ENGINEERING MATERIALS REFERENCE GUIDE

0.05%, it has also proved tough, ductile, and even stronger than steels of conventional purity. "Secondary steel refining techniques using ladle furnaces and vacuum degassing make it possible to melt steel with virtually no impurities and containing only the intended composition," explains Jaffee.

Following successful trials of 34-ton rotors made from the superclean steel, EPRI funded follow-on work at Japan Steel Works to produce the superclean steel for more-uniform yield strength. Meantime, Toshiba has made four production-size rotor forgings of the steel for use in supercritical steam plants in Japan. EPRI is actively seeking to apply the steel as a retrofit material in a U.S. utility plant in advance of the time it could show up in future units. In addition to use as turbine rotor material, superclean steels could also be applied in future chemical and nuclear reactor pressure vessels.

In contrast to advanced structural metallurgy, where much of the focus is on single-crystal metals and manipulating grain orientation, many frontiers in electrical and electronic materials are targeting amorphous (non-crystalline) forms of metals and ceramics for their unique properties. After nearly a decade of EPRI support at one of the leading technology firms in metallic glasses, Allied-Signal has produced a major materials breakthrough for power transformer cores. Ribbons of metallic glass spin off a high-speed casting wheel as molten alloy cools at a rate of a million degrees per second. A utility-scale 500-kVA power transformer with an amorphous steel alloy core is now being tested in actual utility use, as are a thousand 25-kVA distribution transformers, now commercially available. Amorphous metal technology promises to cut typical transformer core losses by 75%.

Amorphous forms of semiconductors offer a route to low-cost, moderately efficient photovoltaic cells. Rather than being grown as a crystal, thin films of silicon only a few atoms thick can be laid down on a substrate by plasma and chemical vapor deposition techniques. Amorphous silicon now forms the thin-film transistor arrays that drive liquid crystal computer displays. Meanwhile, intense materials research in hyperpure, single-crystal silicon is being pursued by EPRI for high-efficiency utility photovoltaics and for very high current power system applications, such as metal oxide thyristor–controlled solid-state switching devices and circuit breakers. In the future, such large-scale semiconductor switching devices will instantly route and control power flows and, possibly under the guidance of superconducting supercomputers and fiber-optic data communications, maintain something of a real-time spot market for electricity across large regions of the country.

Utilities are already installing fiber-optic communications links between generating plants and control centers. Some are entering the telecommunications business, stringing fiber-optic connections along their power transmission rights-of-way. Because virtually any physical phenomenon—heat, radiation, magnetism—can be made to alter the flow of light through an optical fiber, researchers are interested in the possibilities of placing fiber-optic sensors in such diverse places as along transmission lines, inside nuclear reactors, or in electrical gear for detecting minute changes in key conditions.

Data from such sensors might be collected and transmitted almost instantaneously—specifically, at the speed of light—and become part of a larger data/electron flow between and among utility plants, substations, and control centers. EPRI recently initiated an applications study of fiber-optic sensors and last September conducted the first conference for utilities on the possibilities for optical voltage/current measurement technology for power systems.

Successful development of the new high-temperature ceramic superconductor materials could bring profound changes in the utility industry. Power from generating plants may someday be temporarily stored in giant superconducting magnet coils. When it is needed, this electricity could be carried hundreds of miles by superconducting transmission cables with no losses to electrical resistance.

The change promised by superconductors and advanced data transmission are so sweeping that it is easy to lose sight of the many important materials improvements being pursued in smaller components and systems. For example, ceramic-based catalytic combustors similar in concept to those on most automobiles may provide a solution to nitrogen oxide emissions in gas turbines. Porous ceramic filters for hot gas cleanup may be key elements of pressurized fluidized-bed coal plants. And other hot section parts like turbine inlet vanes may also eventually be made of reinforced ceramic matrix composite materials, permitting higher operating temperatures and efficiencies.

MATERIALS RESEARCH

More-stable and longer-lasting anode, cathode, and electrolytic materials are being developed for advanced batteries and fuel cells. Efforts continue in the development of a more efficient molten carbonate–based successor to the phosphoric acid fuel cell, which is approaching full-scale utility application. Sodium-sulfur mixtures and zinc compounds still hold hope of someday becoming the basis for advanced deep-cycle, high-capacity batteries for utility load leveling.

Polymers are also receiving increased attention at EPRI. The emergence of polymer-based concretes in the 1960s led to the development by EPRI of Polysil—a superior insulator now being commercially manufactured as a substitute for ceramic porcelain. Other ceramiclike, chemically bonded polymer concretes may eventually encapsulate nuclear waste with exceedingly low rates of leaching. One of the more recent niches for polypropylene is as a film sandwiched in paper—PPP cable insulation, developed largely under EPRI auspices over the last 15 years—that will cut the cost and size of conventional underground utility lines.

1-6. Future Trends. From the mundane to the exotic, from steel to silicon and beyond, materials help propel science and engineering beyond the present boundaries of understanding to improve or even revolutionize products, systems, and everyday activities. "The dominant materials of a decade ago are being supplemented or replaced by new superior combinations. Computer modeling is improving designs. A materials science and engineering revolution is under way that will be a key factor in determining the outcome of global economic competition," Philip Abelson, deputy editor of *Science* magazine, wrote recently.

MIT's Clark and Flemings sound a similar note. "In addition to meeting needs, materials science and engineering creates opportunities and provides society with new ways to address such problems as the scarcity of resources, the maintenance of economic growth, and the formation of capital. Productivity and the structure of the labor force are also profoundly affected by advances in the field. For industrial, financial, and government leaders, the definition and implementation of strategies that exploit opportunities created by materials science is a central challenge of the last quarter of the century."

Materials research is one of the foundations of modern electric power, and there is every reason to believe that as many materials advances await in the future as have been achieved in the past. The frontiers are much the same, but the new horizons promise developments we have not yet imagined.

1-7. Strength of Materials. Improvements in material strength in the past century have quite literally changed the shape of modern society. Skyscrapers simply could not be built before there were steel superstructures to hang them on. Suspension bridges, a clear triumph of innovative design and material strength, have become leaner and more streamlined over the years as higher metal strength-to-weight ratios have been achieved. But improvements have not been confined to metals; high-strength polymer fibers many times stronger than steel are replacing other materials in dozens of applications, from bullet-proof vests to medical prostheses to the sails used in the America's Cup–winning yacht Stars & Stripes. Utility industry research to develop exceptionally strong superalloys for key components is focused on improved purity and new production techniques, such as directional solidification and the creation of large single metallic crystals.

1-8. Corrosion. Corrosive deterioration is the seemingly inevitable result when moisture meets metal. The threat is particularly extreme in the high-temperature, corrosive environments characteristic of power plants, although milder conditions can also be problematic; the Statue of Liberty suffered considerable corrosion damage over the years just from exposure to the weather. The primary approach to fighting corrosion in industry has been to use corrosion-resistant alloys rich in such metals as chromium or titanium. Nonmetallic materials may take over in many corrosion-prone applications as new composites, high-temperature plastics, and high-strength ceramics are developed. Use of protective coatings, both metal and nonmetal, can also be effective; a number of organic and inorganic coating materials have been tested for use in power plant scrubbers, including sprayed-on silicate cement, fluoroelastomers, vinyl esters, and reinforced epoxies.

1-9. Effect of Heat. The search for higher efficiencies has meant higher temperatures for all types of power-producing devices, from truck engines to utility power plants (Fig. 1-1). Boiler materials were unequal to the temperature and pressure demands of the early supercritical steam plants in the sixties, leading to severe reliability and availability prob-

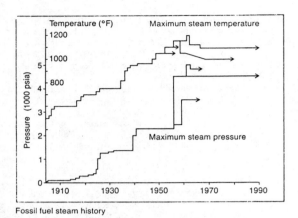

Fossil fuel steam history

FIG. 1-1 Efficiencies increase as temperature and pressure increase.

lems; advanced metal alloys are now reestablishing the supercritical option for coal-fired plants of the future. Low temperatures can be a problem, too. The carbon steel used in the famed Liberty ships of World War II sometimes cracked at the welds when subjected to cold ocean waters. One of the greatest challenges in designing the space shuttle was to develop advanced ceramics and composites that could weather both extremes of temperature—the cold of space and the white heat of reentry.

1-10. Materials for Light and Electronics. The importance of electric power and communications in the modern world has sparked fresh interest in materials that conduct electricity and light. Of particular interest to utilities are advanced solar cells, whose high sunlight-to-electricity efficiencies have resulted in part from advances in crystalline and amorphous silicon materials (Fig. 1-2). Other glassy, amorphous materials are revolutionizing communication by speeding information along transparent optical fibers at the speed of light. Meanwhile, scientists around the world are in a race to engineer and perfect a new family of superconductors made of ceramic oxides that can transmit or store electricity with no resistance losses. Even plastics, long used as electrical insulators, have shown indications of good electrical conductivity under certain conditions—a prospect that could have a tremendous impact on the technology of the future.

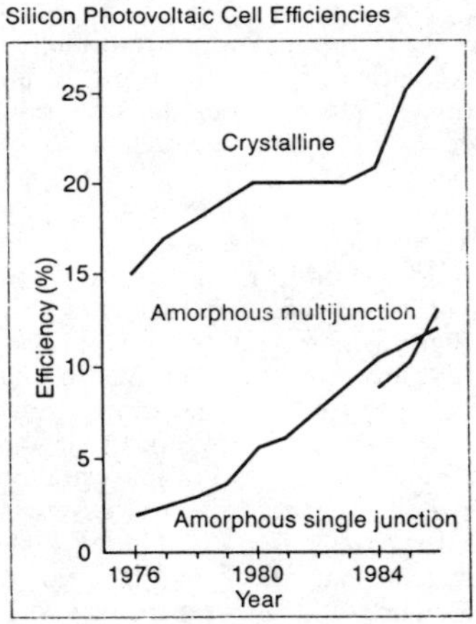

1-11. Materials on a National Scale. An estimated 20,000 pounds of nonrenewable, nonfuel mineral resources are processed on average each year for every person in the United States. Production of primary materials directly involves some 1.5 million workers, plus another 3 million to 4.5 million whose jobs in other sectors also depend on materials. The value of shipments of advanced materials—mainly in the aircraft, aerospace, electronics, and automobile industries—has been estimated at $70 billion annually, or about 14% of the total materials shipments in this country.

FIG. 1-2 Advances in materials science increase efficiencies of solar cells.

That puts the value of the overall materials component of the American economy at about $500 billion, or over 12% of 1985's $4 trillion gross national product.

Increasing per capita consumption of materials has been one of the hallmarks of economic growth since the Industrial Revolution. But analysts note a classical demand cycle in the leveling off of consumption for most basic materials, including steel, cement, and paper.

The trends are attributed to several factors: substitution of a better material for an existing one, design improvements that use less material, saturated traditional markets, and new consumer markets involving products with a relatively low materials content. Together, they suggest a shift is under way toward high-technology products characterized by a lower materials content per dollar of value added, such as high-strength and corrosion-resistant alloys and specialty chemicals.

There are numerous signs that American industry is adapting to the changes, using advanced materials processed and fabricated with state-of-the-art technology into products with special niche markets. In the case of steel, for example, although the total tonnage of steel produced has declined in recent years, the use of specialty alloys designed for specific purposes is in ascendancy.

Issues of materials research policy and funding have attracted increasing attention from Congress and the science policy community in recent years. Approximately $1.2 billion is earmarked for materials R&D in the federal government's proposed 1988 fiscal year budget, down slightly from the current year, according to an analysis by the Federation of Materials Societies.

"The fact that materials undergird the nation's economy and defense is reflected in the range of R&D activities. Examples include programs aimed at developing new materials and improving materials processing for advanced energy systems, the National Aerospace Plane, and the Strategic Defense Initiative; supporting university-based materials research laboratories; increasing research on production and performance characteristics of advanced ceramics, composites, and polymers; expanding the engineering science base in such technology-driven fields as mechanics, structures, and materials engineering; and understanding the effects of gravity on materials processing," the federation notes.

The Department of Commerce has identified a group of emerging technologies in the areas of advanced materials, electronics, biotechnology, automation, and computing that it says American industry must master by the turn of the century if it is to successfully compete with foreign manufacturers. Key materials areas include superconductors, polymer composites, optical electronics, and advanced microelectronics.

Momentum reportedly is building for expanding federal involvement in the development and commercialization of long-term, high-risk technologies through increased funding of cooperative research with industry, universities, and other private organizations. Materials R&D areas likely to be in the focus include superconductors, advanced ceramics, and submicrometer-circuit computer chips. The National Science Foundation has plans for new university-based centers in materials science, along with others in computer and information sciences, biotechnology, and social and behavioral sciences. The emergence of an active Materials Research Society is seen as further evidence that professional as well as corporate and federal attention and resources are converging for a renewed national R&D commitment to materials science and engineering.

1-12. Suggestions for Further Reading

Various authors. "Frontiers in Chemistry: Materials Science." *Science,* Vol. 235, No. 4792 (27 February 1987), pp. 953; 997–1035.

T. W. Eager. "The Real Challenge in Materials Engineering." *Technology Review,* Vol. 90, No. 2 (February-March 1987), pp. 24–35.

P. J. Steinhardt. "Quasicrystals." *American Scentist,* Vol. 74, No. 6 (November-December 1986), pp. 586–597.

Various authors. "Materials for Economic Growth." *Scientific American,* Vol. 255, No. 4 (October 1986).

E. D. Larson, M. H. Ross, and R.H. Williams. "Beyond the Era of Materials." *Scientific American,* Vol. 254, No. 6 (June 1986), pp. 34–41.

R. I. Jaffee. "Materials and Electricity." *Metallurgical Transactions A.* Vol. 17A (May 1986), pp. 755–775.

1-10 ELECTRICAL ENGINEERING MATERIALS REFERENCE GUIDE

Materials for Large Land-Based Gas Turbines. Final report for RP2382-1, prepared by National Research Council, March 1986. EPRI AP-4476.

Ceramics for Electric Power Generating Systems. Final report for RP2608-1, prepared by SRI International, January 1986. EPRI AP-4380.

National Academy of Engineering. *Advancing Materials Research.* Washington, D.C.: National Academy Press, 1986.

M. A. Meyers and O. T. Inal, eds. *Frontiers in Materials Technologies.* Amsterdam: Elsevier, 1985.

G. G. Libowitz and M. S. Whittingham, eds. *Materials Science in Energy Technology.* New York: Academic Press, 1979.

Chapter 2

CONDUCTOR MATERIALS

By MARK A. BAKER

General Properties   2-1
Metal Properties   2-2
Conductor Properties   2-15

Fusible Metals and Alloys   2-44
Miscellaneous Metals and Alloys   2-76

General Properties

2-1. Conducting Materials. A conductor of electricity is any substance or material which will afford continuous passage to an electric current when subjected to a difference of electric potential. The greater the density of current for a given potential difference, the more efficient the conductor is said to be. Virtually all substances in solid or liquid state possess the property of electric conductivity in some degree, but certain substances are relatively efficient conductors, while others are almost totally devoid of this property. The metals, for example, are the best conductors, while many other substances, such as metal oxides and salts, minerals, and fibrous materials, are relatively poor conductors, but their conductivity is beneficially affected by the absorption of moisture. Some of the less efficient conducting materials, such as carbon and certain metal alloys, as well as the efficient conductors such as copper and aluminum, have very useful applications in the electrical arts.Certain other substances possess so little conductivity that they are classed as nonconductors, a better term being insulators or dielectrics. In general, all materials which are used commercially for conducting electricity for any purpose are classed as conductors.

2-2. Definition of Conductor. A conductor is a body so constructed from conducting material that it may be used as a carrier of electric current. In ordinary engineering usage, a conductor is a material of relatively high conductivity.

2-3. Definition of Circuit. An electric circuit is the path of an electric current, or, more specifically, it is a conducting part or a system of parts through which an electric current is intended to flow.

2-4. General Properties of Conductors. Electric circuits in general possess four fundamental electrical properties, consisting of resistance, inductance, capacitance, and leakance. That portion of a circuit which is represented by its conductors will also possess these four properties, but only two of them are related to the properties of the conductor considered by itself. Capacitance and leakance depend in part upon the external dimensions of the con-

ELECTRICAL ENGINEERING MATERIALS REFERENCE GUIDE

ductors and their distances from one another and from other conducting bodies and in part upon the dielectric properties of the materials employed for insulating purposes. The inductance is a function of the magnetic field established by the current in a conductor, but this field as a whole is divisible into two parts, one being wholly external to the conductor and the other being wholly within the conductor; only the latter portion can be regarded as corresponding to the magnetic properties of the conductor material. The resistance is strictly a property of the conductor itself. Both the resistance and the internal inductance of conductors change in effective values when the current changes with great rapidity, as in the case of high-frequency alternating currents; this is termed the "skin effect."

In certain cases, conductors are subjected to various mechanical stresses. Consequently their weight, tensile strength, and elastic properties require consideration in all applications of this character. Conductor materials as a class are affected by changes in temperature and by the conditions of mechanical stress to which they are subjected in service. They are also affected by the nature of the mechanical working and the heat-treatment which they receive in the course of manufacture or fabrication into finished products.

2-5. Types of Conductor. In general, a conductor consists of a solid wire or a multiplicity of wires stranded together, made of a conducting material and used either bare or insulated. Only bare conductors are considered in this subsection. Usually the conductor is made of copper or aluminum, but for applications requiring higher strength, such as overhead transmission lines, bronze, steel, and various composite constructions are used. For conductors having very low conductivity and used as resistor materials, a group of special alloys is available.

Metal Properties

2-6. Specific Gravity and Density. Specific gravity is the ratio of mass of any material to that of the same volume of water at 4°C. Density is the unit weight of material expressed as pounds per cubic inch, grams per cubic centimeter, etc., at some reference temperature, usually 20°C. For all practical purposes, the numerical values of specific gravity and density are the same, expressed in g/cm^3.

2-7. Density and Weight of Copper. Pure copper, rolled, forged, or drawn and then annealed, has a density of 8.89 g/cm^3 at 20°C or 8.90 g/cm^3 at 0°C. Samples of high-conductivity copper will vary usually from 8.87 to 8.91 and occasionally from 8.83 to 8.94. Variations in density may be caused by microscopic flaws or seams or the presence of scale or some other defect; the presence of 0.03% oxygen will cause a reduction of about 0.01 in density. Hard-drawn copper has about 0.02% less density than annealed copper, on the average, but for practical purposes the difference is negligible.

The international standard of density, 8.89 at 20°C, corresponds to a weight of 0.32117 lb/in^3 or 3.0270×10^{-6} lb/(cmil)(ft) or 15.982×10^{-3} lb/(cmil)(mile). Multiplying either of the last two figures by the square of the diameter of the wire in mils will produce the total weight of wire in pounds per foot or per mile, respectively.

2-8. Density and weight of copper alloys varies with the composition. For hard-drawn wire covered by ASTM Specification B105, the density of alloys 85 to 20 is 8.89 g/cm^3 (0.32117 lb/in^3) at 20°C; alloy 15 is 8.54 (0.30853); alloys 13 and 8.5 are 8.78 (0.31720).

2-9. Density and weight of copper-clad steel wire is a mean between the density of copper and the density of steel, which can be calculated readily when the relative volumes or cross sections of copper and steel are known. For practical purposes a value of 8.15 g/cm^3 (0.29444 lb/in^3) at 20°C is used.

2-10. Density and weight of aluminum wire (commercially hard-drawn) is 2.705 g/cm^3 (0.0975 lb/in^3) at 20°C. The density of electrolytically refined aluminum (99.97% Al) and for hard-drawn wire of the same purity is 2.698 at 20°C. With less pure material there is an appreciable decrease in density on cold working. Annealed metal having a density of 2.702 will have a density of about 2.700 when in the hard-drawn or fully cold-worked conditions (see *NBS Circ.* 346, pp. 68 and 69).

2-11. Density and weight of aluminum-clad wire is a mean between the density of aluminum and the density of steel, which can be calculated readily when the relative volumes or cross sections of aluminum and steel are known. For practical purposes a value of 6.59 g/cm^3 (0.23808 lb/in^3) at 20°C is used.

CONDUCTOR MATERIALS
2-3

2-12. Density and weight of aluminum alloys varies with type and composition. For hard-drawn aluminum alloy wire 5005-H19 and 6201-T81, a value of 2.703 g/cm^3 (0.09765 lb/in^3) at 20°C is used.

2-13. Density and weight of pure iron is 7.90 g/cm^3 [2.690 × 10^{-6} lb/(cmil)(ft)] at 20°C.

2-14. Density and weight of galvanized steel wire (EBB, BB, HTL-85, HTL-135, and HTL-195) with Class A weight of zinc coating is 7.83 g/cm^3 (0.283 lb/in^3) at 20°C, with Class B is 7.80 g/cm^3 (0.282 lb/in^3), and with Class C is 7.78 g/cm^3 (0.281 lb/in^3).

2-15. Percent Conductivity. It is very common to rate the conductivity of a conductor in terms of its percentage ratio to the conductivity of chemically pure metal of the same kind as the conductor is primarily constituted or in ratio to the conductivity of the international copper standard. Both forms of the conductivity ratio are useful for various purposes.

This ratio can also be expressed in two different terms, one where the conductor cross sections are equal and therefore termed the *volume-conductivity ratio* and the other where the conductor masses are equal and therefore termed the *mass-conductivity ratio*.

2-16. The International Annealed Copper Standard (IACS) is the internationally accepted value for the resistivity of annealed copper of 100% conductivity. This standard is expressed in terms of mass resistivity as 0.15328 $\Omega \cdot$g/m^2, or the resistance of a uniform round wire 1 m long weight 1 g at the standard temperature of 20°C. Equivalent expressions of the annealed copper standard, in various units of mass resistivity and volume resistivity, are as follows:

0.15328	$\Omega \cdot$g/m^2
875.20	$\Omega \cdot$lb/mi^2
1.7241	$\mu\Omega \cdot$cm
0.67879	$\mu\Omega \cdot$in at 20°C
10.371	$\Omega \cdot$cmil/ft
0.017241	$\Omega \cdot$mm^2/m

The above values are the equivalent of $\frac{1}{58}$ $\Omega \cdot$mm^2/m, so that the volume conductivity can be expressed as 58 S·mm^2/m at 20°C.

2-17. Conductivity of conductor materials varies with chemical composition and processing. For industry specification values, see Table 2-1.

2-18. Electrical resistivity is a measure of the resistance of a unit quantity of a given material. It may be expressed in terms of either mass or volume; mathematically,

Mass resistivity:
$$\delta = \frac{Rm}{l^2} \qquad (2\text{-}1)$$

Volume resistivity:
$$\rho = \frac{RA}{l} \qquad (2\text{-}2)$$

where R = resistance, m = mass, A = cross-sectional area, l = length.

2-19. Electrical resistivity of conductor materials varies with chemical composition and processing. For industry specification values, see Tables 2-1 and 2-2.

2-20. Electrical Conductivity and Resistivity: Nonferrous Conductors. See Table 2-1.

2-21. Electrical Resistivity: Ferrous Conductors. See Table 2-2.

2-22. Effects of Temperature Changes. Within the temperature ranges of ordinary service there is no appreciable change in the properties of conductor materials, except in electrical resistance and physical dimensions. The change in resistance with change in temperature is sufficient to require consideration in many engineering calculations. The change in physical dimensions with change in temperature is also important in certain cases, such as in overhead spans and in large units of apparatus or equipment.

2-23. Temperature Coefficient of Resistance. Over moderate ranges of temperature, such as 100°C, the change of resistance is usually proportional to the change of temperature. Resistivity is always expressed at a standard temperature, usually 20°C (68°F). In general if R_{t_1} is the resistance at a temperature t_1, and α_{t_1} is the temperature coefficient at that tem-

TABLE 2-1 Electrical Conductivity and Resistivity*—Nonferrous Conductors

Specification†	Temper and shape	Size limits, in	Conductivity at 20°C (68°F), IACS, %	Weight resistivity at 20°C (68°F)		Volume resistivity at 20°C (68°F)			
				$\Omega \cdot g/m^2$	$\Omega \cdot lb/mi^2$	$\Omega \cdot cmil/ft$	$\Omega \cdot mm^2/m$	$\mu\Omega \cdot cm$	$\mu\Omega \cdot in$
Copper and copper alloy—specific gravity 8.89:			100.17	0.15302‡	873.75	10.354	0.017213	1.7213	0.67767
B4 B5 B170 B187 B188 B298 B75	Low-resistance Lake wirebar Electrolytic wirebar Oxygen-free wirebar Soft bus bar, rod and shape Soft bus tube Silver-coated soft, round Tube, soft, OF copper	All	100.00	**0.15328§**	875.20	10.371	0.017241	1.7241	0.67879
B49 B3 B48	Hot-rolled rod Soft, round Soft, rectangular	1.375–0.250 All All	100.00	0.15328	**875.20**	10.371	0.017241	1.7241	0.67879
			99.50 99.00 98.50	0.15405 0.15482 0.15561	879.60 884.04 888.53	10.423 10.476 10.529	0.017328 0.017416 0.017504	1.7328 1.7416 1.7504	0.68220 0.68565 0.68913
B187	Hard bus bar, rod and shape	Over 1 in OD Over 0.375 × 4	98.40	**0.15577**	889.42	10.539	0.017521	1.7521	0.68981
B188	Hard bus tube, rectangular or square	Over 6 in	98.16	**0.15614**					
B75	Tube, soft, DLP copper	All							
B188	Hard bus tube, rectangular or square	Up to 6 × 3/16 in wall	98.00 97.80	0.15640 **0.15673**	893.06 894.90	10.583 10.604	0.017593 0.017629	1.7593 1.7629	0.69265 0.69406
B2 B33 B189	Medium hard, round Tinned soft, round Lead-coated soft, round	0.460–0.325 0.460–0.290 0.460–0.290	97.66	0.15694	**896.15**	10.619	0.017654	1.7654	0.69504
			97.50	0.15721	897.64	10.637	0.017683	1.7683	0.69620
B187	Hard bus bar and rod	Up to 1 in OD Up to 0.375 × 4							
B188	Hard bus tube, round Hard bus pipe, IPS and extra strong Hard bus tube, rectangular or square	Over 1 in OD Over 4 in OD Up to 6 in, over 3/16 in wall	97.40	**0.15737**	898.55	10.648	0.017701	1.7701	0.69690
B75 B372	Tube, hard, OF copper Waveguide tube, OF copper	All							

ASTM spec	Description	Size, in	% Cond.						
B1	Hard, round	0.460–0.325	97.30	0.15753	899.49	10.659	0.017720	1.7720	0.69763
B47 and B116	Hard trolley wire	All	97.16	0.15775	**900.77**	10.674	0.017745	1.7745	0.69863
B33 B189	Tinned soft, round Lead-coated soft, round	0.289–0.103							
B2	Medium hard, round	0.324–0.0403	97.00	0.15802	902.27	10.692	0.017775	1.7775	0.69979
B188	Hard bus pipe, SPS and extra strong	Up to 4 in OD	96.66	0.15857	905.44	10.729	0.017837	1.7837	0.70224
			96.60	**0.15865**	905.86	10.734	0.017845	1.7845	0.70257
			96.50	0.15884	906.94	10.747	0.017867	1.7867	0.70341
B1 B33 B189	Hard, round Tinned soft, round Lead-coated soft, round	0.324–0.0403 0.102–0.0201 0.102–0.0201	96.16	0.15940	**910.15**	10.785	0.017930	1.7930	0.70590
B75 B372	Tube, hard, DLP copper Waveguide tube, DLP copper	All	96.16	**0.15940**	910.15	10.785	0.017930	1.7930	0.70590
B355	Nickel-coated soft, round, Class 2	All	96.00	0.15966	**911.67**	10.803	0.017960	1.7960	0.70708
B33 B189	Tinned soft, round Lead-coated soft, round	0.0200–0.0111	94.16	0.16279	**929.52**	11.015	0.018312	1.8312	0.72092
B355	Nickel-coated soft, round, Class 4	All	94.0	0.16306	**931.06**	11.033	0.018342	1.8342	0.72212
B246	Tinned medium hard, round	0.2043–0.103	93.22	0.16443	**938.85**	11.125	0.018495	1.84949	0.72816
B33 B189	Tinned soft, round Lead-coated soft, round	0.0110–0.0030	93.15	0.16454	**939.51**	11.133	0.018508	1.8508	0.72867
B246	Tinned hard, round	0.2043–0.103	92.72	0.16532	**943.92**	11.185	0.018595	1.85947	0.73209
B246	Tinned medium hard, round	0.103–0.0508	92.51	0.16569	**946.06**	11.211	0.018637	1.86369	0.73375
B246	Tinned hard, round	0.103–0.0508	91.96	0.16668	**951.72**	11.278	0.018748	1.87484	0.73814
B355	Nickel-coated soft, round, Class 7	All	91.0	0.16844	**961.76**	11.397	0.018947	1.8947	0.74593
B355	Nickel-coated soft, round, Class 10	All	90.00	0.17031	972.45	11.524	0.019157	1.9157	0.75421
B105	Hard round, alloy 85	All	88.0‖	0.17418	**994.55**	11.785	0.019592	1.9592	0.77136
			85.00‖	0.18039	**1,030**	12.206	0.020291	2.0291	0.79885
B105 B9	Hard round, alloy 80 Trolley wire, alloy 80	All	80.00‖	0.19160	**1,094**	12.964	0.021551	2.1551	0.84849
B355	Nickel-coated soft, round, Class 27	All	71.0	0.21588	**1,232.7**	14.607	0.024284	2.4284	0.95605
B105 B9	Hard round, alloy 65 Trolley wire, alloy 65	All	65.00‖	0.23573	**1,346**	15.950	0.026516	2.6516	1.0439
B105 B9	Hard round, alloy 55 Trolley wire, alloy 55	All	55.00‖	0.27864	1,591	18.854	0.031343	3.1343	1.2340
B9 B9	Hard round, alloy 40 Trolley wire, alloy 40	All	40.00‖	0.38320	2,188	25.928	0.043103	4.3103	1.6970
B105	Hard round, alloy 30	All	30.00‖	0.51086	2,917	34.567	0.057465	5.7465	2.2624
B105	Hard round, alloy 20	All	20.00‖	0.76638	4,376	51.856	0.086207	8.6207	3.3940
Resistivity temperature constant.				0.000597	3.41	0.0409	0.0000681	0.00681	0.00268

TABLE 2-1 Electrical Conductivity and Resistivity*—Nonferrous Conductors (*Continued*)

Specification†	Temper and shape	Size limits, in	Conductivity at 20°C (68°F), IACS, %	Weight resistivity at 20°C (68°F)		Volume resistivity at 20°C (68°F)			
				Ω·g/m²	Ω·lb/mi²	Ω·cmil/ft	Ω·mm²/m	μΩ·cm	μΩ·in
Copper alloy—specific gravity 8.54:									
B105	Hard, round, alloy 15	All	15.00‖	0.98162	**5,605**	69.142	0.11494	11.494	4.5253
Resistivity temperature constant.				0.000597	3.41	0.0409	0.000681	0.00681	0.00268
Copper alloy—specific gravity 8.78:									
B105	Hard, round, alloy 13	All	13.00‖	1.1645	**6,649**	79.778	0.13263	13.263	5.2215
B105	Hard, round, alloy 8.5	All	8.50‖	1.7809	**10,169**	122.01	0.20284	20.284	7.9857
Resistivity temperature constant.				0.000597	3.41	0.0409	0.0000681	0.00681	0.00268
Aluminum and aluminum alloy—specific gravity 2.703:									
B233	Redraw rod 1350-0	1.000	61.8	0.075410	**430.59**	16.782	0.027899	2.7899	1.0983
	Redraw rod 1350-H12 and -H22		61.5	0.075778	**432.69**	16.864	0.028035	2.8035	1.1037
	Redraw rod 1350-H14 and EC-H24		61.4	0.075901	**433.39**	16.891	0.028080	2.8080	1.1055
	Redraw rod 1350-H16 and EC-H26		61.3	0.076025	**434.10**	16.919	0.028126	2.8126	1.1073
B236	Bus bar	All	61	**0.07640**	436.24	17.002	0.028264	2.8624	1.1128
B230	Hard, round								
B262	Three-quarter hard, round	All	61.0	0.076399	436.23	**17.002**	0.028265	2.8625	1.1128
B323	Half hard, round								
B314	Half hard, round								
B324	All tempers, rectangular	All	61.0	0.076397	**436.24**	17.002	0.028264	2.8624	1.1128
B317	T64 temper, extruded alloy 6101	All	59.0	**0.0782**	446.74	17.430	0.028976	2.8976	1.1408
B317	H111 temper, extruded alloy 6101	All	59.0	**0.0789**	450.52	17.578	0.029222	2.9222	1.1505
B317	T61 temper, extruded alloy 6101	All	57.0	**0.0817**	466.33	18.195	0.030247	3.0247	1.1909
B317	T65 temper, extruded alloy 6201	All	56.5	0.082401	470.50	18.356	0.030518	3.0518	1.2015
B317	T63 temper, extruded alloy 6101	All	56.0	**0.0831**	474.66	18.520	0.030788	3.0788	1.2121
B317	T6 temper, extruded alloy 6101	All	55.0	**0.0846**	483.29	18.856	0.031347	3.1347	1.2342
B396	Hard, round, alloy 5005	All	53.5	0.087106	497.38	**19.385**	0.032226	3.2226	1.2687
B398	Hard, round, alloy 6201	All	52.5	0.088764	506.85	**19.754**	0.032839	3.2839	1.2929
Resistivity temperature constant.				0.000305	1.74	0.0689	0.000115	0.0115	0.00451
Copper-clad steel—specific gravity 8.15:									
B227	Hard, round, grades 40 HS and EHS	All	40‖	0.035837	2,046.3	**26.45**	0.043971	4.3971	1.7311
B227	Hard, round, grades 30 HS and EHS	All	30‖	0.047773	2,727.8	**35.26**	0.058617	5.8617	2.3078
Resistivity temperature constant (40%)				0.00135	7.68	0.100	0.000167	0.0167	0.00657
Resistivity temperature constant (30%)				0.00179	10.24	0.134	0.000222	0.0222	0.00875
Aluminum-clad steel—specific gravity 6.59:									
B415	Hard, round	0.204–0.080	20‖	0.55886	3,191.0	**51.01**	0.084805	8.4805	3.3384
Resistivity temperature constant.				0.00200	11.40	0.184	0.000306	0.0306	0.0121

* The value established as standard is indicated in boldface type; other values given are calculated.
† ASTM unless otherwise noted.
‡ Matthiessen's standard.
§ International Annealed Copper Standard (IACS).
‖ Nominal value.
NOTE: 1 in = 2.54 cm.

CONDUCTOR MATERIALS

TABLE 2-2 Electrical Resistivity—Ferrous Conductors

Material	ASTM specification	Weight resistivity at 20°C (68°F)		Volume resistivity at 20°C (68°F)			
		Maximum	Average	Range	Average	Range	Average
		$\Omega \cdot lb/mi^2$		$\Omega \cdot cmil/ft$		$\mu\Omega \cdot cm$	
Pure iron		4,410	....	......	58.83	....	9.78
Contact rails:							
Open hearth			....	100–135	......	16.6–22.4	
Mild to soft steel			....	77–95	......	12.8–15.8	
Telephone and telegraph (galvanized):							
Extra Best Best (EBB)	A111	5,000					
Best Best (BB)	A111	5,600					
Grade 85	A326	5,800					
Grade 135 and 195	A326	6,500					
Commercial galvanized:							
Siemens-Martin			7,280	....	97.9*	....	16.3*
High strength			9,000	....	121*	....	20.1*
Extra-high strength			9,360	....	126*	....	20.9*

*Calculated from average weight resistivity, with average specific gravity 7.83.

NOTE: 1 lb = 0.4536 kg; 1 mi² = 2.589 × 10⁶ m².

perature, the resistance at some other temperature t_2 is expressed by the formula

$$R_{t2} = R_{t1}[1 + \alpha_{t1}(t_2 - t_1)] \tag{2-3}$$

Over wide ranges of temperature the linear relationship of this formula is not usually applicable, and the formula then becomes a series involving higher powers of t, which is unwieldy for ordinary use.

When the temperature of reference t_1 is changed to some other value, the coefficient changes also. Upon assuming the general linear relationship between resistance and temperature previously mentioned, the new coefficient at any temperature t within the linear range is expressed

$$\alpha_t = \frac{1}{(1/\alpha_{t1}) + (t - t_1)} \tag{2-4}$$

The reciprocal of α is termed the inferred absolute zero of temperature. Equation (2-3) takes no account of the change in dimensions with change in temperature and therefore applies to the case of conductors of constant mass, usually met in engineering work. For a more extended discussion of this subject see J. H. Dellinger, The Temperature Coefficient of Resistance of Copper, *NBS Bull.*, 1911, Vol. 8, pp. 71–101; also see *NBS Handbook* 100, Copper Wire Tables.

The coefficient for copper of less than standard (or 100%) conductivity is proportional to the actual conductivity, expressed as a decimal percentage. Thus, if n is the percentage conductivity (95% = 0.95), the temperature coefficient will be $\alpha_t' = n\alpha_t$, where α_t is the coefficient of the annealed copper standard.

The coefficients given in Table 2-3 were computed from the formula

$$\alpha_1 = \frac{1}{[1/n(0.00393)] + (t_1 - 20)} \tag{2-5}$$

The inferred absolute zero of temperature, upon assuming a linear relationship between resistance and temperature, is given as quantity $-T$ in the last column of Table 2-3. At the absolute zero of temperature, the resistance would be zero (see Fig. 2-1).

The coefficient changes with the temperature of reference as shown in Table 2-3.

2-24. Temperature-Resistance Coefficients for Copper. See Table 2-3.

2-25. Temperature-resistance coefficients for copper alloys usually can be approximated by multiplying the corresponding coefficient for copper (100% IACS) by the alloy conduc-

2-8 ELECTRICAL ENGINEERING MATERIALS REFERENCE GUIDE

TABLE 2-3 Temperature-Resistance Coefficients for Aluminum and Copper

Conductivity IACS, %	Temperature, deg C						Temperature $-T$ for inferred-zero resistance,* deg C
	0	15	20	25	30	50	
	Temperature coefficient of resistance, α_t per deg C						
Aluminum							
55	0.00392	0.00370	0.00363	0.00357	0.00351	0.00328	255.2
56	0.00400	0.00377	0.00370	0.00363	0.00357	0.00333	250.3
57	0.00407	0.00384	0.00377	0.00370	0.00363	0.00338	245.6
58	0.00415	0.00391	0.00383	0.00376	0.00369	0.00344	241.0
59	0.00423	0.00398	0.00390	0.00382	0.00375	0.00349	236.6
60	0.00431	0.00404	0.00396	0.00389	0.00381	0.00354	232.3
60.6	0.00435	0.00409	0.00400	0.00393	0.00385	0.00357	229.8
60.97	0.00438	0.00411	0.00403	0.00395	0.00387	0.00359	228.3
61.0	0.00438	0.00411	**0.00403**	0.00395	0.00387	0.00360	228.1
61.2	0.00440	0.00412	0.00404	0.00396	0.00388	0.00360	227.3
61.3	0.00441	0.00413	0.00405	0.00397	0.00389	0.00361	226.9
61.4	0.00441	0.00414	0.00406	0.00398	0.00390	0.00362	226.5
61.5	0.00442	0.00415	0.00406	0.00398	0.00390	0.00362	226.1
61.8	0.00445	0.00417	0.00408	0.00400	0.00392	0.00364	224.9
62.0	0.00446	0.00418	0.00410	0.00401	0.00393	0.00365'	224.1
63	0.00454	0.00425	0.00416	0.00408	0.00400	0.00370	220.3
64	0.00462	0.00432	0.00423	0.00414	0.00406	0.00375	216.5
65	0.00470	0.00439	0.00429	0.00420	0.00412	0.00380	212.9
Copper							
95	0.00403	0.00380	0.00373	0.00367	0.00360	0.00336	247.8
96	0.00408	0.00385	0.00377	0.00370	0.00364	0.00339	245.1
97	0.00413	0.00389	0.00381	0.00374	0.00367	0.00342	242.3
97.5	0.00415	0.00391	0.00383	0.00376	0.00369	0.00344	241.0
98	0.00417	0.00393	0.00385	0.00378	0.00371	0.00345	239.6
99	0.00422	0.00397	0.00389	0.00382	0.00374	0.00348	237.0
100	0.00427	0.00401	**0.00393**	0.00385	0.00378	0.00352	234.5
101	0.00431	0.00405	0.00397	0.00389	0.00382	0.00355	231.9
102	0.00436	0.00409	0.00401	0.00393	0.00385	0.00358	229.5

*See Par. 2-23.

Conductivities 95 to 102 from *NBS Handbook* 100.

Coefficient 0.00403 at 20°C for 61.0% conductivity from ASTM Designation B193; others calculated on same basis.

Boldface type indicates standard values.

tivity expressed as a decimal. For some complex alloys, however, this relation does not hold even approximately, and suitable values should be obtained from the supplier.

2-26. Temperature-resistance coefficient for copper-clad steel wire is 0.00378/°C at 20°C.

2-27. Temperature-Resistance Coefficients for Aluminum. See Table 2-3.

2-28. Temperature-resistance coefficients for aluminum-alloy wires are: for 5005-H19, 0.00353/°C; for 6201-T81, 0.00347/°C at 20°C.

2-29. Temperature-resistance coefficient for aluminum-clad wire is 0.0036/°C at 20°C.

2-30. Temperature-resistance coefficient for pure iron is 0.0064/°C at 20°C. The coefficient, determined by extrapolation from tests on galvanized telephone and telegraph wire, for wire of 100% conductivity at 20°C, is 0.0061/°C.

2-31. Temperature-resistance coefficients for galvanized-steel conductors are as follows:

Commercial grade	Approximate temperature coefficient per °C
Extra Best Best (EBB)	0.0056
Best Best (BB), HTL-85	0.0046
HTL-135, HTL-195	0.0043
High strength	0.0032
Extra-high strength	0.0031

2-32. Temperature-resistance coefficients for typical composite conductors are as follows:

Type	Approximate temperature coefficient per °C at 20°C
Copper—copper-clad steel	0.00381
ACSR (aluminum-steel)	0.00403
Aluminum-aluminum alloy	0.00394
Aluminum—aluminum-clad steel	0.00396

2-33. Reduction of Observations to Standard Temperature. A table of convenient corrections and factors for reducing resistivity and resistance to standard temperature, 20°C, will be found in Copper Wire Tables, *NBS Handbook* 100.

2-34. Resistivity-Temperature Constant. The *change of resistivity per degree* may be readily calculated, taking account of the expansion of the metal with rise of temperature. The proportional relation between temperature coefficient and conductivity may be put in the following convenient form for reducing *resistivity* from one temperature to another: *The change of resistivity of copper per degree Celsius is a constant, independent of the temperature of reference and of the sample of copper.* This "resistivity-temperature constant" may be taken, for general purposes, as $0.00060 \ \Omega$ *(meter, gram)*, or $0.0068 \ \mu\Omega \cdot cm$. More exact values for this constant are given in Table 2-1.

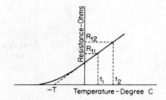

FIG. 2-1 Resistance-temperature relationship.

Details of the calculation of the resistivity-temperature constant will be found in Copper Wire Tables, *NBS Handbook* 100; also see this reference for expressions for the temperature coefficients of resistivity and their derivation.

2-35. Calculation of Percent Conductivity. The percent conductivity of a sample of copper is calculated by dividing the resistivity of the International Annealed Copper Standard at 20°C by the resistivity of the sample at 20°C. Either the mass resistivity or the volume resistivity may be used. Inasmuch as the temperature coefficient of copper varies with the conductivity, it is to be noted that a different value will be found if the resistivity at some other temperature is used. This difference is of practical moment in some cases. In order that such differences may not arise, it is best always to use the 20°C value of resistivity in computing the percent conductivity of copper. When the resistivity of the sample is known at some other temperature t, it is very simply reduced to 20°C by adding the quantity $20 - t$ multiplied by the resistivity-temperature constant, given in Table 2-1.

2-36. Temperature coefficient of expansion (linear) of pure metals over a range of several hundred degrees is not a linear function of the temperature but is well expressed by a quadratic equation

$$\frac{L_{t2}}{L_{t1}} = 1 + [\alpha(t_2 - t_1) + \beta(t_2 - t_1)^2] \tag{2-6}$$

Over the temperature ranges for ordinary engineering work (usually 0 to 100°C), the coefficient can be taken as a constant (assumed linear relationship) and a simplified formula employed:

$$L_{t2} = L_{t1}[1 + \alpha_{t1}(t_2 - t_1)] \tag{2-7}$$

Changes in linear dimensions, superficial area, and volume take place in most materials with changes in temperature. In the case of linear conductors, only the change in length is ordinarily important.

The coefficient for changes in superficial area is approximately twice the coefficient of linear expansion for relatively small changes in temperature. Similarly the volume coefficient is three times the linear coefficient, with similar limitations.

2-37. Temperature-Expansion Coefficients for Conductors. See Table 2-4.

2-38. Specific heat of electrolytic tough pitch copper is 0.092 cal/(g)(°C) at 20°C (see *NBS Circ.* 73).

2-39. Specific heat of aluminum is 0.226 cal/(g)(°C) at room temperature (see *NBS Circ.* C447, Mechanical Properties of Metals and Alloys).

2-40. Specific heat of iron (wrought) or very soft steel from 0 to 100°C is 0.114 cal/(g)(°C); the true specific heat of iron at 0°C is 0.1075 cal/(g)(°C) (see *International Critical Tables,* vol. II, p. 518; also ASM, *Metals Handbook*).

2-41. Thermal conductivity of electrolytic tough pitch copper at 20°C is 0.934 cal/(cm^2)(cm)(s)(°C), adjusted to correspond to an electrical conductivity of 101% (see *NBS Circ.* 73).

2-42. Thermal-Electrical Conductivity Relation of Copper. The Wiedemann-Franz-Lorenz law, which states that the ratio of the thermal and electrical conductivities at a given temperature is independent of the nature of the conductor, holds closely for copper. The ratio $K/\lambda T$ (where K = thermal conductivity, λ = electrical conductivity, T = absolute temperature) for copper is 5.45 at 20°C.

2-43. Thermal Conductivity of Copper Alloys.

ASTM alloy (Spec. B105)	Thermal conductivity (volumetric) at 20 C	
	Btu per sq ft per ft per hr per deg F	Cal per sq cm per cm per sec per deg C
8.5	31	0.13
15	50	0.21
30	84	0.35
55	135	0.56
80	199	0.82
85	208	0.86

2-44. Thermal Conductivity of Aluminum. The determination made by the Bureau of Standards at 50°C for aluminum of 99.66% purity is 0.52 cal/(cm^2)(cm)(s)(°C) (*Circ.* 346; also see *Smithsonian Physical Tables* and *International Critical Tables*).

2-45. Thermal conductivity of iron (mean) from 0 to 100°C is 0.143 cal/(cm^2)(cm)(s)(°C); with increase of carbon and manganese content, it tends to decrease and may reach a figure

TABLE 2-4 Temperature-Expansion Coefficients for Conductors

Conductor type	Temperature coefficient	
	Per °F	Per °C
Copper	9.4 $\times 10^{-6}$	16.92 $\times 10^{-6}$
Copper alloy	9.4 $\times 10^{-6}$	16.92 $\times 10^{-6}$
Copper-clad steel	7.2 $\times 10^{-6}$	12.96 $\times 10^{-6}$
Aluminum	12.8 $\times 10^{-6}$	23.0 $\times 10^{-6}$
Aluminum alloy	12.8 $\times 10^{-6}$	23.0 $\times 10^{-6}$
Aluminum-clad steel	7.2 $\times 10^{-6}$	13.0 $\times 10^{-6}$
Pure iron	6.72 $\times 10^{-6}$	12.1 $\times 10^{-6}$
Galvanized steel:		
Mild steel	6.22 $\times 10^{-6}$	11.2 $\times 10^{-6}$
ACSR core wire	6.4 $\times 10^{-6}$	11.52 $\times 10^{-6}$
HTL-85, HTL-135, HTL-195	5.7 $\times 10^{-6}$	10.26 $\times 10^{-6}$
Composite conductors:		
Copper–copper-clad:		
Type 2A to 6A	8.5 $\times 10^{-6}$	15.30 $\times 10^{-6}$
Type F	9.0 $\times 10^{-6}$	16.20 $\times 10^{-6}$
Type E	8.4 $\times 10^{-6}$	15.12 $\times 10^{-6}$
Type EK	8.8 $\times 10^{-6}$	15.84 $\times 10^{-6}$
Copper–copper alloy	9.4 $\times 10^{-6}$	16.92 $\times 10^{-6}$
Aluminum–aluminum clad:		
AWAC 5/2	10.0 $\times 10^{-6}$	18.00 $\times 10^{-6}$
AWAC 4/3	9.3 $\times 10^{-6}$	16.74 $\times 10^{-6}$
AWAC 3/4	8.6 $\times 10^{-6}$	15.48 $\times 10^{-6}$
AWAC 2/5	8.0 $\times 10^{-6}$	14.40 $\times 10^{-6}$

CONDUCTOR MATERIALS 2-11

of approximately 0.095 with about 1% carbon, or only about half that figure if the steel is hardened by water quenching (see *International Critical Tables,* vol. II, p. 518).

2-46. Influence of Chemical Composition. The resistivity of most metals is very sensitive to changes in chemical composition or constituents. This applies particularly to the case of relatively small amounts of impurities present in metals which approach closely the chemically pure state. Such impurities may consist of oxides, slag, or traces of various foreign ingredients which escape elimination in the process of reducing the original ores to the metallic state, refining these intermediate products, and fabricating the ingots into the form of conductors by various processes of hot and cold working. The combined effects of such impurities are usually summed up in the percentage of conductivity of the conductor expressed as the ratio to the conductivity of chemically pure metal.

2-47. Influence of Mechanical Treatment. The fabrication of conductors, from the ingot to the finished state, normally starts with hot rolling and finishes with cold drawing. Annealing operations may or may not take place at intermediate stages or at the finished state. The cold-working operations tend in general to harden the material, reduce its ductility, increase its tensile strength, and very slightly increase its resistivity. The increase in tensile strength is frequently very useful, and consequently many types of conductor are finished by cold working, in which condition they are usually described as hard-drawn.

2-48. Copper is a highly malleable and ductile metal, of reddish color. It can be cast, forged, rolled, drawn, and machined. Mechanical working hardens it, but annealing will restore it to the soft state. The density varies slightly with the physical state, 8.9 being an average value. It melts at 1083°C (1981°F) and in the molten state has a sea-green color. When heated to a very high temperature, it vaporizes and burns with a characteristic green flame. Copper readily alloys with many other metals. In ordinary atmospheres it is not subject to appreciable corrosion. Its electrical conductivity is very sensitive to the presence of slight impurities in the metal.

Copper when exposed to ordinary atmospheres becomes oxidized, turning to a black color, but the oxide coating is protective, and the oxidizing process is not progressive. When exposed to moist air containing carbon dioxide, it becomes coated with green basic carbonate, which is also protective. At temperatures above 180°C it oxidizes in dry air. In the presence of ammonia it is readily oxidized in air, and it is also affected by sulfur dioxide. Copper is not readily attacked at high temperatures below the melting point by hydrogen, nitrogen, carbon monoxide, carbon dioxide, or steam. Molten copper readily absorbs oxygen, hydrogen, carbon monoxide, and sulfur dioxide, but on cooling, the occluded gases are liberated to a great extent, tending to produce blowholes or porous castings. Copper in the presence of air does not dissolve in dilute hydrochloric or sulfuric acid but is readily attacked by dilute nitric acid. It is also corroded slowly by saline solutions and sea water.

2-49. Commercial grades of copper in the United States are electrolytic, oxygen-free, Lake, fire-refined, and casting. *Electrolytic copper* is that which has been electrolytically refined from blister, converter, black, or Lake copper. *Oxygen-free copper* is produced by special manufacturing processes which prevent the absorption of oxygen during the melting and casting operations or by removing the oxygen by reducing agents. It is used for conductors subjected to reducing gases at elevated temperature where reaction with the included oxygen would lead to the development of cracks in the metal. *Lake copper* is electrolytically or fire-refined from Lake Superior native copper ores and is of two grades, low resistance and high resistance. *Fire-refined copper* is a lower purity grade intended for alloying or for fabrication into products for mechanical purposes; it is not intended for electrical purposes. *Casting copper* is the grade of lowest purity and may consist of furnace-refined copper, rejected metal not up to grade, or melted scrap; it is exclusively a foundry copper.

2-50. Copper content of commercial grades is given in the following table:

Commercial grade	ASTM Designation	Copper content, minimum %
Electrolytic	B5	99.900
Oxygen-free electrolytic	B170	99.95
Lake, low resistance	B4	99.900
Lake, high resistance	B4	99.900
Fire-refined	B216	99.88
Casting	B119	98

2-51. Hardening and Heat-treatment of Copper. There are but two well-recognized methods for hardening copper; one is by mechanically working it, and the other is by the addition of an alloying element. The properties of copper are not affected by a rapid cooling after annealing or rolling, as are those of steel and certain copper alloys.

2-52. Annealing of Copper. Cold-worked copper is softened by annealing, with decrease of tensile strength and increase of ductility. In the case of pure copper hardened by cold reduction of area to one-third of its initial area, this softening takes place with maximum rapidity between 200 and 325°C. However, this temperature range is affected in general by the extent of previous cold reduction and the presence of impurities. The greater the previous cold reduction, the lower is the range of softening temperatures. The effect of iron, nickel, cobalt, silver, cadmium, tin, antimony, and tellurium is to lower the conductivity and raise the annealing range of pure copper in varying degrees (see Effect of Iron, Cobalt and Nickel on Some Properties of High-purity Copper and Effect of Certain Fifth-period Elements on Some Properties of High-Purity Copper, by J. S. Smart, Jr., and A. A. Smith, Jr., *Trans. AIME,* 1942, vol. 147, p. 48, and 1943, vol. 152, p. 103). Oxygen content lowers the annealing range.

Trade-named coppers, such as *Hy-Therm, Tensilok,* and *High Thermo,* are produced by adding minute amounts of hardening agents to pure electrolytic copper. These coppers meet industry specifications for purity and conductivity but have higher annealing characteristics than commercial brands, as illustrated in Fig. 2-2. Their higher resistance to annealing permits higher continuous and short-time emergency overload current-carrying capacity of conductors used in overhead lines (see Hy-Therm Copper—An Improved Overhead-Line Conductor, by L. F. Hickernell, A. A. Jones, and C. J. Snyder, *Trans. AIEE,* 1949, vol. 68, pt. 1, p. 22).

2-53. Alloying of Copper. Elements that are soluble in moderate amounts in a solid solution of copper, such as manganese, nickel, zinc, tin, and aluminum, generally harden it and diminish its ductility but improve its rolling and working properties. Elements that are but slightly soluble, such as bismuth and lead, do not harden it but diminish both the ductility and the toughness and impair its hot-working properties. See *NBS Circ.* 73, 2d ed., 1922, pp. 53–68, for extended discussion. Small additions (up to 1.5%) of manganese, phosphorus, or tin increase the tensile strength and hardness of cold-rolled copper.

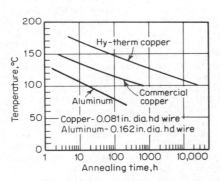

FIG. 2-2 Annealing characteristics of copper and aluminum for 5% loss in strength, no tension.

Brass is usually a binary alloy of copper and zinc, but brasses are seldom employed as electrical conductors, as they have relatively low conductivity though comparatively high tensile strength. In general, brass is not suitable for use where exposed to the weather, owing to the difficulty from stress-corrosion cracking; the higher the zinc content, the more pronounced does this become.

Bronze in its simplest form is a binary alloy of copper and tin, in which the latter element is the hardening and strengthening agent. This material is rather old in the arts and has been used to some extent for electrical conductors for many years past, especially abroad. Modern bronzes are frequently ternary alloys, containing as the third constituent such elements as phosphorus, silicon, manganese, zinc, aluminum, or cadmium; in such cases the third element is usually given in the name of the alloy, as phosphor bronze, silicon bronze. Certain bronzes are quaternary alloys or contain two other elements in addition to copper and tin.

In bronzes for use as electrical conductors the content of tin and other metals is usually less than in bronzes for structural or mechanical applications where physical properties and resistance to corrosion are the governing considerations. High resistance to atmospheric corrosion is always an important consideration in selecting bronze conductors for overhead service.

2-54. Commercial Grades of Bronze. Various bronzes have been developed for use as conductors, and these are now covered by ASTM Specification B105. They all have been

CONDUCTOR MATERIALS 2-13

designed to provide conductors having high resistance to corrosion and tensile strengths greater than hard-drawn copper conductors. The standard specification covers 10 grades of bronze, designated by numbers according to their conductivities.

2-55. Copper-Chromium Alloy. A patented alloy of this type contains from 0.5 to 3.0% of chromium. When cast, rolled, heat-treated, and drawn it may have an electrical conductivity of 80% and a tensile strength of 72,000 to 80,000 lb/in^2.

2-56. Copper-beryllium alloy containing 0.4% of beryllium may have an electrical conductivity of 48% and a tensile strength (in 0.128-in wire) of 86,000 lb/in^2. A content of 0.9% of beryllium may give a conductivity of 28% and a tensile strength of 122,000 lb/in^2. The effect of this element in strengthening copper is about ten times as great as that of tin.

2-57. Copper-clad steel wires have been manufactured by a number of different methods. The general object sought in the manufacture of such wires is the combination of the high conductivity of copper with the high strength and toughness of iron or steel.

The principal manufacturing processes now in commercial use are: (a) coating a steel billet with a special flux, placing it in a vertical mold closed at the bottom, heating the billet and mold to yellow heat, and then casting molten copper around the billet, after which it is hot-rolled to rods and cold-drawn to wire, and (b) electroplating a dense coating of copper on a steel rod and then cold drawing to wire.

2-58. Aluminum is a ductile metal, silver-white in color, which can be readily worked by rolling, drawing, spinning, extruding, and forging. Its specific gravity is 2.703. Pure aluminum melts at 660°C (1220°F). Aluminum has relatively high thermal and electrical conductivities. The metal is always covered with a thin, invisible film of oxide which is impermeable and protective in character. Aluminum, therefore, shows stability and long life under ordinary atmospheric exposure.

Exposure to atmospheres high in hydrogen sulfide or sulfur dioxide does not cause severe attack of aluminum at ordinary temperatures, and for this reason aluminum or its alloys can be used in atmospheres which would be rapidly corrosive to many other metals.

Aluminum parts should, as a rule, not be exposed to salt solutions while in electrical contact with copper, brass, nickel, tin, or steel parts, since galvanic attack of the aluminum is likely to occur. Contact with cadmium in such solutions results in no appreciable acceleration in attack on the aluminum, while contact with zinc (or zinc-coated steel as long as the coating is intact) is generally beneficial, since the zinc is attacked selectively and cathodically protects adjacent areas of the aluminum.

Most organic acids and their water solutions have little or no effect on aluminum at room temperature, although oxalic acid is an exception and is corrosive. Concentrated nitric acid (about 80% by weight) and fuming sulfuric acid can be handled in aluminum containers. However, more dilute solutions of these acids are more active. All but the most dilute (less than 0.1%) solutions of hydrochloric and hydrofluoric acids have a rapid etching action on aluminum.

Solutions of the strong alkalies, potassium, or sodium hydroxides dissolve aluminum rapidly. However, ammonium hydroxide and many of the strong organic bases have little action on aluminum and are successfully used in contact with it (see NBS Circ. 346).

Aluminum in the presence of water and limited air or oxygen rapidly converts into aluminum hydroxide, a whitish powder.

2-59. Commercial grades of aluminum in the United States are designated by their purity, such as 99.99, 99.95, 99.90%.

2-60. Electrical conductor alloy aluminum 1350, having a purity of approximately 99.5% and a minimum conductivity of 61.0% IACS, is used for conductor purposes. Specified physical properties are obtained by closely controlling the kind and amount of certain impurities.

2-61. Annealing of Aluminum. Cold-worked aluminum is softened by annealing, with decrease of tensile strength and increase of ductility. The annealing temperature range is affected in general by the extent of previous cold reduction and the presence of impurities. The greater the previous cold reduction, the lower is the range of softening temperatures. Typical annealing characteristics of 1350 aluminum alloy are shown in Fig. 2-2.

2-62. Alloying of Aluminum. Aluminum can be alloyed with a variety of other elements, with a consequent increase in strength and hardness. With certain alloys, the strength can be further increased by suitable heat-treatment. The alloying elements most generally used are copper, silicon, manganese, magnesium, chromium, and zinc. Some of the aluminum alloys, particularly those containing one or more of the following elements—

copper, magnesium, silicon, and zinc—in various combinations, are susceptible to heat-treatment.

Pure aluminum, even in the hard-worked condition, is a relatively weak metal for construction purposes. Strengthening for castings is obtained by alloying elements. The alloys most suitable for cold-rolling seldom contain less than 90 to 95% aluminum. By alloying, working, and heat-treatment it is possible to produce tensile strengths ranging from 8500 lb/in^2 for pure annealed aluminum up to 82,000 lb/in^2 for special wrought heat-treated alloy, with densities ranging from 2.65 to 3.00.

2-63. Electrical conductor alloys of aluminum are principally alloys 5005 and 6201 covered by ASTM Specifications B396 and B398.

2-64. Aluminum-clad steel wires have a relatively heavy layer of aluminum surrounding and bonded to the high-strength steel core. The aluminum layer can be formed by compacting and sintering a layer of aluminum powder over a steel rod, by electroplating a dense coating of aluminum on a steel rod, or extruding a coating of aluminum on a steel rod, and then cold drawing to wire.

2-65. Magnesium is a ductile metal, silver-white in color, which is distinguished by its light weight (sp. gr. 1.74) and ease of machining.

Pure magnesium has relatively low strength; so its uses are limited to applications where strength is of little importance. However, by alloying magnesium with small amounts of other metals, particularly aluminum, zinc, or manganese or combinations of these, alloys have been developed which show excellent mechanical properties and lead to the high strength-weight ratios of the magnesium products. In a general way the mechanical properties compare favorably with those of the commercial aluminum alloys. The various alloys are divided into those suitable for castings and for wrought products. A number of the standard alloys are susceptible to heat-treatment to improve the properties in general or for the improvement of some particular property.

Magnesium alloys are resistant to attack by alkalies and many organic chemicals. They are attacked by acids of any strength with the exception of pure hydrofluoric and chromic acids. Salt solutions in general corrode the metal, and applications involving continuous contact with saline solutions are not recommended. The alloys are commercially stable against ordinary atmospheric conditions, and for many uses the surface needs no protection.

A number of coatings applied by chemical treatment have been developed to provide a surface stability suitable for more difficult situations.

Pure magnesium is used for ingot, powder, shavings, extruded wire and strip, and rolled ribbon. Sand, permanent-mold, and die castings are available in magnesium alloys. Wrought magnesium alloys are available in extruded round, square, hexagonal, and rectangular bar; as special shapes, moldings, and structural sections; as rolled plate and sheet; and as hammer and hot-pressed forgings.

2-66. Silicon is a light metal having a specific gravity of approximately 2.34. There is lack of accurate data on the pure metal, because its mechanical brittleness bars it from most industrial uses. However, it is very resistant to atmospheric corrosion and to attack by many chemical reagents. Silicon is of fundamental importance in the steel industry, but for this purpose it is obtained in the form of ferrosilicon, which is a coarse granulated or broken product. It is very useful as an alloying element in steel for electrical sheets and substantially increases the electrical resistivity and thereby reduces the core losses. Silicon is peculiar among metals in the respect that its temperature coefficient of resistance may change sign in some temperature range, the exact behavior varying with the impurities.

2-67. Beryllium is a light metal having a specific gravity of approximately 1.84 or nearly the same as magnesium. It is normally hard and brittle and difficult to fabricate. Copper is materially strengthened by the addition of small amounts of beryllium, without very serious loss of electrical conductivity. The principal uses for this metal appear to be as an alloying element with other metals, such as aluminum and copper.

2-68. Sodium is a soft, bright, silvery metal obtained commercially by the electrolysis of absolutely dry fused sodium chloride. It is the most abundant of the alkali group of metals, is extremely reactive, and is never found free in nature. It oxidizes readily and rapidly in air. In the presence of water (it is so light that it floats) it may ignite spontaneously, decomposing the water with evolution of hydrogen and formation of sodium hydroxide. This can be explosive. Sodium should be handled with respect, as it can be dangerous when improperly handled. It melts at 97.8°C, below the boiling point of water, and in the same

CONDUCTOR MATERIALS 2-15

range as many fuse metal alloys. Sodium is approximately one-tenth as heavy as copper and has roughly three-eighths the conductivity; hence 1 lb of sodium is about equal electrically to 3½ lb of copper. Interest in sodium as an electrical conductor recently was renewed with the development of a means of extruding a polyethylene insulating tube, simultaneously filling it with liquid sodium fed through a tube from a closed container, and cooling them together.

2-69. Iron and Steel. Iron is a hard tenacious metal which has a silvery-white luster and takes a high polish. It is strongly attracted by a magnet but retains practically no magnetism. It softens at a red heat and may be readily welded at a white heat. Its melting point is higher than that of steel or wrought iron. Iron is very reactive chemically and dissolves in most dilute acids with liberation of hydrogen. In dry air it undergoes no change, but when exposed to atmospheres containing moisture it corrodes more or less rapidly and forms rust (hydrated ferric oxide); such corrosion is accelerated by the presence of carbon dioxide and sulfur dioxide. There are three oxides of iron: FeO, or ferrous oxide, is a black powder; Fe_2O_3, or ferric oxide, is known also as hematite and is a steel-gray crystalline substance with considerable luster; Fe_3O_4, or ferroso-ferric oxide, is also known as magnetite and is attracted by a magnet but not always magnetic of itself. Iron forms numerous compounds with carbon, sulfur, phosphorus, oxygen, hydrogen, nitrogen, and other elements; it alloys readily with numerous metallic elements such as manganese, silicon, nickel, cobalt, chromium, and tungsten.

The element iron is the base of all commercial iron and steel products, but in order to define these products in reasonably complete terms it is desirable to state both their constituents and the processes by which they were made. Iron and steel are always different substances in the commercial sense and moreover are manufactured in many different commercial varieties. Even the common varieties of steel are somewhat complicated substances; it is insufficient to describe them as alloys of iron and certain other elements, because the alloys take many different forms and impart certain distinctive properties, depending on both the alloy proportions and the processes of manufacture.

2-70. Protective Coatings for Iron and Steel. Iron and steel wires for outdoor service as conductors or guys are protected from corrosion by the application of zinc or aluminum coatings. Such coatings may be applied by the hot-dip process or by electroplating. These coatings themselves are not impregnable against atmospheric attack but afford the best protection which has yet been found practicable.

2-71. Galvanized-iron telephone- and telegraph-line wire is available commercially in two grades, EBB and BB, and with three weights of zinc coating. Characteristics are given in ASTM Specification A111. See Tables 2-2 and 2-25.

2-72. Galvanized high-tensile steel telephone- and telegraph-line wire is available commercially in three grades, 85, 135, and 195, and with three weights of zinc coating. Characteristics are given in ASTM Specification A326. See Tables 2-2 and 2-25.

2-73. Steel Rails. The resistivities of steel rails are given in Table 2-2.

The effective resistance of steel rails to alternating currents is increased on account of skin effect, which in turn is a function of magnetic permeability (see data in "Report of the Electric Railway Test Commission," New York, McGraw-Hill Book Company, 1906; also Experimental Researches on the Skin Effect in Steel Rails, by A. E. Kennelly, F. H. Achard, and A. S. Dana, *J. Franklin Inst.,* August 1916).

Conductor Properties

2-74. Electrical conductors are manufactured in various forms and shapes for various purposes. These may be wires, cables, flat straps, square or rectangular bars, angles, channels, or special designs for particular requirements. The most extensive use of conductors, however, is in the form of round solid wires and stranded conductors. The following terminology describes properly the various terms relating to conductors.

2-75. Definitions of Electrical Conductors

Wire. A rod or filament of drawn or rolled metal whose length is great in comparison with the major axis of its cross section.

The definition restricts the term to what would ordinarily be understood by the term "solid wire." In the definition, the word "slender" is used in the sense that the length is great

2-16   ELECTRICAL ENGINEERING MATERIALS REFERENCE GUIDE

in comparison with the diameter. If a wire is covered with insulation, it is properly called an "insulated wire"; while primarily the term "wire" refers to the metal, nevertheless when the context shows that the wire is insulated, the term "wire" will be understood to include the insulation.

Conductor. A wire or combination of wires not insulated from one another, suitable for carrying an electric current.

The term "conductor" is not to include a combination of conductors insulated from one another, which would be suitable for carrying several different electric currents.

Rolled conductors (such as busbars) are, of course, conductors but are not considered under the terminology here given.

Stranded Conductor. A conductor composed of a group of wires, usually twisted, or any combination of groups of wires.

The wires in a stranded conductor are usually twisted or braided together.

Cable. A stranded conductor (single-conductor cable) or a combination of conductors insulated from one another (multiple-conductor cable).

The component conductors of the second kind of cable may be either solid or stranded, and this kind of cable may or may not have a common insulating covering. The first kind of cable is a single conductor, while the second kind is a group of several conductors. The term "cable" is applied by some manufacturers to a solid wire heavily insulated and lead covered; this usage arises from the manner of the insulation, but such a conductor is not included under this definition of "cable." The term "cable" is a general one and in practice it is usually applied only to the larger sizes. A small cable is called a "stranded wire" or a "cord," both of which are defined below. Cables may be bare or insulated, and the latter may be armored with lead or with steel wires or bands.

Strand. One of the wires of any stranded conductor.

Stranded Wire. A group of small wires used as a single wire.

A wire has been defined as a slender rod or filament of drawn metal. If such a filament is subdivided into several smaller filaments or strands and is used as a single wire, it is called "stranded wire." There is no sharp dividing line of size between a "stranded wire" and a "cable." If used as a wire, for example, in winding inductance coils or magnets, it is called a stranded wire and not a cable. If it is substantially insulated, it is called a "cord," defined below.

Cord. A small cable, very flexible and substantially insulated to withstand wear.

There is no sharp dividing line in respect to size between a cord and a cable, and likewise no sharp dividing line in respect to the character of insulation between a cord and a stranded wire. Usually the insulation of a cord contains rubber.

Concentric Strand. A strand composed of a central core surrounded by one or more layers of helically laid wires or groups of wires.

Concentric-Lay Conductor. Conductor constructed with a central core surrounded by one or more layers of helically laid wires.

Rope-Lay Conductor. Conductor constructed of a bunch-stranded or a concentric-stranded member or members, as a central core, around which are laid one or more helical layers of such members.

N-Conductor Cable. A combination of N conductors insulated from one another.

It is not intended that the name as here given be actually used. One would instead speak of a "3-conductor cable," a "12-conductor cable," etc. In referring to the general case, one may speak of a "multiple-conductor cable."

N-Conductor Concentric Cable. A cable composed of an insulated central conducting core with N-1 tubular-stranded conductors laid over it concentrically and separated by layers of insulation.

This kind of cable usually has only two or three conductors. Such cables are used in carrying alternating currents.

The remark on the expression "N conductor" given for the preceding definition applies here also. (Additional definitions can be found in ASTM B354.)

2-76. Wire sizes have been for many years indicated in commercial practice almost entirely by gage numbers, especially in America and England. This practice is accompanied by some confusion because numerous gages are in common use. The most commonly used gage for electrical wires, in America, is the *American wire gage*. The most commonly used gage for steel wires is the *Birmingham wire gage*.

CONDUCTOR MATERIALS 2-17

There is no legal standard wire gage in this country, although a gage for sheets was adopted by Congress in 1893. In England there is a legal standard known as the *Standard wire gage*. In Germany, France, Austria, Italy, and other Continental countries practically no wire gage is used, but wire sizes are specified directly in millimeters. This system is sometimes called the *Millimeter wire gage*. The wire sizes used in France, however, are based to some extent on the old Paris gage (*jauge de Paris de* 1857) (for a history of wire gages see *NBS Handbook* 100, Copper Wire Tables; also see *Circ.* 67, Wire Gages, 1918).

There is a tendency to *abandon gage numbers* entirely and specify wire sizes by the *diameter in mils* (thousandths of an inch). This practice holds particularly in writing specifications and has the great advantages of being both simple and explicit. A number of the wire manufacturers also encourage this practice, and it was definitely adopted by the U.S. Navy Department in 1911.

2-77. *Mil* is a term universally employed in this country to measure wire diameters and is a unit of length equal to one-thousandth of an inch.

2-78. *Circular mil* is a term universally used to define cross-sectional areas, being a unit of area equal to the area of a circle 1 mil in diameter. Such a circle, however, has an area of 0.7854 (or $\pi/4$) mil^2. Thus a wire 10 mils in diameter has a cross-sectional area of 100 cmils or 78.54 $mils^2$. Hence, a cmil equals 0.7854 mil^2.

2-79. *American wire gage,* also known as the *Brown & Sharpe gage,* was devised in 1857 by J. R. Brown. It is usually abbreviated AWG. This gage has the property, in common with a number of other gages, that its sizes represent approximately the successive steps in the process of wire drawing. Also, like many other gages, its numbers are retrogressive, a larger number denoting a smaller wire, corresponding to the operations of drawing. These gage numbers are not arbitrarily chosen, as in many gages, but follow the mathematical law upon which the gage is founded.

Basis of the AWG is a simple mathematical law. The gage is formed by the specification of two diameters and the law that a given number of intermediate diameters are formed by geometrical progression. Thus, the diameter of No. 0000 is defined as 0.4600 in and of No. 36 as 0.0050 in. There are 38 sizes between these two; hence the ratio of any diameter to the diameter of the next greater number is given by this expression:

$$\sqrt[39]{\frac{0.4600}{0.0050}} = \sqrt[39]{92} = 1.122\ 932\ 2 \tag{2-8}$$

The square of this ratio = 1.2610. The sixth power of the ratio, that is, the ratio of any diameter to the diameter of the sixth greater number, = 2.0050. The fact that this ratio is so nearly 2 is the basis of numerous useful relations or short cuts in wire computations.

There are a number of approximate rules applicable to the AWG which are useful to remember:

1. An increase of three gage numbers (for example, from No. 10 to 7) doubles the area and weight and consequently halves the dc resistance.

2. An increase of six gage numbers (for example, from No. 10 to 4) doubles the diameter.

3. An increase of 10 gage numbers (for example, from No. 10 to 1/0) multiplies the area and weight by 10 and divides the resistance by 10.

4. A No. 10 wire has a diameter of about 0.10 in, an area of about 10,000 cmils, and (for standard annealed copper at 20°C) a resistance of approximately 1.0 Ω/1000 ft.

5. The weight of No. 2 copper wire is very close to 200 lb/1000 ft.

2-80. *Steel wire gage,* also known originally as the *Washburn & Moen gage* and later as the *American Steel & Wire Co.'s gage,* was established by Ichabod Washburn about 1830. This gage also, with a number of its sizes rounded off to thousandths of an inch, is known as the *Roebling gage*. It is used exclusively for steel wire and is frequently employed in wire mills.

2-81. *Birmingham wire gage,* also known as *Stubs' wire gage* and *Stubs' iron wire gage,* is said to have been established early in the eighteenth century in England, where it was long in use. This gage was used to designate the Stubs soft-wire sizes and should not be confused with Stubs' steel-wire gage. The numbers of the Birmingham gage were based upon the reductions of size made in practice by drawing wire from rolled rod. Thus, a wire rod was called "No. 0," "first drawing No. 1," and so on. The gradations of size in this gage are not regular, as will appear from its graph. This gage is generally in commercial use in the United States for iron and steel wires.

2-82. Standard wire gage, which more properly should be designated *(British) Standard wire gage,* is the legal standard of Great Britain for all wires, adopted in 1883. It is also known as the *New British Standard gage,* the *English legal standard gage,* and the *Imperial wire gage.* It was constructed by so modifying the Birmingham gage that the differences between consecutive sizes become more regular. This gage is largely used in England but never has been used extensively in America.

2-83. Old English wire gage, also known as the *London wire gage,* differs very little from the Birmingham gage. It formerly was used to some extent for brass and copper wires but is now nearly obsolete.

2-84. Millimeter wire gage, also known as the *metric wire gage,* is based on giving progressive numbers to the progressive sizes, calling 0.1 mm diameter "No. 1," 0.2 mm "No. 2," etc.

2-85. German wire gage, in which the diameter or thickness is expressed in millimeters, is retrogressive and contains 25 sizes.

German Wire Gage Table

(Diameters in millimeters)

No.	Diam.	No.	Diam.	No.	Diam.	No.	Diam.	No.	Diam.
1	5.50	6	3.75	11	2.50	16	1.375	21	0.750
2	5.00	7	3.50	12	2.25	17	1.250	22	0.625
3	4.50	8	3.25	13	2.00	18	1.125	23	0.562
4	4.25	9	3.00	14	1.75	19	1.000	24	0.500
5	4.00	10	2.75	15	1.50	20	0.875	25	0.438

2-86. Conductor-Size Designation. *America* uses, for sizes up to 4/0, mil, decimals of an inch, or AWG numbers for solid conductors and AWG numbers or circular mils for stranded conductors; for sizes larger than 4/0, circular mils is used throughout. Other countries ordinarily use square millimeter area.

2-87. Conductor-size conversion can be accomplished from the following relation:

$$\text{cmils} = \text{in}^2 \times 1{,}273{,}200 = \text{mm}^2 \times 1973.5 \tag{2-9}$$

2-88. Measurement of wire diameters may be accomplished in many ways, but most commonly by means of a micrometer caliper. Stranded cables usually are measured by means of a circumference tape calibrated directly in diameter readings.

2-89. Comparison of Wire Gages. See Table 2-11.

2-90. Stranded conductors are used generally because of their increased flexibility and consequent ease in handling. The greater the number of wires in any given cross section, the greater will be the flexibility of the finished conductor. Most conductors above 4/0 AWG in size are stranded. Generally, in a given concentric-lay stranded conductor, all wires are of the same size and the same material, although special conductors are available embodying wires of different sizes and of different materials. The former will be found in some insulated cables, and the latter in overhead stranded conductors combining high-conductivity and high-strength wires.

The flexibility of any given size of strand obviously increases as the total number of wires increases. It is common practice to increase the total number of wires as the strand diameter increases, in order to provide reasonable flexibility in handling. So-called *flexible concentric strands* for use in insulated cables have about one to two more layers of wires than the standard type of strand for ordinary use.

2-91. Number of Wires in Stranded Conductors. Each successive layer in a concentri-

CONDUCTOR MATERIALS

cally stranded conductor contains six more wires than the preceding one. The total number of wires in a conductor is

For 1-wire core constructions (1, 7, 19, etc.),

$$N = 3n(n + 1) + 1 \qquad (2\text{-}10)$$

For 3-wire core constructions (3, 12, etc.),

$$N = 3n(n + 2) + 3 \qquad (2\text{-}11)$$

when n is number of layers over core, which is not counted as a layer.

2-92. Wire size in stranded conductors is

$$d = \sqrt{\frac{A}{N}} \qquad (2\text{-}12)$$

where A = total conductor area in circular mils and N = total number of wires.

Copper cables are manufactured usually to certain cross-sectional sizes specified in total circular mils or by gage numbers in AWG. This necessarily requires individual wires drawn to certain prescribed diameters, which are different as a rule from normal sizes in AWG (see Table 2-17).

2-93. Diameter of stranded conductors (circumscribing circle) is

$$D = d(2n + k) \qquad (2\text{-}13)$$

where d = diameter of individual wire, n = number of layers over core which is not counted as a layer, k = 1 for constructions having 1-wire core (1, 7, 19, etc.), and k = 2.155 for constructions having 3-wire core (3, 12, etc.).

For standard concentric-lay stranded conductors, the following rule gives a simple method of determining the outside diameter of a stranded conductor from the known diameter of a solid wire of the same cross-sectional area.

To obtain the diameter of concentric-lay stranded conductor, multiply the diameter of the solid wire of the same cross-sectional area by the appropriate factor as follows:

Number of wires	Factor	Number of wires	Factor
3	1.244	91	1.153
7	1.134	127	1.154
12	1.199	169	1.154
19	1.147	217	1.154
37	1.151	271	1.154
61	1.152		

2-94. Area of stranded conductors is

$$A = Nd^2 \text{ cmils} = \tfrac{1}{4}\pi Nd^2 \times 10^{-6} \text{ in}^2 \qquad (2\text{-}14)$$

where N = total number of wires and d = individual wire diameter in mils.

2-95. Effects of Stranding. All wires in a stranded conductor except the core wire form continuous helices, of slightly greater length than the axis or core. This causes slight increase in weight and electrical resistance and slight decrease in tensile strength and sometimes affects the internal inductance, as compared theoretically with a conductor of equal dimensions but composed of straight wires parallel with the axis.

2-96. Lay, or Pitch. The axial length of one complete turn, or helix, of a wire in a stranded conductor is sometimes termed the lay, or pitch. This is often expressed as the *pitch ratio,* which is the ratio of the length of the helix to its *pitch diameter* (diameter of the helix at the centerline of any individual wire or strand equals the outside diameter of the helix minus the thickness of one wire or strand). If there are several layers, the pitch expressed as an axial length may increase with each additional layer, but when expressed as the ratio of axial length to pitch diameter of helix, it is usually the same for all layers, or nearly so. In commercial practice, the pitch is commonly expressed as the ratio of axial length to outside diameter of helix, but this is an arbitrary designation made for convenience of usage. The *pitch angle* is shown in Fig. 2-3, where *ac* represents the axis of the stranded conductor and *l* is the axial length of one complete turn or helix, *ab* is the length of any individual wire $l + \Delta l$ in one complete turn, and *bc* is equal to the circumference of a circle corresponding to the pitch diameter *d* of the helix. The angle *bac*, or θ, is the pitch angle, and the pitch ratio is expressed by $p = l/d$. There is no standard pitch ratio used by manufacturers generally, since it has been found desirable to vary this depending on the type of service for which the conductor is intended. Applicable lay lengths generally are included in industry specifications covering the various stranded conductors. For bare overhead conductors, a representative commercial value for pitch length is 13.5 times the outside diameter of each layer of strands.

FIG. 2-3 Pitch angle in concentric-lay cable.

2-97. Direction of Lay. The direction of lay is the lateral direction in which the individual wires of a cable run over the top of the cable as they recede from an observer looking along the axis. *Right-hand lay* recedes from the observer in clockwise rotation or like a right-hand screw thread; *left-hand lay* is the opposite. The outer layer of a cable is ordinarily applied with a right-hand lay for bare overhead conductors and left-hand lay for insulated conductors, although the opposite lay can be used if desired.

2-98. Increase in Weight Due to Stranding. Referring to Fig. 2-3, the increase in weight of the spiral members in a cable is proportional to the increase in length:

$$\frac{l + \Delta l}{l} = \sec \theta = \sqrt{1 + \tan^2 \theta}$$

$$= \sqrt{1 + \frac{\pi^2}{p^2}} = 1 + \frac{1}{2}\frac{\pi^2}{p^2} - \frac{1}{8}\left(\frac{\pi^2}{p^2}\right)^2 + \cdots$$

(2-15)

As a first approximation this ratio equals $1 + 0.5(\pi^2/p^2)$, and a pitch of 15.7 produces a ratio of 1.02. This correction factor should be computed separately for each layer if the pitch *p* varies from layer to layer. Practical correction factors for stranded copper conductors are given in Table 2-17.

2-99. Increase in Resistance Due to Stranding. If it were true that no current flows from wire to wire through their lineal contacts, the proportional increase in the total resistance would be the same as the proportional increase in total weight. If all the wires were in perfect and complete contact with each other, the total resistance would decrease in the same proportion that the total weight increases, owing to the slightly increased normal cross section of the cable as a whole. The contact resistances are normally sufficient to make the actual increase in total resistance nearly as much, proportionately, as the increase in total weight, and for practical purposes they are usually assumed to be the same. See Table 2-17.

2-100. Decrease in Strength Due to Stranding. When a concentric-lay cable is subjected to mechanical tension, the spiral members tend to tighten around those layers under them and thus produce internal compression, gripping the inner layers and the core. Consequently the individual wires, taken as a whole, do not behave as they would if they were true linear conductors acting independently. Furthermore, the individual wires are never exactly alike in diameter or in strength or in elastic properties. For these reasons there is ordinarily a loss of about 4 to 11% in total tensile efficiency, depending on the number of layers. This reduction tends to increase as the pitch ratio decreases. Actual tensile tests on cables furnish the most dependable data on their ultimate strength.

2-101. Tensile efficiency of a stranded conductor is the ratio of its breaking strength to the sum of the tensile strengths of all its individual wires. Concentric-lay cables of 12 to 16 pitch ratio have a normal tensile efficiency of approximately 90%; rope-lay cables, approximately 80%.

CONDUCTOR MATERIALS

2-102. Preformed Cable. This type of cable is made by preforming each individual wire (except the core) into a spiral of such length and curvature that the wire will fit naturally into its normal position in the cable instead of being forced into that shape under the usual tension in the stranding machine. This method has the advantage in cable made of the stiffer grades of wire that the individual wires do not tend to spread or untwist if the strand is cut in two without first binding the ends on each side of the cut.

2-103. Weight. A uniform cylindrical conductor of diameter d, length l, and density δ has a total weight expressed by the formula

$$W = \delta l \frac{\pi d^2}{4} \tag{2-16}$$

The weight of any conductor is commonly expressed in pounds per unit of length, such as 1 ft, 1000 ft, or 1 mi. The weight of stranded conductors can be calculated using Eq. (2-16), but allowance must be made for increase in weight due to stranding (see Par. **2-98**). Rope-lay stranding has greater increase in weight because of the multiple stranding operations. As an example, industry standards for increase in weight of *stranded copper conductors* are as follows:

ASTM stranding classes: AA, A, B, C, D	Increase, %	ASTM stranding classes: G, H	Increase, %	ASTM stranding classes: I, J, K, L, M, O, P, Q	Increase, %
1 to 4 AWG (AA only)	1	49 wires or less....	3	Single bunched strand.	2
Up to 2,000 Mcm.....	2	133 wires..........	4	7 ropes of bunched strand............	4
Over 2,000–3,000 Mcm.	3	259 wires..........	4.5	19 ropes of bunched strand............	5
Over 3,000–4,000 Mcm.	4	427 wires..........	5	7 × 7 ropes of bunched strand............	6
Over 4,000–5,000 Mcm.	5	Over 427 wires....	6	19, 87 or 61 × 7 ropes of bunched strand...	7

2-104. Breaking Strength. The maximum load that a conductor attains when tested in tension to rupture.

2-105. Total Elongation at Rupture. When a sample of any material is tested under tension until it ruptures, measurement is usually made of the total elongation in a certain initial test length. In certain kinds of testing, the initial test length has been standardized, but in every case, the total elongation at rupture should be referred to the initial test length of the sample on which it was measured. Such elongation is usually expressed in percentage of original unstressed length and is a general index of the ductility of the material. Elongation is determined on solid conductors or on individual wires before stranding; it is rarely determined on stranded conductors.

2-106. Elasticity. All materials are deformed in greater or lesser degree under application of mechanical stress. Such deformation may be either of two kinds, known, respectively, as "elastic deformation" and "permanent deformation." When a material is subjected to stress and undergoes deformation but resumes its original shape and dimensions when the stress is removed, the deformation is said to be elastic. If the stress is so great that the material fails to resume its original dimensions when the stress is removed, the permanent change in dimensions is termed permanent deformation or "set." In general the stress at which appreciable permanent deformation begins is termed the "working elastic limit." Below this limit of stress the behavior of the material is said to be elastic, and in general the deformation is proportional to the stress.

2-107. Stress and Strain. The stress in a material under load, as in simple tension or compression, is defined as the total load divided by the area of cross section normal to the direction of the load, assuming the load to be uniformly distributed over this cross section. It is commonly expressed in pounds per square inch. The strain in a material under load is defined as the total deformation measured in the direction of the stress, divided by the total unstressed length in which the measured deformation occurs, or the deformation per unit length. It is expressed as a decimal ratio or numeric.

In order to show the complete behavior of any given conductor under tension, it is customary to make a graph in terms of loading or stress as the ordinates and elongation or strain as the abscissas. Such graphs or curves are useful in determining the elastic limit and the yield point if the loading is carried to the point of rupture. Graphs showing the relationship between stress and strain in a material tested to failure are termed load-deformation or stress-strain curves.

2-108. Hooke's law consists of the simple statement that the stress is proportional to the strain. It obviously implies a condition of perfect elasticity, which is true only for stresses less than the elastic limit.

2-109. Stress-Strain Curves. A typical stress-strain diagram of hard-drawn copper wire is shown in Fig. 2-4, which represents No. 9 AWG. The curve *ae* is the actual stress-strain curve; *ab* represents the portion which corresponds to true elasticity, or for which Hooke's law holds rigorously; *cd* is the tangent *ae* which fixes the Johnson elastic limit (see Par. **2-113**); and the curve *af* represents the set, or permanent elongation due to flow of the metal under stress, being the difference between *ab* and *ae*.

A typical stress-strain diagram of hard-drawn aluminum wire, based on data furnished by the Aluminum Company of America is shown in Fig. 2-5.

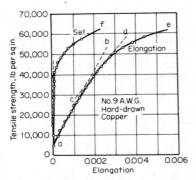

FIG. 2-4 Stress-strain curves of No. 9 AWG hard-drawn copper wire (Watertown Arsenal Test).

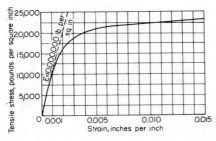

FIG. 2-5 Typical stress-strain curve of hard-drawn aluminum wire.

2-110. Modulus (or coefficient) of elasticity is the ratio of internal stress to the corresponding strain or deformation. It is a characteristic of each material, form (shape or structure), and type of stressing. For deformations involving changes both in volume and in shape, special coefficients are used. For conductors under axial tension, the ratio of stress to strain is called *Young's modulus*.

If F is the total force or load acting uniformly on the cross section A, the stress is F/A. If this magnitude of stress causes an elongation e in an original length l, the strain is e/l. Young's modulus is then expressed

$$M = \frac{Fl}{Ae} \quad (2\text{-}17)$$

If a material were capable of sustaining an elastic elongation sufficient to make e equal to l, or such that the elongated length is double the original length, the stress required to produce this result would equal the modulus. This modulus is very useful in computing the sags of overhead conductor spans under loads of various kinds. It is usually expressed in pounds per square inch.

Stranding usually lowers the Young's modulus somewhat, rope-lay stranding to a greater extent than concentric-lay stranding.

When a new cable is subjected initially to tension and the loading is carried up to the maximum working stress, there is an apparent elongation which is greater than the subsequent elongation under the same loading. This is apparently due to the removal of a very slight slackness in the individual wires, causing them to fit closely together and adjust themselves to the conditions of tension in the strand. When a new cable is loaded to the working limit, unloaded, and then reloaded, the value of Young's modulus determined on initial loading may be on the order of one-half to two-thirds of its true value on reloading. The

CONDUCTOR MATERIALS

2-23

latter figure should approach within a few percent of the modulus determined by test on individual straight wires of the same material.

For those applications where elastic stretching under tension needs consideration, the stress-strain curve should be determined by test, with the precaution not to prestress the cable before test unless it will be prestressed when installed in service.

Commercially used values of Young's modulus for conductors are given in Table 2-5.

2-111. *Young's Moduli for Conductors.* See Table 2-5.

2-112. *Young's Modulus for ACSR.* The permanent modulus of ACSR is dependent upon the proportions of steel and aluminum in the cable and upon the distribution of stress between aluminum and steel. This latter condition is dependent upon temperature, tension, and previous maximum loadings. Because of the interchange of stress between the steel and the aluminum caused by changes of tension and temperature, graphical methods are much superior in accuracy and convenience to analytical methods for sag-tension calculations.

TABLE 2-5 Young's Moduli for Conductors

Conductor	Young's modulus,* lb/in²		Reference
	Final†	Virtual initial‡	
Copper wire, hard-drawn	17.0×10^6	14.5×10^6	Copper Wire Engineering Assoc.
Copper wire, medium hard-drawn	16.0×10^6	14.0×10^6	Anaconda Wire and Cable Co.
Copper cable, hard-drawn, 3 and 12 wire	17.0×10^6	14.0×10^6	Copper Wire Engineering Assoc.
Copper cable, hard-drawn, 7 and 19 wire	17.0×10^6	14.5×10^6	Copper Wire Engineering Assoc.
Copper cable, medium hard-drawn	15.5×10^6	14.0×10^6	Anaconda Wire and Cable Co.
Bronze wire, alloy 15	14.0×10^6	13.0×10^6	Anaconda Wire and Cable Co.
Bronze wire, other alloys	16.0×10^6	14.0×10^6	Anaconda Wire and Cable Co.
Bronze cable, alloy 15	13.0×10^6	12.0×10^6	Anaconda Wire and Cable Co.
Bronze cable, other alloys	16.0×10^6	14.0×10^6	Anaconda Wire and Cable Co.
Copper-clad steel wire	24.0×10^6	22.0×10^6	Copperweld Steel Co.
Copper-clad steel cable	23.0×10^6	20.5×10^6	Copperweld Steel Co.
Copper–copper-clad steel cable, type E	19.5×10^6	17.0×10^6	Copperweld Steel Co.
Copper–copper-clad steel cable, type EK	18.5×10^6	16.0×10^6	Copperweld Steel Co.
Copper–copper-clad steel cable, type F	18.0×10^6	15.5×10^6	Copperweld Steel Co.
Copper–copper-clad steel cable, type 2A to 6A	19.0×10^6	16.5×10^6	Copper Wire Engineering Assoc.
Aluminum wire	10.0×10^6	....	Reynolds Metals Co.
Aluminum cable	9.1×10^6	7.3×10^6	Reynolds Metals Co.
Aluminum-alloy wire	10.0×10^6	....	Reynolds Metals Co.
Aluminum-alloy cable	9.1×10^6	7.3×10^6	Reynolds Metals Co.
Aluminum-steel cable, aluminum wire	10.0×10^6	....	Aluminum Co. of America
Aluminum-steel cable, steel wire	29.0×10^6	....	Aluminum Co. of America
Aluminum-clad steel wire	23.5×10^6	22.0×10^6	Copperweld Steel Co.
Aluminum-clad steel cable	23.0×10^6	21.5×10^6	Copperweld Steel Co.
Aluminum-clad steel–aluminum cable:			
AWAC 5/2	13.5×10^6	12.0×10^6	Copperweld Steel Co.
AWAC 4/3	15.5×10^6	14.0×10^6	Copperweld Steel Co.
AWAC 3/4	17.5×10^6	16.0×10^6	Copperweld Steel Co.
AWAC 2/5	19.0×10^6	18.0×10^6	Copperweld Steel Co.
Galvanized-steel wire, Class A coating	28.5×10^6	....	Indiana Steel & Wire Co.
Galvanized-steel cable, Class A coating	27.0×10^6	....	Indiana Steel & Wire Co.

*For stranded cables the moduli are usually less than for solid wire and vary with number and arrangement of strands, tightness of stranding, and length of lay. Also, during initial application of stress, the stress-strain relation follows a curve throughout the upper part of the range of stress commonly used in transmission-line design.

†Final modulus is the ratio of stress to strain (slope of the curve) obtained after fully prestressing the conductor. It is used in calculating design or final sags and tensions.

‡Virtual initial modulus is the ratio of stress to strain (slope of the curve) obtained during initial sustained loading of new conductor. It is used in calculating initial or stringing sags and tensions.

NOTE: $1 \text{ lb/in}^2 = 6.895 \text{ kPa}$.

Because ACSR is a composite cable made of aluminum and steel wires, additional phenomena occur which are not found in tests of cable composed of a single material. As shown in Fig. 2-6, the part of the curve obtained in the second stress cycle contains a comparatively large "foot" at its base, which is caused by the difference in extension at the elastic limits of the aluminum and steel.

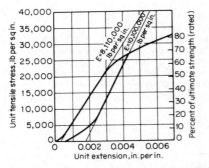

FIG. 2-6 Repeated stress-strain curve, 795,000 cmils ACSR; 54 × 0.1212 aluminum strands; 7 × 0.1212 steel strands.

2-113. Elastic Limit. This is variously defined as the limit of stress beyond which permanent deformation occurs or the stress limit beyond which Hooke's law ceases to apply or the limit beyond which the stresses are not proportional to the strains or the *proportional limit*. In some materials the elastic limit occurs at a point which is readily determined, but in others it is quite difficult to determine because the stress-strain curve deviates from a straight line but very slightly at first, and the point of departure from true linear relationship between stress and strain is somewhat indeterminate.

The late Dean J. B. Johnson of the University of Wisconsin, well-known authority on materials of construction, proposed the use of an arbitrary determination referred to frequently as the "Johnson definition of elastic limit." This proposal, which has been quite largely used, was that an "apparent elastic limit" be employed, defined as that point on the stress-strain curve at which the rate of deformation is 50% greater than at the origin. The apparent elastic limit thus defined is a practical value, which is suitable for engineering purposes, as it involves negligible permanent elongation.

The *Johnson elastic limit* is that point on the stress-strain curve at which the natural tangent is equal to 1.5 times the tangent of the angle of the straight or linear portion of the curve, with respect to the axis of ordinates, or *Y* axis.

2-114. Yield Point. In many materials a point is reached on the stress-strain diagram at which there is a marked increase in strain or elongation without an increase in stress or load. The point at which this occurs is termed the yield point. It is usually quite noticeable in ductile materials but may be scarcely perceptible or possibly not present at all in certain hard-drawn materials such as hard-drawn copper.

2-115. Maximum working stresses of conductor materials must be determined by tests on samples or specimens comparable in size and condition with the shapes or members intended for use under service conditions. The ratio of the safe maximum working stress to the ultimate breaking strength is from 50 to 60% in many classes of materials, but in others it may range as high as 75 to 80%. The maximum working stresses of any material should be determined only from complete knowledge of its properties and the conditions under which it will be used in service.

The working strength of strands is also affected by the mode of attachment to their supports. When gripped in clamping devices, the edges of the clamps should be rounded to prevent injury to the wires, and the grooves in the clamps should be of suitable diameter to fit the strand very closely and of sufficient length to grip all the wire surfaces firmly. If these precautions are not followed, the strength of any strand as a whole may be appreciably impaired at the clamps. In some forms of construction, strands are supported by means of tapered sockets, to which their ends are made fast by having a matrix, such as zinc, cast in the open end of the socket so as to fill it and grip the individual wires.

2-116. Maximum Working Stress for Annealed-Copper Conductors. The stress-strain curves for annealed copper show that it has no definite elastic limit and starts to take a permanent deformation, or set, at comparatively small stresses. It is characteristic of such wire to stretch slowly but permanently under relatively moderate stresses, but in so doing it hardens and tends to increase its own elastic limit. The actual condition of any given wire depends on its previous loading. In overhead spans, where the slack has been pulled up repeatedly, the maximum working stress may approach a high percentage of the original ultimate strength; according to one authority this may become as much as 85%.

CONDUCTOR MATERIALS

2-117. Maximum Working Stress for Medium-hard-drawn Copper Conductors. The average is 50% of the ultimate tensile strength, but where more exact information is desired, stress-strain curves should be determined by test. If the wire will be prestressed at the time it is installed for service, it should be similarly prestressed before determining the stress-strain curve or tested in appropriate manner to show the stress-strain curve after prestressing.

2-118. Maximum Working Stress for Hard-drawn Copper Conductors. For wire sizes from 0.460 to 0.325 in, inclusive, the average is 55% of the ultimate tensile strength, with a minimum of 50%. For sizes from 0.324 to 0.040 in, inclusive, the average is 60%, with a minimum of 55%. For more exact information, stress-strain curves should be determined by test.

2-119. Maximum Working Stress for ACSR. Although the ultimate strength of the steel strands is 200,000 lb/in² and that of the aluminum strands is from 23,000 to 30,000 lb/in², depending on the size, the ultimate stresses do not occur at the same elongation. Since the aluminum strands have less elongation at rupture than the steel strands, it is evident that the breaking strength of ACSR is the sum of two quantities, the first being the cross-sectional area of aluminum multiplied by its ultimate stress and the second being the cross-sectional area of steel multiplied by its stress at the ultimate elongation of the aluminum wires. The ultimate strength of ACSR thus obtained is the tension at which the aluminum strands fail; the steel core will then stretch, thus reducing the tension. The safe maximum working tension may be defined as that value of tension which can be experienced repeatedly without altering the tensile properties which obtain after the first application of that tension.

2-120. Maximum Working Stresses for Copper-clad, Aluminum-clad, and Galvanized-Steel Conductors. For sizes customarily employed, 60% of the breaking strength is used. Under some special conditions, this value may be exceeded.

2-121. Prestressed Conductors. In the case of some materials, especially those of considerable ductility, which tend to show permanent elongation or "drawing" under loads just above the initial elastic limit, it is possible to raise the working elastic limit by loading them to stresses somewhat above the elastic limit as found on initial loading. After such loading, or prestressing, the material will behave according to Hooke's law at all loads less than the new elastic limit. This applies not only to many ductile materials, such as soft or annealed copper wire, but also to cables or stranded conductors, in which there is a slight inherent slack or looseness of the individual wires that can be removed only under actual loading. It is sometimes the practice, when erecting such conductors for service, to prestress them to the working elastic limit or safe maximum working stress and then reduce the stress to the proper value for installation at the stringing temperature without wind or ice.

2-122. Resistance is the property of an electric circuit or of any body that may be used as part of an electric circuit which determines for a given current the average rate at which electrical energy is converted into heat. The term is properly applied only when the rate of conversion is proportional to the square of the current and is then equal to the power conversion divided by the square of the current. A uniform cylindrical conductor of diameter d, length l, and *volume resistivity* ρ has a total resistance to continuous currents expressed by the formula

$$R = \frac{\rho l}{\pi d^2 / 4} \qquad (2\text{-}18)$$

The resistance of any conductor is commonly expressed in ohms per unit of length, such as 1 ft, 1000 ft, or 1 mi. When used for conducting alternating currents, the effective resistance may be higher than the dc resistance defined above. In the latter case it is common practice to apply the proper factor, or ratio of effective ac resistance to dc resistance, sometimes termed the "skin-effect resistance ratio" (see Par. **2-126**). This ratio may be determined by test, or it may be calculated if the necessary data are available.

2-123. Magnetic permeability applies to a field in which the flux is uniformly distributed over a cross section normal to its direction or to a sufficiently small cross section of a nonuniform field so that the distribution can be assumed as substantially uniform. In the case of a cylindrical conductor, the magnetomotive force (mmf) due to the current flowing in the conductor varies from zero at the center or axis to a maximum at the periphery or surface of the conductor and sets up a flux in circular paths concentric with the axis and perpendicular to it but of nonuniform distribution between the axis and the periphery. If

2-26   ELECTRICAL ENGINEERING MATERIALS REFERENCE GUIDE

the permeability is nonlinear with respect to the mmf, as is usually true with magnetic materials, there is no correct single value of permeability which fits the conditions, although an apparent or equivalent average value can be determined. In the case of other forms of cross section, the distribution is still more complex, and the equivalent permeability may be difficult or impossible to determine except by test.

2-124. Internal Inductance. A uniform cylindrical conductor of nonmagnetic material, or of unit permeability, has a constant magnitude of internal inductance per unit length, independent of the conductor diameter. This is commonly expressed in microhenrys or millihenrys per unit of length, such as 1 ft, 1000 ft, or 1 mi. When the conductor material possesses magnetic susceptibility, and when the magnetic permeability μ is constant and therefore independent of the current strength, the internal inductance is expressed in absolute units by the formula

$$L = \frac{\mu l}{2} \tag{2-19}$$

In most cases μ is not constant but is a function of the current strength. When this is true, there is an effective permeability, one-half of which ($\mu/2$) expresses the inductance per centimeter of length, but this figure of permeability is virtually the ratio of the effective inductance of the conductor of susceptible material to the inductance of a conductor of material which has a permeability of unity. When used for conducting alternating currents, the effective inductance may be less than the inductance with direct current; this is also a direct consequence of the same skin effect which results in an increase of effective resistance with alternating currents, but the overall effect is usually included in the figure of effective permeability. It is usually the practice to determine the effective internal inductance by test, but it may be calculated if the necessary data are available.

2-125. Skin effect is a phenomenon which occurs in conductors carrying currents whose intensity varies rapidly from instant to instant but does not occur with continuous currents. It arises from the fact that elements or filaments of variable current at different points in the cross section of a conductor do not encounter equal components of inductance, but the central or axial filament meets the maximum inductance, and in general the inductance offered to other filaments of current decreases as the distance of the filament from the axis increases, becoming a minimum at the surface or periphery of the conductor. This, in turn, tends to produce unequal current density over the cross section as a whole; the density is a minimum at the axis and a maximum at the periphery. Such distribution of the current density produces an increase in effective resistance and a decrease in effective internal inductance; the former is of more practical importance than the latter. In the case of large copper conductors at commercial power frequencies, and in the case of most conductors at carrier and radio frequencies, the increase in resistance should be considered (see Pars. 2-130 and 2-131).

2-126. Skin-Effect Ratios. If R' is the effective resistance of a linear cylindrical conductor to sinusoidal alternating current of given frequency and R is the true resistance with continuous current, then

$$R' = KR \qquad \text{ohms} \tag{2-20}$$

where K is determined from Table 2-6 in terms of x. The value of x is given by

$$x = 2\pi a \sqrt{\frac{2f\mu}{\rho}} \tag{2-21}$$

where a = radius of conductor in centimeters, f = frequency in cycles per second, μ = magnetic permeability of conductor (here assumed to be constant), ρ = resistivity in abohm-centimeters (abohm = $10^{-9}\ \Omega$).

For practical calculation, Eq. (2-21) can be written

$$x = 0.063598 \sqrt{\frac{f\mu}{R}} \tag{2-22}$$

where R = dc resistance at operating temperature in ohms per mile.

CONDUCTOR MATERIALS

TABLE 2-6 Skin-Effect Ratios

(Bur. Std. *Bull.* 169, pp. 226–228)

x	K	K'	x	K	K'	x	K	K'	x	K	K'
0.0	1.00000	1.00000	2.9	1.28644	0.86012	6.6	2.60313	0.42389	17.0	6.26817	0.16614
0.1	1.00000	1.00000	3.0	1.31809	0.84517	6.8	2.67312	0.41171	18.0	6.62129	0.15694
0.2	1.00001	1.00000	3.1	1.35102	0.82975	7.0	2.74319	0.40021	19.0	6.97446	0.14870
0.3	1.00004	0.99998	3.2	1.38504	0.81397	7.2	2.81334	0.38933	20.0	7.32767	0.14128
0.4	1.00013	0.99993	3.3	1.41999	0.79794	7.4	2.88355	0.37902	21.0	7.68091	0.13456
0.5	1.00032	0.99984	3.4	1.45570	0.78175	7.6	2.95380	0.36923	22.0	8.03418	0.12846
0.6	1.00067	0.99966	3.5	1.49202	0.76550	7.8	3.02411	0.35992	23.0	8.38748	0.12288
0.7	1.00124	0.99937	3.6	1.52879	0.74929	8.0	3.09445	0.35107	24.0	8.74079	0.11777
0.8	1.00212	0.99894	3.7	1.56587	0.73320	8.2	3.16480	0.34263	25.0	9.09412	0.11307
0.9	1.00340	0.99830	3.8	1.60314	0.71729	8.4	3.23518	0.33460	26.0	9.44748	0.10872
1.0	1.00519	0.99741	3.9	1.64051	0.70165	8.6	3.30557	0.32692	28.0	10.15422	0.10096
1.1	1.00758	0.99621	4.0	1.67787	0.68632	8.8	3.37597	0.31958	30.0	10.86101	0.09424
1.2	1.01071	0.99465	4.1	1.71516	0.67135	9.0	3.44638	0.31257	32.0	11.56785	0.08835
1.3	1.01470	0.99266	4.2	1.75233	0.65677	9.2	3.51680	0.30585	34.0	12.27471	0.08316
1.4	1.01969	0.99017	4.3	1.78933	0.64262	9.4	3.58723	0.29941	36.0	12.98160	0.07854
1.5	1.02582	0.98711	4.4	1.82614	0.62890	9.6	3.65766	0.29324	38.0	13.68852	0.07441
1.6	1.03323	0.98342	4.5	1.86275	0.61563	9.8	3.72812	0.28731	40.0	14.39545	0.07069
1.7	1.04205	0.97904	4.6	1.89914	0.60281	10.0	3.79857	0.28162	42.0	15.10240	0.06733
1.8	1.05240	0.97390	4.7	1.93533	0.59044	10.5	3.97477	0.26832	44.0	15.80936	0.06427
1.9	1.06440	0.96795	4.8	1.97131	0.57852	11.0	4.15100	0.25622	46.0	16.51634	0.06148
2.0	1.07816	0.96113	4.9	2.00710	0.56703	11.5	4.32727	0.24516	48.0	17.22333	0.05892
2.1	1.09375	0.95343	5.0	2.04272	0.55597	12.0	4.50358	0.23501	50.0	17.93032	0.05656
2.2	1.11126	0.94482	5.2	2.11353	0.53506	12.5	4.67993	0.22567	60.0	21.46541	0.04713
2.3	1.13069	0.93527	5.4	2.18389	0.51566	13.0	4.85631	0.21703	70.0	25.00063	0.04040
2.4	1.15207	0.92482	5.6	2.25393	0.49764	13.5	5.03272	0.20903	80.0	28.53593	0.03535
2.5	1.17538	0.91347	5.8	2.32380	0.48086	14.0	5.20915	0.20160	90.0	32.07127	0.03142
2.6	1.20056	0.90126	6.0	2.39359	0.46521	14.5	5.38560	0.19468	100.0	35.60666	0.02828
2.7	1.22753	0.88825	6.2	2.46338	0.45056	15.0	5.56208	0.18822	∞	∞	0
2.8	1.25620	0.87451	6.4	2.53321	0.43682	16.0	5.91509	0.17649			

If L' is the effective inductance of a linear conductor to sinusoidal alternating current of a given frequency,

$$L' = L_1 + K'L_2 \tag{2-23}$$

where L_1 = external portion of inductance, L_2 = internal portion (due to the magnetic field within the conductor), and K' is determined from Table 2-6 in terms of x. Thus the total effective inductance per unit length of conductor is

$$L' = 2 \ln \frac{d}{a} + K' \frac{\mu}{2} \tag{2-24}$$

The inductance is here expressed in abhenrys per centimeter of conductor, in a linear circuit; a is the radius of the conductor and d is the separation between the conductor and its return conductor, expressed in the same units.

Values of K and K' in terms of x are shown in Table 2-6 and Figs. 2-7 and 2-8 (see *NBS Circ.* 74, pp. 309–311, for additional tables, and *Sci. Paper* 374).

Value of μ for nonmagnetic materials (copper, aluminum, etc.) is 1; for magnetic materials, it varies widely with composition, processing, current density, etc., and should be determined by test in each case.

2-127. Skin-Effect Ratios. See Table 2-6.

2-128. Skin-Effect Ratios—Copper Conductors NOT in Close Proximity. See Table 2-7.

2-129. Alternating-Current Resistance. For small conductors at power frequencies, the frequency has a negligible effect, and dc resistance values can be used. For large conductors, frequency must be taken into account in addition to temperature effects. To do this, first calculate the dc resistance at the operating temperature, then determine the skin-effect ratio K, and finally determine the ac resistance at operating temperature (see Par. **2-126**).

2-130. AC resistance for copper conductors NOT in close proximity can be obtained from

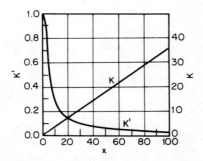
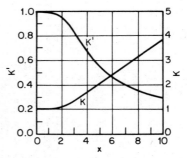

FIG. 2-7 K and K' for values of x from 0 to 100.   **FIG. 2-8** K and K' for values of x from 1 to 10.

the skin-effect ratios given in Tables 2-6 and 2-7. Practical values for usual conductor sizes are given in Table 2-27.

2-131. AC resistance for copper conductors in close proximity or for insulated copper conductors installed in conduit may be calculated from Table 2-8.

2-132. AC/DC Resistance Ratios: Copper Conductors in Close Proximity. See Table 2-8.

2-133. AC resistance for copper-clad and aluminum-clad steel conductors is dependent on several variables and should be determined by test. Practical values for usual conductor sizes are given in Tables 2-28, 2-29, and 2-31.

2-134. AC Resistance for Aluminum Conductors. The increase in resistance and decrease in internal inductance of cylindrical aluminum conductors can be determined from the data in Par. **2-126**. It is not the same as for copper conductors of equal diameter but is slightly less because of the higher volume resistivity of aluminum.

2-135. AC/DC Resistance Ratios: Aluminum Conductors in Close Proximity. See Table 2-8.

TABLE 2-7 Skin-Effect Ratios—Copper Conductors NOT in Close Proximity
(See also Table 4-8; AC/DC Resistance Ratios—Copper and Aluminum Conductors IN Close Proximity)

Conductor size, Mcm	Skin-effect ratio K at 60 cycles and 65 C (149 F)														
	Inside conductor diameter, in.														
	0*		0.25		0.50		0.75		1.00		1.25		1.50		2.00	
	Outside diameter, in	K	Outside diameter, in	K	Outside diameter, in	K	Outside diameter, in	K	Outside diameter, in	K	Outside diameter, in	K	Outside diameter, in	K	Outside diameter, in	K
3000	1.998	1.439	2.02	1.39	2.08	1.36	2.15	1.29	2.27	1.23	2.39	1.19	2.54	1.15	2.87	1.08
2500	1.825	1.336	1.87	1.28	1.91	1.24	2.00	1.20	2.12	1.16	2.25	1.12	2.40	1.09	2.75	1.05
2000	1.631	1.239	1.67	1.20	1.72	1.17	1.80	1.12	1.94	1.09	2.09	1.06	2.25	1.05	2.61	1.02
1500	1.412	1.145	1.45	1.12	1.52	1.09	1.63	1.06	1.75	1.04	1.91	1.03	2.07	1.02	2.47	1.01
1000	1.152	1.068	1.19	1.05	1.25	1.03	1.39	1.02	1.53	1.01	1.72	1.01				
800	1.031	1.046	1.07	1.04	1.16	1.02	1.28	1.01	1.45	1.01						
600	0.893	1.026	0.94	1.02	1.04	1.01										
500	0.814	1.018	0.86	1.01	0.97	1.01										
400	0.728	1.012	0.78	1.01												
300	0.630	1.006														

*For standard concentric-stranded conductors (i.e., inside diameter = 0).
NOTE: 1 in = 2.54 cm.

CONDUCTOR MATERIALS

TABLE 2-8 AC/DC Resistance Ratios—Copper and Aluminum Conductors IN Close Proximity
(*IPCEA Publ.* P-34-359)

Conductor size, Mcm or AWG	A-c/d-c resistance ratio at 60 cycles and 65°C (149°F)							
	Single-conductor cable in air or separate nonmetallic conduit						5–15-kV nonleaded shielded power cable, 3 single conductor cables in same metallic conduit	
	Concentric		Segmental		Annular		Concentric	
	Copper	Aluminum	Copper	Aluminum	Copper	Aluminum	Copper	Aluminum
5,000	1.77	1.42	....	....	1.12	1.05		
4,500	1.69	1.36	....	....	1.12	1.05		
4,000	1.61	1.31	1.18	1.08	1.12	1.05		
3,500	1.52	1.25	1.15	1.06	1.11	1.04		
3,000	1.43	1.20	1.11	1.04	1.11	1.04		
2,500	1.33	1.14	1.08	1.03	1.07	1.03		
2,250	1.28	1.12	1.06	1.03	....	....		
2,000	1.24	1.10	1.05	1.02	1.05	1.02		
1,750	1.19	1.08	1.04	1.02	1.04	1.02		
1,500	1.14	1.06	1.03	1.01	1.04	1.01		
1,250	1.10	1.04	1.02	1.01	1.04	1.01		
1,000	1.07	1.03	1.01	1.01	1.03	1.01	1.36	1.17
900	1.06	1.02	....	....	1.03	1.01	1.30	1.14
800	1.04	1.02	....	....	1.02	1.01	1.24	1.11
750	1.04	1.01	....	....	1.02	1.01	1.22	1.10
700	1.03	1.01	....	....	....	....	1.19	1.09
600	1.03	1.01	....	....	....	....	1.14	1.07
500	1.02	1.01	....	....	....	....	1.10	1.05
400	1.01	1.01*	....	....	....	....	1.07	1.03
350	1.01	....	....	....	....	....	1.05	1.03
300	1.01	....	....	....	....	....	1.04	1.02
250	1.01*	....	....	....	....	....	1.03	1.01
4/0	1.01*	....	....	....	....	....	1.02	1.01
3/0	....	....	....	....	....	....	1.01	
2/0	....	....	....	....	....	....	1.01	

*Conductor skin effect less than 1%.

2-136. AC Resistance for ACSR. In the case of ACSR conductors, the steel core is of relatively high resistivity, and therefore its conductance is usually neglected in computing the total resistance of such strands. The effective permeability of the grade of steel employed in the core is also relatively small. It is approximately correct to assume that such a strand is hollow and consists exclusively of its aluminum wires; in this case the laws of skin effect in tubular conductors will be applicable. Conductors having a single layer of aluminum wires over the steel core have higher ac/dc ratios than those having multiple layers of aluminum wires. Practical values for usual conductor sizes are given in Table 2-30.

2-137. AC Resistance for Steel Conductors. The increase in effective resistance with alternating currents depends fundamentally upon the circular permeability; there is also an increase, of much smaller proportions, caused by hysteresis. The permeability is very sensitive to variations in composition, heat-treatment, and working of the metal; for this reason it is not usually feasible to compute the skin effect except as an approximation. The results of tests show great variations depending upon the factors just mentioned. Typical test results are shown in Figs. 2-9 and 2-10. Practical values for small conductors are given in Table 2-26.

2-138. Inductive Reactance. Present practice is to consider inductive reactance as split into two components: (1) that due to flux within a radius of 1 ft including the internal reactance within the conductor of radius r and (2) that due to flux between 1 ft radius and the equivalent conductor spacing D_s or geometric mean distance (GMD).

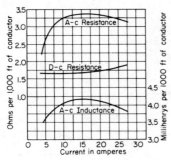
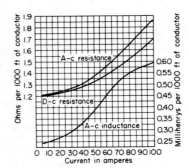

FIG. 2-9 Effective resistance and internal inductance at 60 Hz of No. 6 AWG galvanized BB wire.

FIG. 2-10 Effective resistance and internal inductance at 60 Hz of ⅜-in 7-wire Class A galvanized high-strength steel strand.

The fundamental inductance formula is

$$L = 2 \ln \frac{D_s}{r} + \frac{\mu}{2} \quad \text{abH/(cm)(conductor)} \tag{2-25}$$

This can be rewritten

$$L = 2 \ln \frac{D_s}{1} + 2 \ln \frac{1}{r} + \frac{\mu}{2} \tag{2-26}$$

where the term $2 \ln (D_s/1)$ represents inductance due to flux between 1 ft radius and the equivalent conductor spacing and $2 \ln (1/r) + (\mu/2)$ represents the inductance due to flux within 1 ft radius [$2 \ln (1/r)$ represents inductance due to flux between conductor surface and 1 ft radius, and $\mu/2$ represents internal inductance due to flux within the conductor].

By definition, geometric mean radius (GMR) of a conductor is the radius of an infinitely thin tube having the same internal inductance as the conductor. Therefore,

$$L = 2 \ln \frac{D_s}{1} + 2 \ln \frac{1}{\text{GMR}} \tag{2-27}$$

Since inductance reactance $= 2\pi f L$, for practical calculation Eq. (2-27) can be written

$$X = 0.004657 f \log \frac{D_s}{1} + 0.004657 f \log \frac{1}{\text{GMR}} \quad \Omega/\text{(mi)(conductor)} \tag{2-28}$$

In the conductor tables in this section, inductive reactance is calculated from Eq. (2-28), considering that

$$X = x_a + x_d \tag{2-29}$$

2-139. Inductive reactance for conductors using steel varies in a manner similar to ac resistance (see Pars. **133** to **137**). Practical values for usual conductor sizes are given in Tables 2-30, 2-31, and 2-32.

2-140. Capacitive Reactance. By the same reasoning used in Par. **138**, the capacitive reactance can be considered in two parts also, giving

$$X = \frac{4.099}{f} \log \frac{D_s}{1} + \frac{4.099}{f} \log \frac{1}{r} \quad \text{M}\Omega/\text{(mi)(conductor)} \tag{2-30}$$

In the conductor tables in this section, capacitive reactance is calculated from Eq. (2-30), it being considered that

$$X' = x'_a + x'_d \tag{2-31}$$

It is important to note that in capacitance calculations the conductor radius used is the actual physical radius of the conductor.

2-141. Capacitive Susceptance

$$B = \frac{1}{x'_a + x'_d} \quad \mu S/(mi)(conductor) \tag{2-32}$$

2-142. Charging Current

$$I_C = eB \times 10^{-3} \quad A/(mi)(conductor) \tag{2-33}$$

where e = voltage to neutral in kilovolts.

2-143. Current-Carrying Capacity of Bare Conductors. The IEEE publication titled *Calculation of Bare Overhead Conductor Temperature and Ampacity for Steady-State Conditions*, 1984, has been accepted generally by the industry for the calculation of the current-carrying capacity of conductors for overhead power-transmission lines. For other methods in use, see the following:

Schurig, O. R., and Frick, C. W.: Heating and Current-Carrying Capacity of Bare Conductors for Outdoor Service; *Gen. Electr. Rev.*, March 1930, vol. 33, no. 3, p. 141.

Frick, C. W.: Current-Carrying Capacity of Bare Cylindrical Conductors for Indoor and Outdoor Service; *Gen. Electr. Rev.*, August 1934, vol. 34, no. 8, p. 464.

Kidder, A. H., and Woodward, C. B.: Ampere Load Limits for Copper in Overhead Lines; *Trans. AIEE*, 1943, March section, vol. 62, p. 149.

House, H. E., and Tuttle, P. D.: Current Carrying Capacity of ACSR; *Trans. AIEE*, 1958, vol. 77, pt. III, p. 1169.

2-144. Current-Carrying Capacity of Insulated Conductors. For *power transmission* and distribution cables in air, in enclosed and exposed conduit and in underground ducts, the current-carrying capacities sponsored by the Insulated Power Cable Engineers Association (IPCEA) are in general use. See Power Cable Ampacities, *AIEE Publ.* S-135-1, 1962, vol. 1, Copper Conductors, vol. 2, Aluminum Conductors (available from IEEE, 345 E. 47th St., N.Y. 10017).

For *interior wiring* under jurisdiction of the National Electrical Code, the latest issue of the Standard of the National Board of Fire Underwriters for Electric Wiring and Apparatus, *NBFU Pamphlet 70*, should be consulted.

2-145. Contact (Trolley) Wires. The special requirements for trolley contact wires have resulted in the development of materials having special characteristics. For general use, hard-drawn copper wire manufactured under particular specifications gives satisfactory service. Under unusual conditions of service, however, harder materials are needed to avoid constant wire replacements.

FIG. 2-11   Grooved section.

Trolley contact wires are designed and manufactured for the dual purpose of providing current-carrying capacity and resisting the constant abrasive effect of the trolley wheel, shoe, or pantograph. They must be of good-quality material with a hard-wearing surface, and special care is necessary in manufacture to obtain a surface smooth and free from imperfections.

FIG. 2-12   Figure 8 section.

Because of the need for special methods of supporting trolley contact wires, the use of grooved wires is very common. These as well as other special designs are illustrated in Figs. 2-11 to 2-13.

2-146. Bus conductors require that greater attention be given to certain physical

FIG. 2-13   Figure 9 deep section.

and electrical characteristics of the metals than is usually necessary in designing line conductors. These characteristics are current-carrying capacity, emissivity, skin effect, expansion, and mechanical deflection. To obtain the most satisfactory and economical designs for busbars in power stations and substations, where they are used extensively, consideration must be given to choice not only of material but also of shape. Both copper and aluminum are used for busbars, and in certain outdoor substations, steel has proved satisfactory. The most common busbar form for carrying heavy current, especially indoors, is flat copper bar. Busbars in the form of angles, channels, and tubing have been developed for heavy currents and, because of better distribution of the conducting material, make more efficient use of the metal both electrically and mechanically. All such designs are based upon the need for proper current-carrying capacity without excess busbar temperatures and upon the necessity for adequate mechanical strength.

2-147. Solid conductors (wire) for bare power conductors are usually made of copper or copper alloy, although for particular applications aluminum, aluminum alloy, copper-clad steel, aluminum-clad steel, or steel is used.

2-148. Copper solid conductors for overhead line applications are supplied normally in hard-drawn or medium hard-drawn tempers in sizes No. 10 to 4 AWG. Number 2 AWG is occasionally used, but for this size and larger, stranded conductors are preferable. Soft or annealed conductors are available in all sizes for use in insulated conductors, grounding or bonding wires, weather-resisting (weatherproof) wire, etc.

2-149. Hard-drawn copper wire is drawn through dies from rod to finished product without intermediate annealing. This results in a wire of high strength and low elongation. Both these characteristics can be controlled to some extent by choice of size of rod and by modification of certain details of the drawing process.

2-150. Medium-hard-drawn copper wire is essentially and necessarily a special product, because when wire has once started on its course through the drawing operations, it can finish only as a hard-drawn wire to be used as such or to be annealed and become soft or annealed wire. Medium-hard-drawn wire is annealed wire drawn to a slightly smaller diameter.

Medium-hard-drawn wire approaches hard-drawn wire in its characteristics, but from the very nature of the product, exact uniformity in tensile strength cannot be obtained; hence the necessity for establishing a range of tensile strength within which standard medium-hard-drawn wire must be expected to be found.

2-151. Soft or annealed copper wire is drawn by customary operations and annealed, finished by cleaning when necessary to remove scale or oxide. The wire is so soft and ductile that it is easily marred and even stretched by careless handling in the operations of winding or cabling; hence the necessity for confining specifications and inspection to wire in packages as it leaves the manufacturer and before being put through processes incident to its use by the purchaser.

2-152. Copper-alloy solid conductors for overhead line applications normally are supplied only in hard-drawn temper. The usual range of sizes is No. 10 to 4 AWG. Wires are available in any of the several alloys designated in ASTM B105 as alloys 8.5 to 85 in accordance with their conductivities.

2-153. Copper-clad steel solid conductors are rarely used for purposes other than signal or communication wire (weatherproofed or insulated) or telephone drop wire. Usual sizes are No. 17 to 10 AWG.

Aluminum-clad steel solid conductors are used primarily for signal and communication wire (either bare or insulated). The usual sizes employed are No. 8 through No. 12 AWG.

2-154. Aluminum solid conductors are practically never used for overhead-line applications because of their relatively poor physical characteristics. Intermediate temper or hard-drawn conductors are used in all permitted sizes of insulated solid conductors.

2-155. Iron and steel solid conductors are used occasionally for telephone and telegraph lines in Nos. 16, 10, and 8 BWG and Nos. 14, 13, and 12 SWG (see Table 2-11).

2-156. Concentric-lay stranded conductors are made up of successive layers of helically laid solid members (wires). Usual strandings have 6 wires more in each added layer and start with either a 1- or a 3-wire core.

2-157. Bunch-stranded conductors are made up of any number of wires grouped together without regard to their accurate geometric arrangement. Usually they are twisted together; if not, they are generally referred to as "parallel-strand."

TABLE 2-9 Industry Specifications for Conductors

Sponsoring organization	Specification designation	Specification title	Corresponding ANSI specification
		General definitions	
AIEE	Book	Definitions of electrical terms	C42
AIEE/IEEE	No. 30	Definitions and general standards for wire and cable	C8.1
ASTM	B354	Definitions of terms relating to uninsulated metallic electrical conductors	
		Conductor materials	
ASTM	B258	Standard nominal diameters and cross-sectional areas of AWG sizes of solid round wires used as electrical conductors	C7.36
ASTM	B5	Electrolytic copper wire, bars, cakes, slabs, billets, ingots, and ingot bars	H17.2
ASTM	B4	Lake copper wire, bars, cakes, slabs, billets, ingots, and ingot bars	H17.1
ASTM	B49	Hot-rolled copper rods for electrical purposes	C7.7
ASTM	B170	Oxygen-free electrolytic copper wire bars, billets, and cakes	H23.11
ASTM	B224	Classification of coppers	
ASTM	B263	Method for determination of cross-sectional area of stranded conductors	C7.29
ASTM	B233	Aluminum-alloy 1350 redraw rod for electrical purposes	C7.23
ASTM	B296	Recommended practice for temper designation of aluminum and magnesium alloys, cast and wrought	
		Bare solid copper wire	
ASTM	B1	Hard-drawn copper wire	C7.2
ASTM	B2	Medium hard-drawn copper wire	C7.3
ASTM	B3	Soft or annealed copper wire	C7.1
ASTM	B48	Soft rectangular and square bare copper wire for electrical conductors	C7.9
ASTM	B47	Copper trolley wire	C7.6
ASTM	B116	ASTM figure 9 deep-section grooved and figure 8 copper trolley wire for industrial haulage	C711
		Coated-copper wire	
ASTM	B33	Tinned soft or annealed copper wire for electrical purposes	C7.4
ASTM	B246	Tinned hard-drawn and medium-hard-drawn copper wire for electrical purposes	C7.37
ASTM	B189	Lead-coated and lead-alloy-coated soft copper wire for electrical purposes	C7.15
ASTM	B298	Silver-coated soft or annealed copper wire	C7.38
ASTM	B355	Nickel-coated soft or annealed copper wire	C7.48
		Copper-clad-steel wire	
ASTM	B227	Hard-drawn copper-clad steel wire	C7.17
		Copper-alloy wires	
ASTM	B9	Bronze trolley wire	C7.5
ASTM	B105	Hard-drawn copper-alloy wires for electrical conductors	C7.10
		Bare solid aluminum wire	
ASTM	B230	Aluminum wire, 1350-H19, for electrical purposes	C7.20

2-33

TABLE 2-9 Industry Specifications for Conductors (*Continued*)

Sponsoring organization	Specification designation	Specification title	Corresponding ANSI specification
		General definitions	
ASTM	B262	Aluminum wire, 1350-H16 or -H26 for electrical purposes	C7.35
ASTM	B314	Aluminum wire for communication cable	C7.40
ASTM	B323	Aluminum wire, 1350-H14 or -H24, for electrical purposes	C7.42
ASTM	B324	Aluminum rectangular and square wire for electrical purposes	C7.43
		Aluminum-alloy wire	
ASTM	B396	Aluminum alloy 5005-H19 wire for electrical purposes	C7.49
ASTM	B398	Aluminum alloy 6201-T81 wire for electrical purposes	C7.51
		Aluminum-clad steel wire	
ASTM	B415	Hard-drawn aluminum-clad steel wire	C7.55
		Copper cable	
ASTM	B8	Concentric-lay-stranded copper conductors, hard, medium hard, or soft	C7.8
ASTM	B172	Rope-lay-stranded copper conductors having bunch-stranded members for electrical conductors	C7.12
ASTM	B173	Rope-lay-stranded copper conductors having concentric-stranded members for eletrical conductors	C7.13
ASTM	B174	Bunch-stranded copper conductors for electrical conductors	C7.14
ASTM	B226	Cored, annular, concentric-lay-stranded copper conductors	C7.16
ASTM	B286	Copper conductors for use in hookup wire for electronic equipment	C7.39
		Copper-clad steel and composite cables	
ASTM	B228	Concentric-lay-stranded copper-clad steel conductors	C7.18
ASTM	B229	Concentric-lay-stranded copper and copper-clad steel composite conductors	C7.19
		Aluminum and ACSR cables	
ASTM	B231	Aluminum conductors, concentric-lay-stranded 1350	C7.21
ASTM	B232	Aluminum conductors, concentric-lay-stranded conductors, coated steel-reinforced (ACSR)	C7.22
ASTM	B341	Aluminum-coated (aluminized) steel-core wire for aluminum conductors, steel-reinforced (ACSR/AZ)	C7.47
ASTM	B400	Compact round concentric-lay-stranded aluminum 1350 conductors	C7.53
ASTM	B401	Compact round concentric-lay-stranded aluminum conductors, steel-reinforced (ACSR/COMP)	C7.54
ASTM	B498	Zinc-coated (galvanized) steel-core wire for aluminum conductors, steel-reinforced (ACSR)	
ASTM	B502	Aluminum-clad steel-core wire for aluminum conductors, aluminum clad steel reinforced.	
ASTM	B524	Concentric-lay stranded aluminum conductors, aluminum alloy reinforced (ACAR)	

TABLE 2-9 Industry Specifications for Conductors (*Continued*)

Sponsoring organization	Specification designation	Specification title	Corresponding ANSI specification
ASTM	B549	Aluminum conductors, concentric-lay stranded aluminum-clad, steel-reinforced (ACSR/AW)	
ASTM	B701	Concentric-lay stranded self-damping aluminum conductors, steel-reinforced (ACSR/SD)	

		Aluminum-alloy cables	
ASTM	B397	Concentric-lay-stranded 5005-H19 aluminum alloy conductors	C7.50
ASTM	B399	Concentric-lay-stranded 6201-T81 aluminum alloy conductors	C7.52

		Aluminum-clad steel cables	
ASTM	B416	Concentric-lay-stranded aluminum-clad steel conductors	

		Bus conductors	
ASTM	B187	Copper busbar, rod, and shapes	C7.25
ASTM	B188	Seamless copper bus pipe and tube	C7.26
ASTM	B236	Aluminum bars for electrical purposes (bus bars)	C7.27
ASTM	B75	Seamless copper tube	H23.3
ASTM	B317	Aluminum-alloy extruded bar, rod, pipe, and structural shapes for electrical purposes (bus conductos)	C7.45

		Waveguide tube	
ASTM	B372	Seamless copper and copper-alloy rectangular waveguide tube	H37.1

		Stainless-steel strand	
ASTM	A368	Stainless-steel wire strand	

		Galvanized steel wire and strand	
ASTM	A111	Zinc-coated (galvanized) "iron" telephone and telegraph line wire	C7.31
ASTM	A326	Zinc-coated (galvanized) high tensile steel telephone- and telegraph-line wire	C7.30
ASTM	A363	Zinc-coated (galvanized) steel overhead ground wire strand	
ASTM	A411	Zinc-coated (galvanized) low-carbon steel armor wire	
ASTM	A475	Zinc-coated steel wire strand	C7.46

		Methods of test	
ASTM	B193	Resistivity of electrical conductor materials	C7.24
ASTM	E8	Tension testing of metallic materials	
ASTM	B279	Stiffness of bare soft square and rectangular copper wire for magnet wire fabrication	C7.41
ASTM	A90	Weight of coating or zinc-coated (galvanized) iron or steel articles	G8.12
ASTM	A239	Uniformity of coating by the Preece test (copper sulfate dip) on zinc-coated (galvanized) iron or steel articles	
ASTM	B342	Electrical conductivity by use of eddy currents	C7.44

TABLE 2-10 Conductors—Physical and Electrical Properties

	Conductivity % IACS, min. %	Resistivity, Ω				Temp. coefficient of resistance per °C		Weight at 20°C			Coefficient of linear expansion per		Modulus of elasticity	
		cmil/ft		mm²/m						lb/million cmil/1000 ft				
	20°C	20°C	25°C	20°C	25°C	20°C	25°C	g/cm³	lb/in³		°F	°C	lb/in²	kg/mm²
Commercial 1350 aluminum wire	61.0	17.002	17.345	0.028265	0.028834	0.00403	0.00395	2.705	0.09750	920.3	0.0000128	0.000023	10×10^6	7,030
Aluminum alloy wire 6201	52.5	19.754	20.097	0.032840	0.033373	0.00347	0.00340	2.703	0.09765	920.3	0.0000128	0.000023	10×10^6	7,030
Commercial hard-drawn copper wire	97.0	10.692	10.895	0.017774	0.018113	0.00381	0.00374	8.89	0.32	3027	0.0000094	0.0000169	17×10^6	11,950
Standard annealed copper wire	100.0	10.371	10.575	0.017241	0.017579	0.00393	0.00385	8.89	0.321	3027	0.0000094	0.0000169	17×10^6	11,950
Aluminum-coated steel-core wire	9.0*	115.23*		0.19157*				7.78	0.281	2649	0.0000064	0.0000115	29×10^6	20,400
Zinc-coated steel-core wire	9.0*	115.23*		0.19157*				7.78	0.281	2649	0.0000064	0.0000115	29×10^6	20,400
Aluminum-clad steel-core wire	20.33	51.01	51.52	0.0848	0.08563	0.0036	0.00356	6.59	0.2380	2243	0.0000072	0.0000130	23.5×10^6	16,500

*Typical.

2-158. Rope-lay stranded conductors are made up of successive layers of helically laid stranded members. The members may be concentric-lay or bunch-stranded.

2-159. Special Stranded Conductor Shapes. For use in multiconductor insulated cables whose finished cross section must be round, special shapes such as D shape (hemispherical), sector shape (triangular), and semisector (oval) are commonly used.

Annular conductors are formed by stranding helically laid wires over a central core, which may be (1) rope or fibrous material, (2) copper helix, or (3) twisted copper I beam. This construction reduces the skin-effect ratio and is desirable in order to obtain economical use of the copper at high currents (see Table 2-8).

Segmental conductors are single conductors composed of either four or three segments which are combined to give a substantially circular cross section. The segments are electrically separated (usually by means of paper tape), and each strand of the individually stranded segments is alternately transposed between the inner and outer positions in the complete conductor due to its concentric lay in its segment. This construction reduces the skin-effect ratio and is desirable where high current-carrying capacity must be combined with small diameter (see Table 2-8).

2-160. Copper stranded conductors for overhead-line applications are normally supplied in hard-drawn or medium-hard-drawn tempers in No. 4 AWG and larger. Soft or annealed conductors are used in all sizes for insulated conductors and to some extent for weather-resisting (weatherproof) conductors in overhead distribution systems.

2-161. Copper-alloy stranded conductors are available in the same grades as copper-alloy solid conductors (see Par. **2-153**). Generally they are used for high strength together with conductance, where corrosion conditions do not permit the use of cheaper constructions; for applications such as overhead ground wires, messengers, railway catenaries; etc.

2-162. Copper-clad steel stranded conductors are used in the same manner as copper-alloy stranded conductors, where their higher strength and limited range of conductance are suitable.

Aluminum-clad steel stranded conductors are used where high strength, limited conductance, and good corrosion resistance are required. They are widely used for overhead ground wires, neutral messenger and messenger strands, antenna conductors, and guy wires.

2-163. Aluminum stranded conductors for bare and weatherproof overhead-line applications are normally supplied hard-drawn in No. 6 AWG and larger. Hard-drawn or three-quarter hard-drawn conductors are used in all sizes for insulated conductors.

2-164. Iron and steel stranded conductors are used for overhead ground wires, messengers, and guy wires.

2-165. Hollow (expanded) conductors are used on high-voltage transmission lines when, in order to reduce corona loss, it is desirable to increase the outside diameter without increasing the area beyond that needed for maximum line economy. Not only is the initial corona voltage considerably higher than for conventional conductors of equal cross section, but the current-carrying capacity for a given temperature rise is also greater because of the larger surface area available for cooling and the better disposition of the metal with respect to skin effect when carrying alternating currents.

Air-expanded ACSR is a conductor whose diameter has been increased by aluminum skeletal wires between the steel core and the outer layers of aluminum strands creating air spaces. A conductor having the necessary diameter to minimize corona effects on lines operating above 300 kV will, many times, have more metal than is economical if the conductor is made conventionally.

2-166. Composite conductors are those made up of usually two different types of wire having differing characteristics. They are generally designed for a ratio of physical and electrical characteristics different from those found in homogeneous materials.

Cables of this type are particularly adaptable to long-span construction or other service conditions requiring more than average strength combined with liberal conductance. They lend themselves readily to economical, dependable use on transmission lines, rural distribution lines, railroad electrification, river crossings, and many kinds of special construction.

2-167. Self-damping ACSR conductors are used to limit aeolian vibration to a safe level regardless of conductor tension or span length. They are concentrically stranded conductors composed of two layers of trapezoidal-shaped wires or two layers of trapezoidal-shaped

TABLE 2-11 Copper Wire—Wire Gages, Diameter, Area, Weight

Gage name						Diameter at 20 C (68 F)		Area at 20 C (68 F)			Weight at 20 C (68 F) bare copper wire		
American (B & S) wire gage, AWG	Steel wire gage, Stl. wg	Birmingham (Stubs' iron) wire gage, BWG	Old English (London) wire gage	(British) standard wire gage, SWG	Metric wire gage	Mils	Mm	Sq mils	Cir mils	Sq mm	Lb per 1,000 ft	Lb per mile	Kg per km
				7/0		500.0	12.70	196,300	250,000	126.7	756.7	3996	1126
	7/0					490.0	12.45	188,600	240,100	121.7	726.8	3837	1082
				6/0		464	11.79	169,100	215,000	109.1	651.7	3441	969.8
	6/0					461.5	11.72	167,300	213,000	107.9	644.7	3404	959.4
4/0						460.0	11.68	166,200	211,600	107.2	640.5	3382	953.2
		4/0	4/0			454	11.53	161,900	206,100	104.4	623.9	3294	928.5
				5/0		432	10.97	146,600	186,600	94.56	564.9	2983	840.7
	5/0					430.5	10.93	145,600	185,300	93.91	561.0	2962	834.8
		3/0	3/0			425	10.80	141,900	180,600	91.52	546.8	2887	813.7
3/0						409.6	10.40	131,800	167,800	85.01	507.8	2681	755.7
				4/0		400	10.16	125,700	160,000	81.07	484.3	2557	720.8
	4/0					393.8	10.00	121,800	155,100	78.58	469.4	2479	698.6
					100	393.7	10.00	121,700	155,000	78.54	469.2	2477	698.2
		2/0	2/0			380	9.652	113,400	144,400	73.17	437.1	2308	650.5
				3/0		372	9.449	108,700	138,400	70.12	418.9	2212	623.4
2/0						364.8	9.266	104,500	133,100	67.43	402.8	2127	599.5
	3/0					362.5	9.208	103,200	131,400	66.58	397.8	2100	591.9
					90	354.3	9.0	98,589	125,500	63.62	380.0	2007	565.6
				2/0		348	8.839	95,115	121,100	61.36	366.6	1936	545.5
		1/0	1/0			340	8.636	90,790	115,600	58.58	349.9	1848	520.7
	2/0					331.0	8.407	86,050	109,600	55.52	331.6	1751	493.5
1/0						324.9	8.252	82,891	105,600	53.49	319.5	1687	477.6
				1/0		324	8.230	82,450	105,000	53.19	317.8	1678	472.9
					80	314.96	8.0	77,931	99,200	50.27	300.3	1585	446.9
	1/0					306.5	7.785	73,780	93,940	47.60	284.4	1501	423.2
		1	1	1		300	7.620	70,690	90,000	45.60	272.4	1438	405.4
1						289.3	7.348	65,730	83,690	42.41	253.3	1338	377.0
		2	2			284	7.214	63,350	80,660	40.87	244.4	1289	363.3
	1					283.0	7.188	62,900	80,090	40.58	242.4	1280	360.8
				2		276	7.010	59,830	76,180	38.60	230.6	1217	343.1
					70	275.59	7.0	59,650	75,950	38.49	229.9	1214	342.1
	2					262.5	6.668	54,120	68,910	34.92	208.6	1101	310.4
		3	3			259	6.579	52,690	67,080	33.99	203.1	1072	302.2
2						257.6	6.543	52,120	66,360	33.62	200.9	1061	298.9
				3		252	6.401	49,880	63,500	32.18	192.2	1015	286.1

267.5	949.2	179.8	30.09	59,390	46,640	6.190	243.7
255.2	905.3	171.5	28.70	56,640	44,490	6.045	238
251.4	881.8	168.9	28.27	55,800	43,830	6.000	236.2
242.5	860.3	162.9	27.27	53,820	42,270	5.893	232
237.1	841.1	159.3	26.67	52,620	41,330	5.827	229.4
228.7	811.3	153.7	25.72	50,760	39,870	5.723	225.3
218.0	773.6	146.5	24.52	48,400	38,010	5.588	220
202.5	718.3	136.0	22.77	44,940	35,300	5.385	212
193.0	684.0	129.7	21.71	42,850	33,605	5.258	207.0
188.0	667.1	126.3	21.15	41,740	32,780	5.189	204.3
185.6	658.6	124.7	20.88	41,210	32,370	5.156	203
174.6	619.3	117.3	19.63	38,750	30,430	5.0	196.8
166.1	589.2	111.6	18.68	36,860	28,950	4.877	192.0
149.0	528.8	100.2	16.77	33,090	25,990	4.620	181.9
146.0	517.8	98.07	16.42	32,400	25,450	4.572	180
141.4	501.7	95.01	15.90	31,390	24,650	4.500	177.2
141.1	500.7	94.83	15.87	31,330	24,610	4.496	177.0
139.5	495.1	93.76	15.70	30,980	24,330	4.470	176
122.6	435.1	82.41	13.80	27,220	21,380	4.191	165
118.2	419.4	79.44	13.30	26,240	20,610	4.115	162.0
115.3	409.2	77.49	12.97	25,600	20,110	4.064	160
111.7	396.4	75.09	12.57	24,800	19,480	4.000	157.5
99.07	351.5	66.57	11.14	21,990	17,270	3.767	148.3
98.67	350.1	66.30	11.10	21,900	17,200	3.759	148
93.80	332.8	63.03	10.55	20,820	16,350	3.665	144.3
93.41	331.4	62.77	10.51	20,740	16,290	3.658	144
85.53	303.5	57.47	9.622	18,990	14,910	3.5	137.8
82.10	291.0	55.17	9.235	18,220	14,310	3.429	135.0
80.89	287.0	54.35	9.098	17,960	14,100	3.404	134.0
74.38	263.9	49.98	8.367	16,510	12,970	3.264	128.5
73.80	261.9	49.59	8.302	16,380	12,870	3.251	128
65.41	232.1	43.95	7.358	14,520	11,400	3.061	120.5
64.87	230.2	43.59	7.297	14,400	11,310	3.048	120
62.83	222.9	42.22	7.068	13,950	10,960	3.0	118.1
60.61	215.1	40.73	6.818	13,460	10,570	2.946	116
58.95	209.2	39.61	6.631	13,090	10,280	2.906	114.4
53.52	189.9	35.96	6.020	11,880	9,331	2.769	109
50.14	177.9	33.69	5.640	11,130	8,742	2.680	105.5
48.72	172.9	32.74	5.481	10,820	8,495	2.642	104
46.77	166.0	31.43	5.261	10,380	8,155	2.588	101.9
43.63	154.8	29.32	4.908	9,687	7,609	2.5	98.42
40.65	144.2	27.32	4.573	9,025	7,088	2.413	95
38.13	135.3	25.62	4.289	8,464	6,648	2.337	92
37.71	133.8	25.34	4.242	8,372	6,576	2.324	91.5
37.1	131	24.90	4.17	8,230	6,460	2.304	90.7

2-39

TABLE 2-11 Copper Wire—Wire Gages, Diameter, Area, Weight (*Continued*)

American (B & S) wire gage, AWG	Steel wire gage, Stl. wg	Birmingham (Stubs' iron) wire gage, BWG	Old English (London) wire gage	(British) standard wire gage, SWG	Metric wire gage	Mils	Mm	Sq mils	Cir mils	Sq mm	Lb per 1,000 ft	Lb per mile	Kg per km
						Diameter at 20 C (68 F)		**Area at 20 C (68 F)**			**Weight at 20 C (68 F) bare copper wire**		
...	...	14	14	...	.....	83	2.108	5,411	6,889	3.491	20.85	110.1	31.03
12	...	...	...	...	.....	80	2.05	5,130	6,530	3.310	19.8	104	29.4
...	14	...	...	14	.....	80.0	2.032	5,029	6,400	3.243	19.37	102.3	28.83
...	...	...	...	...	20	78.74	2.0	4,869	6,200	3.142	18.77	99.09	27.93
...	15	15	15	15	.....	72.0	1.829	4,072	5,184	2.627	15.69	82.85	23.35
13	...	...	...	...	.....	72.0	1.83	4,070	5,180	2.63	15.7	82.9	23.4
...	...	...	...	...	18	70.87	1.8	3,944	5,022	2.545	15.20	80.27	22.62
...	...	16	16	...	.....	65	1.651	3,318	4,225	2.141	12.79	67.53	19.03
14	...	...	...	...	.....	64.1	1.63	3,230	4,110	2.08	12.4	65.7	18.5
...	...	...	...	16	.....	64	1.626	3,217	4,096	2.075	12.40	65.46	18.45
...	...	...	...	...	16	62.99	1.6	3,116	3,968	2.011	12.01	63.41	17.87
...	16	...	...	...	.....	62.5	1.588	3,068	3,906	1.979	11.82	62.43	17.60
...	...	17	17	...	.....	58	1.473	2,642	3,364	1.705	10.18	53.77	15.15
15	...	...	...	...	.....	57.1	1.45	2,560	3,260	1.650	9.87	52.1	14.7
...	...	...	...	17	.....	56	1.422	2,463	3,136	1.589	9.493	50.12	14.13
...	...	...	...	...	14	55.12	1.4	2,386	3,038	1.539	9.196	48.56	13.69
...	17	...	...	...	.....	54.0	1.372	2,290	2,916	1.478	8.827	46.60	13.14
16	...	...	...	...	.....	50.8	1.29	2,030	2,580	1.31	7.81	41.2	11.6
...	...	18	18	...	.....	49	1.245	1,886	2,401	1.217	7.268	38.37	10.82
...	...	...	...	18	.....	48	1.219	1,810	2,304	1.167	6.974	36.82	10.38
...	18	...	...	...	.....	47.5	1.207	1,772	2,256	1.143	6.830	36.06	10.16
...	...	...	...	...	12	47.24	1.200	1,753	2,232	1.131	6.756	35.67	10.05
17	...	...	...	...	.....	45.3	1.150	1,610	2,050	1.040	6.21	32.8	9.24
...	...	19	...	...	.....	42	1.067	1,385	1,764	0.8938	5.340	28.19	7.946
...	19	...	...	...	.....	41.0	1.041	1,320	1,681	0.8518	5.088	26.87	7.572
18	...	...	...	...	.....	40.3	1.02	1,280	1,620	0.823	4.92	26.0	7.32
...	...	...	19	19	.....	40	1.016	1,257	1,600	0.8107	4.843	25.57	7.207
...	...	...	...	...	10	39.37	1	1,217	1,550	0.7854	4.692	24.77	6.982
...	...	...	...	20	.....	36	0.9144	1,018	1,296	0.6567	3.923	20.71	5.838
19	...	...	...	...	.....	35.9	0.912	1,010	1,290	0.653	3.90	20.6	5.81

(A)	(B)	(C)	(D)	(E)	(F)	(G)	(H)						
5.656	20.07	3.800	0.6362	1,255	986.1	0.90	35.43						
5.520	19.58	3.708	0.6207	1,225	962.1	0.8839	35			20			
5.455	19.36	3.666	0.6136	1,211	951.1	0.8128	34.8	9	21		20	20	
4.613	16.37	3.100	0.5189	1,024	804.2	0.813	32.0						20
4.61	16.4	3.10	0.519	1,020	804.0		32.0			21	21		
4.527	16.06	3.042	0.5092	1,005	789.2	0.8052	31.7					21	
4.469	15.85	3.003	0.5027	992	779.1	0.80	31.5	8					21
3.920	13.91	2.634	0.4410	870	683.3	0.7493	29.50		22	22			
3.685	13.07	2.476	0.4145	818.0	642.4	0.7264	28.6				22	22	
3.66	13.0	2.46	0.412	812.0	638.0	0.724	28.5						22
3.532	12.53	2.373	0.3973	784.0	615.8	0.7112	28			23			
3.421	12.14	2.299	0.3848	759.5	596.5	0.70	27.56	7	23		23	23	
3.284	11.65	2.207	0.3694	729.1	572.6	0.6858	27.00						23
2.998	10.64	2.015	0.3373	665.6	522.8	0.6553	25.8						
2.88	10.2	1.94	0.324	640.0	503.0	0.643	25.3			24			
2.815	9.989	1.892	0.3167	625.0	490.9	0.6350	25				24	24	
2.595	9.206	1.744	0.2919	576.0	452.4	0.60	24	6	24				24
2.514	8.918	1.689	0.2827	558.0	438.3	0.5842	23.62						
2.383	8.455	1.601	0.2680	529.0	415.5	0.574	23.0			25			
2.30	8.16	1.55	0.259	511.0	401.0		22.6				25	25	
2.180	7.736	1.465	0.2452	484.0	380.1	0.5588	22						25
1.893	6.717	1.272	0.2129	420.3	330.1	0.5207	20.5	5	25	26			
1.875	6.651	1.260	0.2109	416.2	326.9	0.5182	20.4				26	26	
1.82	6.46	1.22	0.205	404.0	317.0	0.511	20.1						26
1.802	6.393	1.211	0.2027	400.0	314.2	0.5080	20			27			
1.746	6.193	1.173	0.1963	387.5	304.3	0.50	19.68						
1.584	5.619	1.064	0.1781	351.6	276.1	0.4763	18.75	4.5	26		27	27	
1.476	5.236	0.9916	0.1660	327.6	257.3	0.4597	18.1			28			27
1.460	5.178	0.9807	0.1642	324.0	254.5	0.4572	18						
1.44	5.12	0.970	0.162	320.0	252.0	0.455	17.9						
1.414	5.016	0.9501	0.1590	313.9	246.5	0.45	17.72				28	28	
1.348	4.783	0.9059	0.1517	299.3	235.1	0.4394	17.3	4	27	29			28
1.226	4.351	0.8241	0.1380	272.3	213.8	0.4191	16.50						
1.212	4.299	0.8141	0.1363	269.0	211.2	0.4166	16.4						
1.182	4.194	0.7944	0.1330	262.4	206.1	0.4115	16.2						
1.153	4.092	0.7749	0.1297	256.0	201.1	0.4064	16				29	29	
1.14	4.04	0.765	0.128	253.0	199.0	0.404	15.9	3.5	28	30			
1.117	3.964	0.7507	0.1257	248.0	194.8	0.3937	15.75						
1.082	3.840	0.7272	0.1217	254.1	188.7	0.3810	15.50						
1.014	3.596	0.6811	0.1140	225.0	176.7		15.0						
0.9867	3.501	0.6630	0.1110	219.0	172.0	0.3759	14.8				30	30	
0.908	3.22	0.610	0.102	202.0	158.0	0.361	14.2						
0.8829	3.133	0.5933	0.09932	196.0	153.9	0.3556	14.0						
0.8553	3.035	0.5747	0.09621	189.9	149.1	0.35	13.78						
0.8517	3.022	0.5723	0.09580	189.1	148.5	0.3493	13.75						

TABLE 2-11 Copper Wire—Wire Gages, Diameter, Area, Weight (*Continued*)

American (B & S) wire gage, AWG	Steel wire gage, Stl. wg	Birmingham (Stubs' iron) wire gage, BWG	Old English (London) wire gage	(British) standard wire gage, SWG	Metric wire gage	Mils	Mm	Sq mils	Cir mils	Sq mm	Lb per 1,000 ft	Lb per mile	Kg per km
				Gage No.									
...	...	...	...		29	13.6	0.3454	145.3	185.0	0.09372	0.5599	2.956	0.8332
...	31	...	...	...	.....	13.2	0.3353	136.8	174.2	0.08829	0.5274	2.785	0.7849
...	..	29	...	...		13	0.3302	132.7	169.0	0.08563	0.5116	2.701	0.7613
...	32	...	...	...	.....	12.8	0.3251	128.7	163.8	0.08302	0.4959	2.619	0.7380
28	...	...	...	...		12.6	0.320	125	159.0	0.0804	0.481	2.54	0.715
...	...	...	...		30	12.4	0.3150	120.8	153.8	0.07791	0.4654	2.457	0.6926
...	...	...	31	...	.....	12.25	0.3112	117.9	150.1	0.07604	0.4542	2.398	0.6760
...	...	30	...	...		12	0.3048	113.1	144.0	0.07297	0.4359	2.301	0.6487
...	...	...	...		3	11.81	0.30	109.6	139.5	0.07069	0.4223	2.230	0.6284
...	33	...	...	...		11.8	0.2997	109.4	139.2	0.07055	0.4215	2.225	0.6272
...	...	...	...	31		11.6	0.2946	105.7	134.6	0.06818	0.4073	2.151	0.6061
29	...	...	...	...		11.3	0.287	100.0	128.0	0.0647	0.387	2.04	0.575
...	...	...	32	...		11.25	0.2858	99.40	126.6	0.06413	0.3831	2.023	0.5701
...	...	...	...	32		10.8	0.2743	91.61	116.6	0.05910	0.3531	1.864	0.5254
...	34	...	...	...		10.4	0.2642	84.95	108.2	0.05481	0.3274	1.729	0.4872
...	...	...	33	...		10.25	0.2604	82.52	105.1	0.05324	0.3180	1.679	0.4733
30	...	...	...	...	.....	10.0	0.254	78.50	100.0	0.0507	0.303	1.60	0.450
...	...	31	...	33		10.0	0.2540	78.54	100.0	0.05067	0.3027	1.598	0.4505
...	...	...	...		2.5	9.842	0.25	76.09	96.87	0.04909	0.2932	1.548	0.4364
...	35	...	34	...		9.5	0.2413	70.88	90.25	0.04573	0.2732	1.442	0.4065
...	...	...	...	34		9.2	0.2337	66.48	84.64	0.04289	0.2562	1.353	0.3813
...	36	32	35	...	.....	9.0	0.2286	63.62	81.00	0.04104	0.2452	1.295	0.3649
31	..	...	...	...	.....	8.9	0.226	62.20	79.2	0.0401	0.240	1.27	0.357 .
...	37	...	...	...	.....	8.5	0.2159	56.75	72.25	0.03661	0.2187	1.155	0.3255
...	...	...	...	35		8.4	0.2134	55.42	70.56	0.03575	0.2139	1.128	0.3178
...	38	33	...	...		8.0	0.2032	50.27	64.00	0.03243	0.1937	1.023	0.2883
32	...	...	...	...		8.0	0.203	50.30	64.0	0.0324	0.194	1.02	0.288
...	...	...	...		2	7.874	0.20	48.69	62.00	0.03142	0.1877	0.9909	0.2793
...	...	...	...	36		7.6	0.1930	45.36	57.76	0.02927	0.1748	0.9231	0.2602
...	39	...	36	...		7.5	0.1905	44.18	56.25	0.02850	0.1703	0.8990	0.2534
...	...	...	...		1.8	7.087	0.1800	39.44	50.22	0.02545	0.1520	0.8026	0.2262
33	...	...	...	...	.....	7.1	0.180	39.60	50.40	0.0255	0.153	0.806	0.227
...	40	34	...	...	.....	7.0	0.1778	38.48	49.00	0.02483	0.1483	0.7831	0.2207
...	..	...	...	37	.....	6.8	0.1727	36.32	46.24	0.02343	0.1400	0.7390	0.2083
...	41	...	...	...	.....	6.6	0.1676	34.21	43.56	0.02207	0.1319	0.6962	0.1962

								Gauge
0.1903	0.6753	0.1279	0.02141	42.25	33.18	0.1651	6.50	34
0.179	0.634	0.120	0.0201	39.7	31.20	0.160	6.3	
0.1787	0.6342	0.1201	0.02011	39.68	31.16	0.16	6.299	
0.1732	0.6144	0.1164	0.01948	38.44	30.19	0.1575	6.2	
0.1622	0.5754	0.1090	0.01824	36.00	28.27	0.1524	6.0	
0.1571	0.5575	0.1056	0.01767	34.87	27.39	0.15	5.906	35
0.1515	0.5377	0.1018	0.01705	33.64	26.42	0.1473	5.8	
0.1489	0.5284	0.1001	0.01675	33.06	25.97	0.1461	5.75	
0.141	0.501	0.0949	0.0159	31.4	24.60	0.142	5.6	
0.1369	0.4855	0.09196	0.01539	30.38	23.86	0.14	5.512	
0.1363	0.4835	0.09157	0.01533	30.25	23.76	0.1397	5.5	36
0.1218	0.4322	0.08185	0.01370	27.04	21.24	0.1321	5.2	
0.113	0.400	0.0757	0.0127	25.00	19.60	0.127	5.0	
0.1038	0.3682	0.06974	0.01167	23.04	18.10	0.1219	4.8	
0.1005	0.3567	0.06756	0.01131	22.32	17.53	0.12	4.724	
0.09532	0.3382	0.06405	0.01072	21.16	16.62	0.1168	4.6	37
0.09122	0.3236	0.06130	0.01026	20.25	15.90	0.1143	4.50	
0.0912	0.324	0.0613	0.0103	20.20	15.90	0.114	4.50	
0.08721	0.3094	0.05860	0.009810	19.36	15.21	0.1118	4.4	38
0.07207	0.2557	0.04843	0.008107	16.00	12.57	0.1016	4.0	
0.0721	0.256	0.0484	0.00811	16.0	12.60	0.102	4.0	39
0.06982	0.2477	0.04692	0.007854	15.50	12.17	0.10	3.937	
0.05838	0.2071	0.03923	0.006567	12.96	10.18	0.09144	3.6	40
0.0552	0.196	0.0371	0.00621	12.2	9.62	0.0889	3.5	41
0.04613	0.1637	0.0310	0.005189	10.24	8.042	0.08128	3.2	42
0.0433	0.154	0.0291	0.00487	9.61	7.55	0.0787	3.1	43
0.0353	0.125	0.0237	0.00397	7.84	6.16	0.0711	2.8	
0.03532	0.1253	0.02373	0.003973	7.840	6.158	0.07112	2.8	44
0.0282	0.0999	0.0189	0.00317	6.25	4.91	0.0635	2.5	45
0.02595	0.09206	0.01744	0.002919	5.760	4.524	0.06096	2.4	46
0.0218	0.0774	0.0147	0.00245	4.84	3.80	0.0559	2.2	47
0.01802	0.06393	0.01211	0.002027	4.000	3.142	0.05080	2.0	
0.0180	0.0639	0.0121	0.00203	4.00	3.14	0.0508	2.0	
0.01746	0.06193	0.01173	0.001963	3.875	3.043	0.05	1.969	48
0.0146	0.0519	0.00981	0.00164	3.24	2.54	0.0457	1.8	
0.01153	0.04092	0.007749	0.001297	2.560	2.011	0.04064	1.6	
0.0115	0.0409	0.00775	0.00130	2.56	2.01	0.0406	1.6	49
0.00883	0.0313	0.00593	0.000993	1.96	1.54	0.0356	1.4	
0.00649	0.0230	0.00436	0.000730	1.44	1.13	0.0305	1.2	50
0.006487	0.02301	0.004359	0.0007297	1.440	1.131	0.03048	1.2	
0.00545	0.0193	0.00366	0.000613	1.21	0.950	0.0279	1.1	
0.004505	0.01598	0.003027	0.0005067	1.000	0.7854	0.02540	1.0	
0.00450	0.0160	0.00303	0.000507	1.00	0.785	0.0254	1.0	

wires and one layer of round wires of 1350 (EC) alloy with a high-strength, coated steel core (Fig. 2-14).

FIG. 2-14   Self-damping ACSR conductor.

2-168. Conductor Data Tables
Table 2-9 Industry Specifications for Conductors
Table 2-10 Conductors—Physical and Electrical Properties
Table 2-11 Copper Wire—Wire Gages, Diameter, Area, Weight
Table 2-12 Copper Wire—Tensile Strength, Elongation
Table 2-13 Copper Wire—Weight, Breaking Strength, DC Resistance
Table 2-14 Copper Cable—Stranding Classes, Uses
Table 2-15 Copper Cable, Concentric Lay—Dimensions, Weight
Table 2-16 Copper Cable, Rope Lay—Dimensions, Weight
Table 2-17 Copper Cable, Classes AA, A, B—Weight, Breaking Strength, DC Resistance
Table 2-18 Copper-Clad Steel Wire and Cable—Weight, Breaking Strength, DC Resistance
Table 2-19 Copper-Clad Steel-Copper Cable—Weight, Breaking Strength, DC Resistance
Table 2-20 Aluminum Wire—Dimensions, Weight, DC Resistance
Table 2-21 Aluminum Wire—Stranding Classes, Uses
Table 2-22 Aluminum Conductor—Physical Characteristics (EC-H19)
Table 2-23 ACSR Conductor—Physical Characteristics
Table 2-24 Aluminum-Clad Steel Wire and Cable—Weight, Breaking Strength, DC Resistance
Table 2-25 Galvanized-Steel Wire—Weight, Breaking Strength, DC Resistance
Table 2-26 Galvanized-Steel Strand—Dimensions, Weight, Breaking Strength
Table 2-27 Copper Wire and Cable—Electrical Characteristics
Table 2-28 Copper-Clad-Steel Cable—Electrical Characteristics
Table 2-29 Copper-Clad-Steel Copper Cable—Electrical Characteristics
Table 2-30 ACSR Conductor—Electrical Characteristics
Table 2-31 Aluminum-Clad Steel Cable—Electrical Characteristics
Table 2-32 Inductive-Reactance Spacing Factors
Table 2-33 Capacitive-Reactance Spacing Factors
Table 2-34 Properties of Resistance Metals and Alloys

Fusible Metals and Alloys

2-169. Fusible alloys having melting points in the range from about 60 to 200°C are made principally of bismuth, cadmium, lead, and tin in various proportions. Many of these alloys have been known under the names of their inventors (see index of alloys in "International Critical Tables," vol. 2), but typical compositions and melting points are shown in Table 2-35.

2-170. Compositions and Melting Points of Fusible Alloys. See Table 2-35.

2-171. Fuse metals for electric fuses of the open-link enclosed and expulsion types are ordinarily made of some low-fusible alloy; aluminum also is used to some extent. The resistance of the fuse causes dissipation of energy, liberation of heat, and rise of temperature. Sufficient current will obviously melt the fuse and thus open the circuit if the resultant arc is self-extinguishing. Metals which volatilize readily in the heat of the arc are to be preferred to those which leave a residue of globules of hot metal. The rating of any fuse depends critically upon its shape, dimensions, mounting, enclosure, and any other factors which affect its heat-dissipating capacity.

(Numbered paragraphs resume on page 2-73.)

TABLE 2-12 Copper Wire—Tensile Strength, Elongation

(ASTM Specifications B1, B2, B3)

Size,* AWG	Diameter at 20 C (68 F),† in	Area at 20 C (68 F)		Tensile strength, psi				Elongation, min %				
				Hard	Medium		Soft§	Hard		Medium		Soft
		Cir mils	Sq in	Min	Min	Max		In 10 in	In 60 in	In 10 in	In 60 in	In 10 in
4/0	0.4600	211,600	0.1662	49,000	42,000	49,000	......	3.75	....	3.75	....	35
3/0	0.4096	167,800	0.1318	51,000	43,000	50,000	......	3.25	....	3.60	....	35
2/0	0.3648	133,100	0.1045	52,800	44,000	51,000	......	2.80	....	3.25	....	35
1/0	0.3249	105,600	0.08291	54,500	45,000	52,000	......	2.40	....	3.00	....	35
1	0.2893	83,690	0.06573	56,100	46,000	53,000	......	2.17	....	2.75	....	30
2	0.2576	66,360	0.05212	57,600	47,000	54,000	......	1.98	....	2.50	....	30
3	0.2294	52,620	0.04133	59,000	48,000	55,000	......	1.79	....	2.25	....	30
4	0.2043	41,740	0.03278	60,100	48,330	55,330	......	....	1.24	....	1.25	30
5	0.1819	33,090	0.02599	61,200	48,660	55,660	......	....	1.18	....	1.20	30
...	0.1650‡	27,220	0.02138	62,000	......	......	......	....	1.14			
6	0.1620	26,240	0.02061	62,100	49,000	56,000	......	....	1.14	....	1.15	30
7	0.1443	20,820	0.01635	63,000	49,330	56,330	......	....	1.09	....	1.11	30
...	0.1340‡	17,960	0.01410	63,400				....	1.07			
8	0.1285	16,510	0.01297	63,700	49,660	56,660	......	....	1.06	....	1.08	30
9	0.1144	13,090	0.01028	64,300	50,000	57,000	......	....	1.02	....	1.06	30
...	0.1040‡	10,820	0.008495	64,800				....	1.00			
10	0.1019	10,380	0.008155	64,900	50,330	57,300	......	....	1.00	....	1.04	25
...	0.0920‡	8,460	0.00665	65,400				....	0.97			
11	0.0907	8,230	0.00646	65,400	50,660	57,660	......	....	0.97	....	1.02	25
12	0.0808	6,530	0.00513	65,700	51,000	58,000	......	....	0.95	....	1.00	25
...	0.0800‡	6,400	0.00503	65,700				....	0.94			
13	0.0720	5,180	0.00407	65,900	51,330	58,330	......	....	0.92	....	0.98	25
...	0.0650‡	4,220	0.00332	66,200				....	0.91			
14	0.0641	4,110	0.00323	66,200	51,660	58,660	......	....	0.90	....	0.96	25
15	0.0571	3,260	0.00256	66,400	52,000	59,000	......	....	0.89	....	0.94	25
16	0.0508	2,580	0.00203	66,600	52,330	59,330	......	....	0.87	....	0.92	25
17	0.0453	2,050	0.00161	66,800	52,660	59,660	......	....	0.86	....	0.90	25
18	0.0403	1,620	0.00128	67,000	53,000	60,000	......	....	0.85	....	0.88	25
19	0.0359	1,290	0.00101	......	......	......	......	....	....	....	....	25
20	0.0320	1,020	0.000804	......	......	......	......	....	....	....	....	25
21	0.0285	812	0.000638									25
22	0.0253	640	0.000503									25
23	0.0226	511	0.000401									25
24	0.0201	404	0.000317									20
25	0.0179	320	0.000252									20
26	0.0159	253	0.000199									20
27	0.0142	202	0.000158									20
28	0.0126	159	0.000125									20
29	0.0113	128	0.000100									20
30	0.0100	100	0.0000785									15
31	0.0089	79.2	0.0000622									15
32	0.0080	64.0	0.0000503									15
33	0.0071	50.4	0.0000396									15
34	0.0063	39.7	0.0000312									15
35	0.0056	31.4	0.0000246									15
36	0.0050	25.0	0.0000196									15
37	0.0045	20.2	0.0000159									15
38	0.0040	16.0	0.0000126									15
39	0.0035	12.2	0.00000962									15
40	0.0031	9.61	0.00000755									15
ASTM Specification Designation.............				B1	B2		B3	B1		B2		B3

*The use of gage numbers to specify wire sizes is not recognized in these specifications, because of the possibility of confusion.

†The value of wire diameters in this table which correspond to gage numbers of the AWG are in agreement with the standard nominal diameters prescribed in ASTM Specification B258. For wire whose nominal diameter is more than 0.001 in (1 mil) greater than a size listed in the table and less than that of the next larger size, the requirements of the next larger size shall apply.

‡Diameters often employed by purchasers for communication lines, but not in the AWG (B. & S. wire gage) series. They correspond to certain of the numbers of the Birmingham wire gage or of the (British) standard wire gage.

§No requirements for tensile strength are specified.

NOTE: 1 in = 2.54 cm; 1 in^2 = 64.5 cm^2; 1 lb/in^2 = 6.895 kPa.

TABLE 2-13 Copper Wire—Weight, Breaking Strength, DC Resistance

(Based on ASTM Specifications B1, B2, B3)

Size, AWG	Diameter, in.	Area		Weight		Hard		Medium		Soft	
		Cir mils	Sq in.	Lb per 1,000 ft	Lb per mile	Breaking strength, minimum,* lb	D-c resistance at 20 C (68 F) maximum,† ohms per 1,000 ft	Breaking strength, minimum,* lb	D-c resistance at 20 C (68 F) maximum,† ohms per 1,000 ft	Breaking strength, maximum,‡ lb	D-c resistance at 20 C (68 F) maximum,† ohms per 1,000 ft
4/0	0.4600	211,600	0.1662	640.5	3382	8143	0.05045	6980	0.05019	5983	0.04901
3/0	0.4096	167,800	0.1318	507.8	2681	6720	0.06362	5666	0.06330	4744	0.06182
2/0	0.3648	133,100	0.1045	402.8	2127	5519	0.08021	4599	0.07980	3763	0.07793
1/0	0.3249	105,600	0.08291	319.5	1687	4518	0.1022	3731	0.1016	2985	0.09825
1	0.2893	83,690	0.06573	253.3	1338	3688	0.1289	3024	0.1282	2432	0.1239
2	0.2576	66,360	0.05212	200.9	1061	3002	0.1625	2450	0.1617	1928	0.1563
3	0.2294	52,620	0.04133	159.3	841.1	2439	0.2050	1984	0.2039	1529	0.1971
4	0.2043	41,740	0.03278	126.3	667.1	1970	0.2584	1584	0.2571	1213	0.2485
5	0.1819	33,090	0.02599	100.2	528.8	1590	0.3260	1265	0.3243	961.5	0.3135
6	0.1620	26,240	0.02061	79.44	419.4	1280	0.4110	1010	0.4088	762.6	0.3952
7	0.1443	20,820	0.01635	63.03	332.8	1030	0.5180	806.7	0.5153	605.1	0.4981
8	0.1285	16,510	0.01297	49.98	263.9	826.1	0.6532	644.0	0.6498	479.8	0.6281
9	0.1144	13,090	0.01028	39.61	209.2	660.9	0.8241	513.9	0.8199	380.3	0.7925
10	0.1019	10,380	0.008155	31.43	166.0	529.3	1.039	410.5	1.033	314.0	0.9988
11	0.0907	8,230	0.00646	24.9	131	423	1.31	327	1.30	249	1.26
12	0.0808	6,530	0.00513	19.8	104	337	1.65	262	1.64	197	1.59
13	0.0720	5,180	0.00407	15.7	82.9	268	2.08	209	2.07	157	2.00
14	0.0641	4,110	0.00323	12.4	65.7	214	2.63	167	2.61	124	2.52
15	0.0571	3,260	0.00256	9.87	52.1	170	3.31	133	3.29	98.6	3.18
16	0.0508	2,580	0.00203	7.81	41.2	135	4.18	106	4.16	78.0	4.02
17	0.0453	2,050	0.00161	6.21	32.8	108	5.26	84.9	5.23	62.1	5.05
18	0.0403	1,620	0.00128	4.92	26.0	85.5	6.64	67.6	6.61	49.1	6.39
19	0.0359	1,290	0.00101	3.90	20.6	68.0	8.37	54.0	8.33	39.0	8.05
20	0.0320	1,020	0.000804	3.10	16.4	54.2	10.5	43.2	10.5	31.0	10.1
21	0.0285	812	0.000638	2.46	13.0	43.2	13.3	34.4	13.2	24.6	12.8
22	0.0253	640	0.000503	1.94	10.2	34.1	16.9	27.3	16.8	19.4	16.2
23	0.0226	511	0.000401	1.55	8.16	27.3	21.1	21.9	21.0	15.4	20.3
24	0.0201	404	0.000317	1.22	6.46	21.7	26.7	17.5	26.6	12.7	25.7
25	0.0179	320	0.000252	0.970	5.12	17.3	33.7	13.9	33.5	10.1	32.4
26	0.0159	253	0.000199	0.765	4.04	13.7	42.7	11.1	42.4	7.94	41.0

Size, AWG	Diameter, in.	Area		Weight		Hard		Medium		Soft	
		Cir mils	Sq in.	Lb per 1,000 ft	Lb per mile	Breaking strength, minimum,* lb	D-c resistance at 20 C (68 F) maximum,† ohms per 1,000 ft	Breaking strength, minimum,* lb	D-c resistance at 20 C (68 F) maximum,† ohms per 1,000 ft	Breaking strength, maximum,‡ lb	D-c resistance at 20 C (68 F) maximum,† ohms per 1,000 ft
27	0.0142	202	0.000158	0.610	3.22	10.9	53.5	8.87	53.2	6.33	51.4
28	0.0126	159	0.000125	0.481	2.54	8.64	67.9	7.02	67.6	4.99	65.3
29	0.0113	128	0.000100	0.387	2.04	6.97	84.5	5.68	84.0	4.01	81.2
30	0.0100	100	0.0000785	0.303	1.60	5.47	108	4.48	107	3.14	104
31	0.0089	79.2	0.0000622	0.240	1.27	4.35	136	3.6	135	2.49	131
32	0.0080	64.0	0.0000503	0.194	1.02	3.53	169	2.90	168	2.01	162
33	0.0071	50.4	0.0000396	0.153	0.806	2.79	214	2.30	213	1.58	206
34	0.0063	39.7	0.0000312	0.120	0.634	2.20	272	1.82	270	1.25	261
35	0.0056	31.4	0.0000246	0.0949	0.501	1.75	344	1.44	342	0.985	331
36	0.0050	25.0	0.0000196	0.0757	0.400	1.40	431	1.16	429	0.785	415
37	0.0045	20.2	0.0000159	0.0613	0.324	1.13	533	0.944	530	0.636	512
38	0.0040	16.0	0.0000126	0.0484	0.256	0.898	674	0.750	671	0.503	648
39	0.0035	12.2	0.00000962	0.0371	0.196	0.691	880	0.577	876	0.385	847
40	0.0031	9.61	0.00000755	0.0291	0.154	0.543	1120	0.455	1120	0.302	1080
41	0.0028	7.84	0.00000616	0.0237	0.125	....	1380	....	1370	0.246	1320
42	0.0025	6.25	0.00000491	0.0189	0.0999	....	1730	....	1720	0.196	1660
43	0.0022	4.84	0.00000380	0.0147	0.0774	....	2230	....	2220	0.152	2140
44	0.0020	4.00	0.00000314	0.0121	0.0639	....	2700	....	2680	0.126	2590
ASTM Specification Designation..						B1		B2		B3	

*No. 19 AWG and smaller, based on Anaconda data.
†Based on nominal diameter and ASTM resistivities.
‡No requirements for tensile strength are specified in ASTM B3. Values given here based on Anaconda data.
NOTE: 1 in = 2.54 cm; 1 in^2 = 64.5 cm^2; 1 lb = 0.4536 kg; 1 ft = 0.3048 m; 1 mi = 1.61 km.

TABLE 2-14 Copper Cable—Stranding Classes, Uses
(ASTM Specifications)

ASTM Designation	Construction	Class	Application		
		AA	For bare conductors usually used in overhead lines		
		A	For weather-resistant (weatherproof), slow-burning conductors. For bare conductors where greater flexibility than is afforded by Class AA is required		
B8	Concentric lay	B	For conductors insulated with various materials such as rubber, paper, varnished-cambric, etc. For the conductors indicated under Class A where greater flexibility is required		
		C D	For conductors where greater flexibility is required than is provided by Class B		
B173	Rope lay with concentric-stranded members	G	Conductor constructions having a range of areas from 5,000,000 cir mils and employing 61 stranded members of 19 wires each down to No. 14 AWG containing 7 stranded members of 7 wires each (Typical uses are for rubber-sheathed conductors, apparatus conductors, portable conductors, and similar applications)		
		H	Conductor constructions having a range of areas from 5,000,000 cir mils and employing 91 stranded members of 19 wires each down to No. 9 AWG containing 19 stranded members of 7 wires each (Typical uses are for rubber-sheathed cords and conductors where greater flexibility is required, such as for use on take-up reels over sheaves and extra-flexible apparatus conductors)		
B226	Annular stranded		For bare conductors, or covered with weather-resistant (weatherproof) materials or insulated with rubber, varnished-cambric or solid-type impregnated paper		

ASTM Designation	Construction	Class	Conductor size, AWG	Individual wire size		Application
				In.	AWG	
		I	7, 8, 9, 10	0.0201	24	Rubber-covered, varnished-cambric, and paper-insulated conductors
		J	10, 12, 14, 16, 18, 20	0.0126	28	Fixture wire
		K	10, 12, 14, 16, 18, 20, 22	0.0100	30	Fixture wire, flexible cord, and portable cord
B174	Bunch stranded	L	10, 12, 14, 16, 18, 20, 24	0.0080	32	Fixture wire and portable cord with greater flexibility than Class K
		M	14, 16, 18, 20, 22, 26	0.0063	34	Heater cord and light portable cord
		O	16, 18, 20, 24, 28	0.0050	36	Heater cord with greater flexibility than Class M
		P	16, 18, 20	0.0040	38	More flexible conductors than provided in preceding classes
		Q	18, 20	0.0031	40	Oscillating fan cord. Very great flexibility

ASTM Designation	Construction	Class	Conductor size, cir mils	Individual wire size		Application
				In.	AWG	
B172	Rope lay with bunched-stranded members	I	Up to 2,000,000	0.0201	24	Typical use is for special apparatus cable
		K	Up to 1,000,000	0.0100	30	Typical use, special portable cord and conductors
		M	Up to 1,000,000	0.0063	34	Typical use is for welding conductor

NOTE: 1 in = 2.54 cm.

TABLE 2-15 Copper Cable, Concentric Lay—Dimensions, Weight
(ASTM Specification B8)

Conductor size, Mcm or Awg	Class AA			Class A			Class B			Class C†		Class D†		Approximate weight, lb per 1000 ft
	No. of wires	Diameter each wire, mils	Nominal conductor diam, in.	No. of wires	Diameter each wire, mils	Nominal conductor diam, in.	No. of wires	Diameter each wire, mils	Nominal conductor diam, in.	No. of wires	Diameter each wire, mils	No. of wires	Diameter each wire, mils	
5,000*	..	.....	.....	169	172.0	2.580	217	151.8	2.581	271	135.8	271	135.8	15,890
4,500	..	.....	.....	169	163.2	2.448	217	144.0	2.448	271	128.9	271	128.9	14,300
4,000	..	.....	.....	169	153.8	2.307	217	135.8	2.309	271	121.5	271	121.5	12,590
3,500	..	.....	.....	127	166.0	2.158	169	143.9	2.159	217	127.0	271	113.6	11,020
3,000*	..	.....	.....	127	153.7	1.998	169	133.2	1.998	217	117.6	271	105.2	9,353
2,500*	..	.....	.....	91	165.7	1.823	127	140.3	1.824	169	121.6	217	107.3	7,794
2,000*	..	.....	.....	91	148.2	1.630	127	125.5	1.632	169	108.8	217	96.0	6,175
1,900	..	.....	.....	91	144.5	1.590	127	122.3	1.590	169	106.0	217	93.6	5,866
1,800	..	.....	.....	91	140.6	1.547	127	119.1	1.548	169	103.2	217	91.1	5,558
1,750*	..	.....	.....	91	138.7	1.526	127	117.4	1.526	169	101.8	217	89.8	5,403
1,700	..	.....	.....	91	136.7	1.504	127	115.7	1.504	169	100.3	217	88.5	5,249
1,600	..	.....	.....	91	132.6	1.459	127	112.2	1.459	169	97.3	217	85.9	4,940
1,500*	..	.....	.....	61	156.8	1.411	91	128.4	1.412	127	108.7	169	94.2	4,631
1,400	..	.....	.....	61	151.5	1.364	91	124.0	1.364	127	105.0	169	91.0	4,323
1,300	..	.....	.....	61	146.0	1.314	91	119.5	1.315	127	101.2	169	87.7	4,014
1,250*	..	.....	.....	61	143.1	1.288	91	117.2	1.289	127	99.2	169	86.0	3,859
1,200	..	.....	.....	61	140.3	1.263	91	114.8	1.263	127	97.2	169	84.3	3,705
1,100	..	.....	.....	61	134.3	1.209	91	109.9	1.209	127	93.1	169	80.7	3,396
1,000*	37	164.4	1.151	61	128.0	1.152	61	128.0	1.152	91	104.8	127	88.7	3,088
900	37	156.0	1.092	61	121.5	1.094	61	121.5	1.094	91	99.4	127	84.2	2,779
800*	37	147.0	1.029	61	114.5	1.031	61	114.5	1.031	91	93.8	127	79.4	2,470
750*	37	142.4	0.997	61	110.9	0.998	61	110.9	0.998	91	90.8	127	76.8	2,316
700*	37	137.5	0.963	61	107.1	0.964	61	107.1	0.964	91	87.7	127	74.2	2,161
650	37	132.5	0.928	61	103.2	0.929	61	103.2	0.929	91	84.5	127	71.5	2,007
600*	37	127.3	0.891	37	127.3	0.891	61	99.2	0.893	91	81.2	127	68.7	1,853
550	37	121.9	0.853	37	121.9	0.853	61	95.0	0.855	91	77.7	127	65.8	1,698
500*	19	162.2	0.811	37	116.2	0.813	37	116.2	0.813	61	90.5	91	74.1	1,544
450	19	153.9	0.770	37	110.3	0.772	37	110.3	0.772	61	85.9	91	70.3	1,389
400*	19	145.1	0.726	19	145.1	0.726	37	104.0	0.728	61	81.0	91	66.3	1,235
350*	12	170.8	0.710	19	135.7	0.679	37	97.3	0.681	61	75.7	91	62.0	1,081

TABLE 2-15 Copper Cable, Concentric Lay—Dimensions, Weight (*Continued*)

Conductor size, Mcm or Awg	Class AA			Class A			Class B			Class C†		Class D†		Approximate weight, lb per 1000 ft
	No. of wires	Diameter each wire, mils	Nominal conductor diam, in.	No. of wires	Diameter each wire, mils	Nominal conductor diam, in.	No. of wires	Diameter each wire, mils	Nominal conductor diam, in.	No. of wires	Diameter each wire, mils	No. of wires	Diameter each wire, mils	
300*	12	158.1	0.657	19	125.7	0.629	37	90.0	0.630	61	70.1	91	57.4	926.3
250*	12	144.3	0.600	19	114.7	0.574	37	82.2	0.575	61	64.0	91	52.4	771.9
4/0*	7	173.9	0.522	7	173.9	0.552	19	105.5	0.528	37	75.6	61	58.9	653.3
3/0*	7	154.8	0.464	7	154.8	0.464	19	94.0	0.470	37	67.3	61	52.4	518.1
2/0*	7	137.9	0.414	7	137.9	0.414	19	83.7	0.419	37	60.0	61	46.7	410.9
1/0*	7	122.8	0.368	7	122.8	0.368	19	74.5	0.373	37	53.4	61	41.6	325.8
1	3	167.0	0.360	...	.....	.....	...	.....	.....	...	.....	...	.....	255.9
1*	..	.....	.....	7	109.3	0.328	19	66.4	0.332	37	47.6	61	37.0	258.4
2	3	148.7	0.320	...	.....	.....	...	.....	.....	...	.....	...	.....	202.9
2*	..	.....	.....	7	97.4	0.292	7	97.4	0.292	19	59.1	37	42.4	204.9
3	3	132.5	0.285	...	.....	.....	...	.....	.....	...	.....	...	.....	160.9
3*	..	.....	.....	7	86.7	0.260	7	86.7	0.260	19	52.6	37	37.7	162.5
4	3	118.0	0.254	...	.....	.....	...	.....	.....	...	.....	...	.....	127.6
4*	..	.....	.....	7	77.2	0.232	7	77.2	0.232	19	46.9	37	33.6	128.9
5*	..	.....	.....	...	.....	.....	7	68.8	0.206	19	41.7	37	29.9	102.2
6*	..	.....	.....	...	.....	.....	7	61.2	0.184	19	37.2	37	26.6	81.05
7*	..	.....	.....	...	.....	.....	7	54.5	0.164	19	33.1	37	23.7	64.28
8*	..	.....	.....	...	.....	.....	7	48.6	0.146	19	29.5	37	21.1	50.97
9*	..	.....	.....	...	.....	.....	7	43.2	0.130	19	26.2	37	18.8	40.42
10*	..	.....	.....	...	.....	.....	7	38.5	0.116	19	23.4	37	16.7	32.06
12*	..	.....	.....	...	.....	.....	7	30.5	0.0915	19	18.5	37	13.3	20.16
14*	..	.....	.....	...	.....	.....	7	24.2	0.0726	19	14.7	37	10.5	12.68
16*	..	.....	.....	...	.....	.....	7	19.2	0.0576	19	11.7	...	.....	7.974
18*	..	.....	.....	...	.....	.....	7	15.2	0.0456	19	9.2	...	.....	5.015
20*	..	.....	.....	...	.....	.....	7	12.1	0.0363	19	7.3	...	.....	3.154

* The sizes of conductors which have been marked with an asterisk provide for one or more schedules of preferred series and are commonly used in the industry.   Those not marked are given simply as a matter of reference, and it is suggested that their use be discouraged.

† To calculate the nominal diameters of Class C or Class D conductors or of any concentric-lay-stranded conductors made from round wires of uniform diameters, multiply the diameter of an individual wire by that one of the following factors which applies:

Number of wires in conductor	3	7	12	19	37	61	91	127	169	217	271
Diameter calculation factor	2.155	3	4.155	5	7	9	11	13	15	17	19

NOTE: 1 in = 2.54 cm; 1 lb = 0.4536 kg; 1 ft = 0.3048 m.

TABLE 2-16 Copper Cable, Rope Lay—Dimensions, Weight
(ASTM Specification B173)

Conductor size, Mcm or Awg	Class G						Class H					
	No. of wires	No. of members	No. of wires in each member	Diameter each wire, mils	Nominal conductor diameter, in.	Approx weight, lb per 1000 ft	No. of wires	No. of members	No. of wires in each member	Diameter each wire, mils	Nominal conductor diameter, in.	Approx weight, lb per 1000 ft
5000	1159	61	19	65.7	2.957	16,050	1729	91	19	53.8	2.959	16,060
4500	1159	61	19	62.3	2.804	14,435	1729	91	19	51.0	2.805	14,430
4000	1159	61	19	58.7	2.642	12,820	1729	91	19	48.1	2.646	12,840
3500	1159	61	19	55.0	2.475	11,255	1729	91	19	45.0	2.475	11,235
3000	1159	61	19	50.9	2.291	9,635	1729	91	19	41.7	2.294	9,650
2500	703	37	19	59.6	2.086	8,015	1159	61	19	46.4	2.088	8,010
2000	703	37	19	53.3	1.866	6,415	1159	61	19	41.5	1.868	6,400
1900	703	37	19	52.0	1.820	6,100	1159	61	19	40.5	1.823	6,100
1800	703	37	19	50.6	1.771	5,775	1159	61	19	39.4	1.773	5,770
1750	703	37	19	49.9	1.747	5,620	1159	61	19	38.9	1.751	5,625
1700	703	37	19	49.2	1.722	5,460	1159	61	19	38.3	1.724	5,455
1600	703	37	19	47.7	1.670	5,130	1159	61	19	37.2	1.674	5,145
1500	427	61	7	59.3	1.601	4,775	703	37	19	46.2	1.617	4,815
1400	427	61	7	57.3	1.547	4,460	703	37	19	44.6	1.561	4,485
1300	427	61	7	55.2	1.490	4,135	703	37	19	43.0	1.505	4,170
1250	427	61	7	54.1	1.461	3,975	703	37	19	42.2	1.477	4,015
1200	427	61	7	53.0	1.431	3,810	703	37	19	41.3	1.446	3,845
1100	427	61	7	50.8	1.372	3,500	703	37	19	39.6	1.386	3,535
1000	427	61	7	48.4	1.307	3,180	703	37	19	37.7	1.320	3,205
900	427	61	7	45.9	1.239	2,860	703	37	19	35.8	1.253	2,895
800	427	61	7	43.3	1.169	2,545	703	37	19	33.7	1.180	2,560
750	427	61	7	41.9	1.131	2,385	703	37	19	32.7	1.145	2,410
700	427	61	7	40.5	1.094	2,230	703	37	19	31.6	1.106	2,255
650	427	61	7	39.0	1.053	2,070	703	37	19	30.4	1.064	2,085
600	427	61	7	37.5	1.013	1,910	703	37	19	29.2	1.022	1,920
550	427	61	7	35.9	0.969	1,750	703	37	19	28.0	0.980	1,770
500	259	37	7	43.9	0.922	1,585	427	61	7	34.2	0.923	1,590
450	259	37	7	41.7	0.876	1,425	427	61	7	32.5	0.878	1,435
400	259	37	7	39.3	0.825	1,265	427	61	7	30.6	0.826	1,270
350	259	37	7	36.8	0.773	1,110	427	61	7	28.6	0.772	1,110
300	259	37	7	34.0	0.714	945	427	61	7	26.5	0.716	953
250	259	37	7	31.1	0.653	795	427	61	7	24.2	0.653	795
4/0	133	19	7	39.9	0.599	668	259	37	7	28.6	0.601	670
3/0	133	19	7	35.5	0.533	529	259	37	7	25.5	0.536	533
2/0	133	19	7	31.6	0.474	419	259	37	7	22.7	0.477	422
1/0	133	19	7	28.2	0.423	334	259	37	7	20.2	0.424	334
1	133	19	7	25.1	0.377	264	259	37	7	18.0	0.378	266
2	49	7	7	36.8	0.331	207	133	19	7	22.3	0.335	208
3	49	7	7	32.8	0.295	164	133	19	7	19.9	0.299	167
4	49	7	7	29.2	0.263	130	133	19	7	17.7	0.266	132
5	49	7	7	26.0	0.234	103	133	19	7	15.8	0.237	105
6	49	7	7	23.1	0.208	82	133	19	7	14.0	0.210	82
7	49	7	7	20.6	0.185	65	133	19	7	12.5	0.188	65
8	49	7	7	18.4	0.166	51	133	19	7	11.1	0.167	52
9	49	7	7	16.4	0.148	40.8	133	19	7	9.9	0.149	41
10	49	7	7	14.6	0.131	32.3						
12	49	7	7	11.6	0.104	20.3						
14	49	7	7	9.2	0.083	12.8						

TABLE 2-17 Copper Cable, Classes AA, A, B—Weight, Breaking Strength, DC Resistance
(ASTM Specifications B1, B2, B3, B8)

Conductor size, Mcm or Awg	No. of wires (ASTM stranding class)	Wire diameter, in.	Conductor diameter, in.	Conductor area, sq in.	Conductor weight, lb Per 1000 ft	Per mile	Hard Breaking strength, minimum,* lb	Hard D-c resistance at 20 C (68 F), ohms per 1000 ft	Medium Breaking strength, minimum,* lb	Medium D-c resistance at 20 C (68 F), ohms per 1000 ft	Soft Breaking strength, maximum,† lb	Soft D-c resistance at 20 C (68 F), ohms per 1000 ft
5000	169 (A)	0.1720	2.580	3.927	15,890	83,910	216,300	0.002265	172,000	0.002253	145,300	0.002178
5000	217 (B)	0.1518	2.581	3.927	15,890	83,910	219,500	0.002265	173,200	0.002253	145,300	0.002178
4500	169 (A)	0.1632	2.448	3.534	14,300	75,520	197,200	0.002517	154,800	0.002504	130,800	0.002420
4500	217 (B)	0.1440	2.448	3.534	14,300	75,520	200,400	0.002517	156,900	0.002504	130,800	0.002420
4000	169 (A)	0.1538	2.307	3.142	12,590	66,490	175,600	0.002804	138,500	0.002790	116,200	0.002697
4000	217 (B)	0.1358	2.309	3.142	12,590	66,490	178,100	0.002804	139,500	0.002790	116,200	0.002697
3500	127 (A)	0.1660	2.158	2.749	11,020	58,180	153,400	0.003205	120,400	0.003188	101,700	0.003082
3500	169 (B)	0.1439	2.159	2.749	11,020	58,180	155,900	0.003205	122,000	0.003188	101,700	0.003082
3000	127 (A)	0.1537	1.998	2.356	9,353	49,390	131,700	0.003703	103,900	0.003684	87,180	0.003561
3000	169 (B)	0.1332	1.998	2.356	9,353	49,390	134,400	0.003703	104,600	0.003684	87,180	0.003561
2500	91 (A)	0.1657	1.823	1.963	7,794	41,150	109,600	0.004444	85,990	0.004421	72,650	0.004273
2500	127 (B)	0.1403	1.824	1.963	7,794	41,150	111,300	0.004444	87,170	0.004421	72,650	0.004273
2000	91 (A)	0.1482	1.630	1.571	6,175	32,600	87,790	0.005501	69,270	0.005472	58,120	0.005289
2000	127 (B)	0.1255	1.632	1.571	6,175	32,600	90,050	0.005501	70,210	0.005472	58,120	0.005289
1750	91 (A)	0.1387	1.526	1.374	5,403	28,530	77,930	0.006286	61,020	0.006254	50,850	0.006045
1750	127 (B)	0.1174	1.526	1.374	5,403	28,530	78,800	0.006286	61,430	0.006254	50,850	0.006045
1500	61 (A)	0.1568	1.411	1.178	4,631	24,450	65,840	0.007334	51,950	0.007296	43,590	0.007052
1500	91 (B)	0.1284	1.412	1.178	4,631	24,450	67,540	0.007334	52,650	0.007296	43,590	0.007052
1250	61 (A)	0.1431	1.288	0.9817	3,859	20,380	55,670	0.008801	43,590	0.008755	36,320	0.008463
1250	91 (B)	0.1172	1.289	0.9817	3,859	20,380	56,280	0.008801	43,880	0.008755	36,320	0.008463
1000	37 (AA)	0.1644	1.151	0.7854	3,088	16,300	43,830	0.01100	34,400	0.01094	29,060	0.01058
1000	61 (A-B)	0.1280	1.152	0.7854	3,088	16,300	45,030	0.01100	35,100	0.01094	29,060	0.01058
900	37 (AA)	0.1560	1.092	0.7069	2,779	14,670	39,510	0.01222	31,170	0.01216	26,150	0.01175
900	61 (A-B)	0.1215	1.094	0.7069	2,779	14,670	40,520	0.01222	31,590	0.01216	26,150	0.01175
850	37 (AA)	0.1516	1.061	0.6676	2,624	13,860	37,310	0.01294	29,440	0.01288	24,700	0.01245
850	61 (A-B)	0.1180	1.062	0.6676	2,624	13,860	38,270	0.01294	29,840	0.01288	24,700	0.01245
800	37 (AA)	0.1470	1.029	0.6283	2,470	13,040	35,120	0.01375	27,710	0.01368	23,250	0.01322
800	61 (A-B)	0.1145	1.031	0.6283	2,470	13,040	36,360	0.01375	28,270	0.01368	23,250	0.01322
750	37 (AA)	0.1424	0.997	0.5890	2,316	12,230	33,400	0.01467	26,150	0.01459	21,790	0.01410
750	61 (A-B)	0.1109	0.998	0.5890	2,316	12,230	34,090	0.01467	26,510	0.01459	21,790	0.01410

C1	C2	C3	C4	C5	C6	C7	C8	C9	C10	C11	Strands	Size
0.01511	20,340	0.01563	24,410	0.01572	31,170	11,410	2,161	0.5498	0.963	0.1375	37 (AA)	700
0.01511	20,340	0.01563	24,740	0.01572	31,820	11,410	2,161	0.5498	0.964	0.1071	61 (A-B)	700
0.01627	18,890	0.01684	22,670	0.01692	29,130	10,600	2,007	0.5105	0.928	0.1325	37 (AA)	650
0.01627	18,890	0.01684	22,970	0.01692	29,770	10,600	2,007	0.5105	0.929	0.1032	61 (A-B)	650
0.01763	17,440	0.01824	21,060	0.01834	27,020	9,781	1,853	0.4712	0.891	0.1273	37 (AA-A)	600
0.01763	18,140	0.01824	21,350	0.01834	27,530	9,781	1,853	0.4712	0.893	0.0992	61 (B)	600
0.01923	15,980	0.01990	19,310	0.02000	24,760	8,966	1,698	0.4320	0.855	0.1219	37 (AA-A)	550
0.01923	16,630	0.01990	19,570	0.02000	25,230	8,966	1,698	0.4320		0.0950	61 (B)	550
0.02116	14,530	0.02189	17,320	0.02200	21,950	8,151	1,544	0.3927	0.811	0.1622	19 (AA)	500
0.02116	14,530	0.02189	17,550	0.02200	22,510	8,151	1,544	0.3927	0.813	0.1162	37 (A-B)	500
0.02351	13,080	0.02432	15,590	0.02445	19,750	7,336	1,389	0.3534	0.770	0.1539	19 (AA)	450
0.02351	13,080	0.02432	15,900	0.02445	20,450	7,336	1,389	0.3534	0.772	0.1103	37 (A-B)	450
0.02645	11,620	0.02736	13,950	0.02750	17,810	6,521	1,235	0.3142	0.726	0.1451	19 (AA)	400
0.02645	11,620	0.02736	14,140	0.02750	18,320	6,521	1,235	0.3142	0.728	0.1040	37 (B)	400
0.03022	10,170	0.03127	12,040	0.03143	15,140	5,706	1,081	0.2749	0.710	0.1708	12 (AA)	350
0.03022	10,170	0.03127	12,200	0.03143	15,590	5,706	1,081	0.2749	0.679	0.1357	19 (A)	350
0.03022	10,580	0.03127	12,450	0.03143	16,060	5,706	1,081	0.2749	0.681	0.0973	37 (B)	350
0.03526	8,718	0.03648	10,390	0.03667	13,170	4,891	926.3	0.2356	0.657	0.1581	12 (AA)	300
0.03526	8,718	0.03648	10,530	0.03667	13,510	4,891	926.3	0.2356	0.629	0.1257	19 (A)	300
0.03526	9,071	0.03648	10,740	0.03667	13,870	4,891	926.3	0.2356	0.630	0.0900	37 (B)	300
0.04231	7,265	0.04378	8,717	0.04400	11,130	4,076	771.9	0.1963	0.600	0.1443	12 (AA)	250
0.04231	7,265	0.04378	8,836	0.04400	11,360	4,076	771.9	0.1963	0.574	0.1147	19 (A)	250
0.04231	7,559	0.04378	8,952	0.04400		4,076	771.9	0.1963	0.575	0.0822	37 (AA-A)	250
0.04999	6,149	0.05172	7,278	0.05199	9,154	3,450	653.3	0.1662	0.522	0.1739	7 (AA-A)	4/0
0.04999	6,149	0.05172	7,378	0.05199	9,483	3,450	653.3	0.1662	0.552	0.1328	12 —	4/0
0.04999	6,149	0.05172	7,479	0.05199	9,617	3,450	653.3	0.1662	0.528	0.1055	19 (B)	4/0
0.06304	4,876	0.06522	5,812	0.06556	7,366	2,736	518.1	0.1318	0.464	0.1548	7 (AA-A)	3/0
0.06304	4,876	0.06522	5,890	0.06556	7,556	2,736	518.1	0.1318	0.492	0.1183	12 —	3/0
0.06304	5,074	0.06522	5,970	0.06556	7,698	2,736	518.1	0.1318	0.470	0.0940	19 (B)	3/0
0.07949	3,867	0.08224	4,640	0.08267	5,926	2,169	410.9	0.1045	0.414	0.1379	7 (AA-A)	2/0
0.07949	3,867	0.08224	4,703	0.08267	6,048	2,169	410.9	0.1045	0.438	0.1053	12 —	2/0
0.1002	4,024	0.1037	4,765	0.1042	6,152	1,720	325.8	0.08289	0.419	0.0837	19 (B)	1/0
0.1002	3,067	0.1037	3,755	0.1042	4,841	1,720	325.8	0.08289	0.368	0.1228	7 (AA-A)	1/0
0.1002	3,191	0.1037	3,805	0.1042	4,901	1,720	325.8	0.08289	0.390	0.0938	19 (B)	1/0
0.1252	2,432	0.1295	2,879	0.1302	3,621	1,351	255.9	0.06573	0.360	0.1670	3 (AA)	1
0.1264	2,432	0.1308	2,958	0.1314	3,804	1,364	258.4	0.06573	0.328	0.1093	7 (A)	1
0.1264	2,531	0.1308	3,037	0.1314	3,899	1,364	258.4	0.06573	0.332	0.0664	19 (AA)	1
0.1578	1,929	0.1633	2,299	0.1641	2,913	1,071	202.9	0.05213	0.320	0.1487	3 (AA)	2
0.1594	2,007	0.1649	2,361	0.1657	3,045	1,082	204.9	0.05213	0.292	0.0974	7 (A-B)	2

TABLE 2-17 Copper Cable, Classes AA, A, B—Weight, Breaking Strength, DC Resistance
(*Continued*)

Conductor size, Mcm or Awg	No. of wires (ASTM stranding class)	Wire diameter, in.	Conductor diameter, in.	Conductor area, sq in.	Conductor weight, lb		Hard		Medium		Soft	
					Per 1000 ft	Per mile	Breaking strength, minimum,* lb	D-c resistance at 20 C (68 F), ohms per 1000 ft	Breaking strength, minimum,* lb	D-c resistance at 20 C (68 F), ohms per 1000 ft	Breaking strength, maximum,† lb	D-c resistance at 20 C (68 F), ohms per 1000 ft
3	3 (AA)	0.1325	0.285	0.04134	160.9	849.6	2,359	0.2070	1,835	0.2059	1,530	0.1990
3	7 (A-B)	0.0867	0.260	0.04134	162.5	858.0	2,433	0.2090	1,885	0.2079	1,592	0.2010
4	3 (AA)	0.1180	0.254	0.03278	127.6	673.8	1,879	0.2610	1,465	0.2596	1,213	0.2509
4	7 (A-B)	0.0772	0.232	0.03278	128.9	680.5	1,938	0.2636	1,505	0.2622	1,262	0.2534
5	7 (B)	0.0688	0.206	0.02600	102.2	539.6	1,542	0.3323	1,201	0.3306	1,001	0.3196
6	7 (B)	0.0612	0.184	0.02062	81.05	427.9	1,288	0.4191	958.6	0.4169	793.8	0.4030
7	7 (B)	0.0545	0.164	0.01635	64.28	339.4	977.1	0.5284	765.2	0.5257	629.5	0.5081
8	7 (B)	0.0486	0.146	0.01297	50.97	269.1	777.2	0.6663	610.7	0.6629	499.2	0.6408
9	7 (B)	0.0432	0.130	0.01028	40.42	213.4	618.2	0.8402	487.4	0.8359	395.9	0.8080
10	7 (B)	0.0385	0.116	0.008155	32.06	169.3	491.7	1.060	388.9	1.054	314.0	1.019
12	7 (B)	0.0305	0.0915	0.005129	20.16	106.5	311.1	1.685	247.7	1.676	197.5	1.620
14	7 (B)	0.0242	0.0726	0.003225	12.68	66.95	197.1	2.679	157.7	2.665	124.2	2.576
16	7 (B)	0.0192	0.0576	0.002028	7.974	42.10	124.7	4.259	100.4	4.237	81.14	4.096
18	7 (B)	0.0152	0.0456	0.001276	5.015	26.48	78.99	6.773	63.91	6.738	51.03	6.513
20	7 (B)	0.0121	0.0363	0.0008023	3.154	16.65	50.04	10.77	40.67	10.71	32.09	10.36
ASTM Designation	B8						B1 & B8		B2 & B8		B3 & B8	

*No. 10 AWG and smaller, based on Anaconda data.
†No requirements for tensile strength are specified in ASTM B3. Values given here based on Anaconda data.

Weight and Resistance

Stranding class	Conductor size, Mcm or Awg	Increment of resistance and weight, %
AA............	{4–1	1
	{1/0–1000	2
	{2000 and under	2
A, B, C, D......	Over 2000–3000	3
	Over 3000–4000	4
	{Over 4000–5000	5

Resistance
(ASTM requirements)

Temper	Conductivity at 20 C (68 F), IACS, %	Resistivity at 20 C (68 F), ohms (mile, lb)
Hard.......	96.16	910.15
Medium....	96.66	905.44
Soft........	100	875.20

The resistance values in this table are trade maximums and are higher than the average values for commercial cable.

NOTE: 1 in = 2.54 cm; 1 in² = 64.5 cm²; 1 ft = 0.3048 m; 1 lb = 0.4536 kg; 1 mi = 1.61 km.

TABLE 2-18 Copper-Clad Steel Wire and Cable—Weight, Breaking Strength, DC Resistance

(Based on ASTM Specifications B227 and B228)

Conductor size,* AWG or in.	Conductor stranding		Conductor diam., in.	Conductor area		Conductor weight, lb		Breaking strength, min., lb			D-c resistance at 20 C (68 F), ohms per 1,000 ft	
	No. of wires	Wire size, AWG		Cir mils	Sq in.	Per 1,000 ft	Per mile	High strength		Extra-high strength	Conductivity, IACS	
								Conductivity, IACS				
								40%	30%	30%	40%	30%
Solid (B227)												
4	..	..	0.2043	41,740	0.03278	115.8	611.6	3,541	3,934	4,672	0.6337	0.8447
5	..	..	0.1819	33,090	0.02599	91.86	485.0	2,938	3,250	3,913	0.7990	1.065
0.165	..	..	0.1650	27,230	0.02138	75.55	398.9	2,523	2,780	3,368	0.9715	1.295
6	..	..	0.1620	26,240	0.02061	72.85	384.6	2,433	2,680	3,247	1.008	1.343
7	..	..	0.1443	20,820	0.01635	57.77	305.0	2,011	2,207	2,681	1.270	1.694
8	..	..	0.1285	16,510	0.01297	45.81	241.9	1,660	1,815	2,204	1.602	2.136
0.128	..	..	0.1280	16,380	0.01287	45.47	240.1	1,647	1,802	2,188	1.614	2.152
9	..	..	0.1144	13,090	0.01028	36.33	191.8	1,368	1,491	1,790	2.020	2.693
0.104	..	..	0.1040	10,820	0.008495	30.01	158.5	1,177	1,283	1,487	2.445	3.260
10	..	..	0.1019	10,380	0.008155	28.81	152.1	1,130	1,231	1,460	2.547	3.396
12	..	..	0.0808	6,530	0.005129	18.12	95.68	785	......	......	4.051	
0.080	..	..	0.0800	6,400	0.005027	17.76	93.77	770	......	900	4.133	5.509
Stranded (B228)												
7/8	19	5	0.910	628,900	0.4940	1770	9344	50,240	55,570	66,910	0.04264	0.05685
13/16	19	6	0.810	498,800	0.3917	1403	7410	41,600	45,830	55,530	0.05377	0.07168
23/32	19	7	0.721	395,500	0.3107	1113	5877	34,390	37,740	45,850	0.06780	0.09039
21/32	19	8	0.642	313,700	0.2464	882.7	4660	28,380	31,040	37,690	0.08550	0.1140
9/16	19	9	0.572	248,800	0.1954	700.0	3696	23,390	25,500	30,610	0.1078	0.1437
5/8	7	4	0.613	292,200	0.2295	818.9	4324	22,310	24,780	29,430	0.09143	0.1219
9/16	7	5	0.546	231,700	0.1820	649.4	3429	18,510	20,470	24,650	0.1153	0.1537
1/2	7	6	0.486	183,800	0.1443	515.0	2719	15,330	16,890	20,460	0.1454	0.1938
7/16	7	7	0.433	145,700	0.1145	408.4	2157	12,670	13,910	16,890	0.1833	0.2444
3/8	7	8	0.385	115,600	0.09077	323.9	1710	10,460	11,440	13,890	0.2312	0.3081
11/32	7	9	0.343	91,650	0.07198	256.9	1356	8,616	9,393	11,280	0.2915	0.3886
5/16	7	10	0.306	72,680	0.05708	203.7	1076	7,121	7,758	9,196	0.3676	0.4900
.....	3	5	0.392	99,310	0.07800	277.8	1467	8,373	9,262	11,860	0.2685	0.3579
.....	3	6	0.349	78,750	0.06185	220.3	1163	6,934	7,639	9,754	0.3385	0.4513
.....	3	7	0.311	62,450	0.04905	174.7	922.4	5,732	6,291	7,922	0.4269	0.5691
.....	3	8	0.277	49,530	0.03890	138.5	731.5	4,730	5,174	6,282	0.5383	0.7176
.....	3	9	0.247	39,280	0.03085	109.9	580.1	3,898	4,250	5,129	0.6788	0.9049
.....	3	10	0.220	31,150	0.02446	87.13	460.0	3,221	3,509	4,160	0.8559	1.141
.....	3	12	0.174	19,590	0.01539	54.80	289.3	2,236	....	......	1.361	

*To determine copper equivalent of copper-clad steel conductor, multiply circular-mil area by percent conductivity expressed as a decimal.

NOTE: 1 in = 2.54 cm; 1 in² = 64.5 cm²; 1 ft = 0.3048 m; 1 mi = 1.61 km; 1 lb = 0.4536 kg.

TABLE 2-19 Copper-Clad Steel-Copper Cable—Weight, Breaking Strength, DC Resistance
(ASTM Specification B229)

Hard-drawn copper equiva-lent,* Mcm or AWG	Con-ductor type	Conductor stranding				Con-ductor diam., in	Con-ductor area, in²	Conductor weight, lb		Break-ing strength, min., lb	D-c re-sistance at 20°C (68°F), Ω/1,000 ft
		EHS 30% copper-clad wires		Hard-drawn copper wires				Per 1,000 ft	Per mi		
		No.	Diam., in	No.	Diam, in						
350	E	7	0.1576	12	0.1576	0.788	0.3706	1403	7409	32,420	0.03143
350	EK	4	0.1470	15	0.1470	0.735	0.3225	1238	6536	23,850	0.03143
300	E	7	0.1459	12	0.1459	0.729	0.3177	1203	6351	27,770	0.03667
300	EK	4	0.1361	15	0.1361	0.680	0.2764	1061	5602	20,960	0.03667
250	E	7	0.1332	12	0.1332	0.666	0.2648	1002	5292	23,920	0.04400
250	EK	4	0.1242	15	0.1242	0.621	0.2302	884.2	4669	17,840	0.04400
4/0	E	7	0.1225	12	0.1225	0.613	0.2239	848.3	4479	20,730	0.05199
4/0	EK	4	0.1143	15	0.1143	0.571	0.1950	748.4	3951	15,370	0.05199
4/0	F	1	0.1833	6	0.1833	0.550	0.1847	710.2	3750	12,290	0.05199
3/0	E	7	0.1091	12	0.1091	0.545	0.1776	672.7	3552	16,800	0.06556
3/0	EK	4	0.1018	15	0.1018	0.509	0.1546	593.5	3134	12,370	0.06556
3/0	F	1	0.1632	6	0.1632	0.490	0.1464	563.2	2974	9,980	0.06556
2/0	F	1	0.1454	6	0.1454	0.436	0.1162	446.8	2359	8,094	0.08265
1/0	F	1	0.1294	6	0.1294	0.388	0.09206	354.1	1870	6,536	0.1043
1	F	1	0.1153	6	0.1153	0.346	0.07309	280.9	1483	5,266	0.1315
2†	A	1	0.1699	2	0.1699	0.366	0.06801	256.8	1356	5,876	0.1658
2	F	1	0.1026	6	0.1026	0.308	0.05787	222.8	1176	4,233	0.1658
4†	A	1	0.1347	2	0.1347	0.290	0.04275	161.5	852	3,938	0.2636
6†	A	1	0.1068	2	0.1068	0.230	0.02688	101.6	536.3	2,585	0.4150
8†	A	1	0.1127	2	0.07969	0.199	0.01995	74.27	392.2	2,233	0.6598

*Area of hard-drawn copper cable having the same dc resistance as that of the composite cable.
†Sizes commonly used for rural distribution.
NOTE: 1 in = 2.54 cm; 1 in² = 64.5 cm²; 1 ft = 0.3048 m; 1 lb = 0.4536 kg; 1 mi = 1.61 km.

TABLE 2-20 Aluminum Wire—Dimensions, Weight, DC Resistance

(Based on ASTM Specifications B230, B262, and B323)

Conductor size, AWG	Diam. at 20 C (68 F), mils	Area at 20 C (68 F)		D-c resistance at 20 C (68 F),* ohms per 1,000 ft	Weight at 20 C (68 F),† lb		Length at 20 C (68 F), ft per ohm
		Cir mils	Sq in.		Per 1,000 ft	Per ohm	
2	257.6	66,360	0.05212	0.2562	61.07	238.4	3903
3	229.4	52,620	0.04133	0.3231	48.43	149.9	3095
4	204.3	41,740	0.03278	0.4074	38.41	94.30	2455
5	181.9	33,090	0.02599	0.5139	30.45	59.26	1946
6	162.0	26,240	0.02061	0.6479	24.15	37.28	1544
7	144.3	20,820	0.01635	0.8165	19.16	23.47	1225
8	128.5	16,510	0.01297	1.030	15.20	14.76	971.2
9	114.4	13,090	0.01028	1.299	12.04	9.272	769.7
10	101.9	10,380	0.008155	1.637	9.556	5.836	610.7
11	90.7	8,230	0.00646	2.07	7.57	3.66	484
12	80.8	6,530	0.00513	2.60	6.01	2.31	384
13	72.0	5,180	0.00407	3.28	4.77	1.45	305
14	64.1	4,110	0.00323	4.14	3.78	0.914	242
15	57.1	3,260	0.00256	5.21	3.00	0.575	192
16	50.8	2,580	0.00203	6.59	2.38	0.361	152
17	45.3	2,050	0.00161	8.29	1.89	0.228	121
18	40.3	1,620	0.00128	10.5	1.49	0.143	95.5
19	35.9	1,290	0.00101	13.2	1.19	0.0899	75.8
20	32.0	1,020	0.000804	16.6	0.942	0.0568	60.2
21	28.5	812	0.000638	20.9	0.748	0.0357	47.8
22	25.3	640	0.000503	26.6	0.589	0.0222	37.6
23	22.6	511	0.000401	33.3	0.470	0.0141	30.0
24	20.1	404	0.000317	42.1	0.372	0.00884	23.8
25	17.9	320	0.000252	53.1	0.295	0.00556	18.8
26	15.9	253	0.000199	67.3	0.233	0.00346	14.9
27	14.2	202	0.000158	84.3	0.186	0.00220	11.9
28	12.6	159	0.000125	107	0.146	0.00136	9.34
29	11.3	128	0.000100	133	0.118	0.000883	7.51
30	10.0	100	0.0000785	170	0.0920	0.000541	5.88

*Conductivity = 61.0% IACS.
†Density = 2.703 g per cu cm (0.09765 lb per cu in).
NOTE: 1 in^2 = 64.5 cm^2; 1 ft = 0.3048 m; 1 lb = 0.4536 kg.

TABLE 2-21 Aluminum Cable—Stranding Classes, Uses

(ASTM Specification B231)

Construction	Class	Application
Concentric lay	AA	For bare conductors usually used in **overhead lines**
	A	For conductors to be covered with **weather-resistant** (weatherproof), slow-burning materials and for bare conductors where greater flexibility than is afforded by Class AA is required. Conductors intended for further fabrication into tree wire or to be insulated and laid helically with or around aluminum or ACSR messengers shall be regarded as Class A conductors with respect to direction of lay only
	B	For conductors to be **insulated** with various materials such as rubber, paper, varnished cloth, etc., and for the conductors indicated under Class A where greater flexibility is required
	C, D	For conductors where greater flexibility is required than is provided by Class B conductors

2-57

TABLE 2-22 Aluminum Conductor—Physical Characteristics 1350-H19 Classes AA and A

Cable code word	Conductor size		Current-carrying capacity* A	Stranding		Conductor diam, in	Rated strength, lb	Nominal weight, lb†	
	cmils or AWG	in²		Class	No. and diam of wires, in			Per 1000 ft	Per mile
Peachbell	6	0.0206	95	A	7 × 0.0612	0.184	563	24.6	130
Rose	4	0.0328	130	A	7 × 0.0772	0.232	881	39.2	207
Iris	2	0.0522	175	AA, A	7 × 0.0974	0.292	1,350	62.3	329
Pansy	1	0.0657	200	AA, A	7 × 0.1093	0.328	1,640	78.5	414
Poppy	1/0	0.0829	235	AA, A	7 × 0.1228	0.368	1,990	99.1	523
Aster	2/0	0.1045	270	AA, A	7 × 0.1379	0.414	2,510	124.9	659
Phlox	3/0	0.1317	315	AA, A	7 × 0.1548	0.464	3,040	157.5	832
Oxlip	4/0	0.1663	365	AA, A	7 × 0.1739	0.522	3,830	198.7	1,049
Sneezewort	250,000	0.1964	405	AA	7 × 0.1890	0.567	4,520	234.7	1,239
Valerian	250,000	0.1963	405	A	19 × 0.1147	0.574	4,660	234.6	1,239
Daisy	266,800	0.2097	420	AA	7 × 0.1953	0.586	4,830	250.6	1,323
Laurel	266,800	0.2095	425	A	19 × 0.1185	0.593	4,970	250.4	1,322
Peony	300,000	0.2358	455	A	19 × 0.1257	0.629	5,480	281.8	1,488
Tulip	336,400	0.2644	495	A	19 × 0.1331	0.666	6,150	316.0	1,668
Daffodil	350,000	0.2748	506	A	19 × 0.1357	0.679	6,390	328.4	1,734
Canna	397,500	0.3124	550	AA, A	19 × 0.1447	0.724	7,110	373.4	1,972
Goldentuft	450,000	0.3534	545	AA	19 × 0.1539	0.770	7,890	422.4	2,230
Cosmos	477,000	0.3744	615	AA	19 × 0.1584	0.793	8,360	447.5	2,363
Syringa	477,000	0.3743	615	A	37 × 0.1135	0.795	8,690	447.4	2,362
Zinnia	500,000	0.3926	635	AA	19 × 0.1622	0.811	8,760	469.2	2,477
Hyacinth	500,000	0.3924	635	A	37 × 0.1162	0.813	9,110	469.0	2,476
Dahlia	556,500	0.4369	680	AA	19 × 0.1711	0.856	9,750	522.1	2,757
Mistletoe	556,500	0.4368	680	AA, A	37 × 0.1226	0.858	9,940	522.0	2,756
Meadowsweet	600,000	0.4709	715	AA, A	37 × 0.1273	0.891	10,700	562.8	2,972
Orchid	636,000	0.4995	745	AA, A	37 × 0.1311	0.918	11,400	596.9	3,152
Heuchera	650,000	0.5102	755	AA	37 × 0.1325	0.928	11,600	609.8	3,220
Verbena	700,000	0.5494	790	AA	37 × 0.1375	0.963	12,500	656.6	3,467
Flag	700,000	0.5495	790	A	61 × 0.1071	0.964	12,900	656.8	3,468
Violet	715,500	0.5622	800	AA	37 × 0.1391	0.974	12,800	672.0	3,548
Nasturtium	715,500	0.5619	800	A	61 × 0.1083	0.975	13,100	671.6	3,546
Petunia	750,000	0.5892	825	AA	37 × 0.1424	0.997	13,100	704.3	3,719
Cattail	750,000	0.5892	825	A	61 × 0.1109	0.998	13,500	704.2	3,718
Arbutus	795,000	0.6245	855	AA	37 × 0.1466	1.026	13,900	746.4	3,941
Lilac	795,000	0.6248	855	A	61 × 0.1142	1.028	14,300	746.7	3,943

Cockscomb	900,000	0.7072	925	AA	37 × 0.1560	1.092	15,400	845.2	4,463
Snapdragon	900,000	0.7072	925	A	61 × 0.1215	1.094	15,900	845.3	4,463
Magnolia	954,000	0.7495	960	AA	37 × 0.1606	1.124	16,400	895.8	4,730
Goldenrod	954,000	0.7498	960	A	61 × 0.1251	1.126	16,900	896.1	4,731
Hawkweed	1,000,000	0.7854	990	AA	37 × 0.1644	1.151	17,200	938.7	4,956
Camellia	1,000,000	0.7849	990	A	61 × 0.1280	1.152	17,700	938.2	4,954
Bluebell	1,033,500	0.8124	1015	AA	37 × 0.1672	1.170	17,700	970.9	5,126
Larkspur	1,033,500	0.8122	1015	A	61 × 0.1302	1.172	18,300	970.6	5,125
Marigold	1,113,000	0.8744	1040	AA, A	61 × 0.1351	1.216	19,700	1,045	5,518
Hawthorn	1,192,500	0.9363	1085	AA, A	61 × 0.1398	1.258	21,100	1,119	5,908
Narcissus	1,272,000	0.999	1130	AA, A	61 × 0.1444	1.300	22,000	1,194	6,304
Columbine	1,351,500	1.062	1175	AA, A	61 × 0.1489	1.340	23,400	1,269	6,700
Carnation	1,431,000	1.124	1220	AA, A	61 × 0.1532	1.379	24,300	1,344	7,096
Galdiolus	1,510,500	1.187	1265	AA, A	61 × 0.1574	1.417	25,600	1,419	7,492
Coreopsis	1,590,000	1.250	1305	AA	61 × 0.1615	1.454	27,000	1,493	7,883
Jessamine	1,750,000	1.375	1385	AA	61 × 0.1694	1.525	29,700	1,643	8,675
Cowslip	2,000,000	1.570	1500	A	91 × 0.1482	1.630	34,200	1,876	9,911
Sagebrush	2,250,000	1.766	1600	A	91 × 0.1572	1.729	37,700	2,132	11,257
Lupine	2,500,000	1.962	1700	A	91 × 0.1657	1.823	41,800	2,368	12,503
Bitterroot	2,750,000	2.159	1795	A	91 × 0.1738	1.912	46,100	2,606	13,760
Trillium	3,000,000	2.356	1885	A	127 × 0.1537	1.996	50,300	2,844	15,016
Bluebonnet	3,500,000	2.749	2035	A	127 × 0.1660	2.158	58,700	3,350	17,688

Class of stranding. The class of stranding must be specified on all orders. Class AA stranding is usually specified for bare conductors used on overhead lines. Class A stranding is usually specified for conductors to be covered with weather-resistant (weatherproof) materials and for bare conductors where greater flexibility than afforded by Class AA is required.

Lay. The direction of lay of the outside layer of wires with Class AA and Class A stranding will be right hand unless otherwise specified.

*Ampacity for conductor temperature rise of 40°C over 40°C ambient with a 2 ft/s crosswind and an emissivity factor of 0.5 without sun.

†Nominal conductor weights are based on ASTM standard stranding increments. Actual weights will vary with lay lengths. Invoicing will be based on actual weights.

NOTE: 1 in^2 = 64.5 cm^2; 1 in = 2.54 cm; 1 lb = 0.4536 kg; 1 ft = 0.3048 m; 1 mi = 1.61 km.

TABLE 2-23 ACSR Conductor—Physical Characteristics

	ACSR	Cross section		Current-carrying capacity,* A	Stranding No. and diam of strand, in		Diameter, in		Nominal weight, lb† Per 1000 ft			Rated strength, lb			
		Aluminum	Total									Zinc-coated core			Aluminum-coated core
Code word	cmils or AWG	in²	in²		Aluminum	Steel	Complete cond.	Steel core	Total	Al	Steel	Standard weight coating	Class B coating	Class C coating	
Turkey	6	0.0206	0.0240	95	6 × 0.0661	1 × 0.0661	0.198	0.0661	36.1	24.5	11.6	1,190	1,160	1,120	1,120
Swan	4	0.0328	0.0382	130	6 × 0.0834	1 × 0.0834	0.250	0.0834	57.4	39.0	18.4	1,860	1,810	1,760	1,760
Swanate	4	0.0328	0.0411	130	7 × 0.0772	1 × 0.1029	0.257	0.1029	67.0	39.0	28.0	2,360	2,280	2,200	2,160
Sparrow	2	0.0522	0.0608	175	6 × 0.1052	1 × 0.1052	0.316	0.1052	91.3	62.0	29.3	2,830	2,760	2,680	2,640
Sparate	2	0.0522	0.0654	175	7 × 0.0974	1 × 0.1299	0.325	0.1299	106.7	62.0	44.7	3,640	3,510	3,390	3,260
Robin	1	0.0657	0.0767	200	6 × 0.1181	1 × 0.1181	0.355	0.1182	115.1	78.2	36.9	3,550	3,450	3,340	3,290
Raven	1/0	0.0830	0.0968	230	6 × 0.1327	1 × 0.1327	0.398	0.1327	145.3	98.7	46.6	4,380	4,250	4,120	3,980
Quail	2/0	0.1046	0.1221	265	6 × 0.1490	1 × 0.1490	0.447	0.1490	183.2	124.4	58.8	5,310	5,130	5,050	4,720
Pigeon	3/0	0.1317	0.1537	310	6 × 0.1672	1 × 0.1672	0.502	0.1672	230.8	156.7	74.1	6,620	6,410	6,300	5,880
Penguin	4/0	0.1662	0.1939	350	6 × 0.1878	1 × 0.1878	0.563	0.1878	291.1	197.7	93.4	8,350	8,080	7,950	7,420
Waxwing	266,800	0.2094	0.2210	410	18 × 0.1217	1 × 0.1217	0.609	0.1217	289.5	250.3	39.2	6,880	6,770	6,650	6,540
Owl	266,800	0.2096	0.2368	430	6 × 0.2109	7 × 0.0703	0.633	0.2109	342.4	250.5	91.9	9,680	9,420	9,160	9,160
Partridge	266,800	0.2095	0.2436	440	26 × 0.1013	7 × 0.0788	0.642	0.2364	367.3	251.7	115.6	11,300	11,000	10,600	10,640
Merlin	336,400	0.2642	0.2789	500	18 × 0.1367	1 × 0.1367	0.684	0.1367	365.2	315.7	49.5	8,690	8,540	8,400	8,260
Linnet	336,400	0.2640	0.3070	510	26 × 0.1137	7 × 0.0884	0.720	0.2652	462.5	317.0	145.5	14,100	13,700	13,300	13,300
Oriole	336,400	0.2642	0.3259	515	30 × 0.1059	7 × 0.1059	0.741	0.3177	527.1	318.1	209.0	17,300	16,700	16,200	15,900
Chickadee	397,500	0.3121	0.3295	555	18 × 0.1486	1 × 0.1486	0.743	0.1486	431.6	373.1	58.5	9,940	9,780	9,690	9,530
Brant	397,500	0.3122	0.3527	565	24 × 0.1287	7 × 0.0858	0.772	0.2574	512.1	375.0	137.1	14,600	14,300	13,900	13,900
Ibis	397,500	0.3119	0.3627	570	26 × 0.1236	7 × 0.0961	0.783	0.2883	546.6	374.7	171.9	16,300	15,800	15,300	15,100
Lark	397,500	0.3121	0.3849	575	30 × 0.1151	7 × 0.1151	0.806	0.3453	622.7	375.8	246.9	20,300	19,600	18,900	18,600
Pelican	477,000	0.3747	0.3955	625	18 × 0.1628	1 × 0.1628	0.814	0.1628	518.0	447.8	70.2	11,800	11,600	11,500	11,100
Flicker	477,000	0.3747	0.4233	635	24 × 0.1410	7 × 0.0940	0.846	0.2820	614.6	450.1	164.5	17,200	16,700	16,200	16,000
Hawk	477,000	0.3744	0.4354	640	26 × 0.1354	7 × 0.1053	0.858	0.3159	656.0	449.6	206.4	19,500	18,900	18,400	18,100

Code Word	Size (cmil)			Ampacity	Stranding Al	Stranding St									
Hen	477,000	0.3747	0.4621	645	30 × 0.1261	7 × 0.1261	0.883	0.3783	747.4	451.1	296.3	23,800	23,000	22,100	21,300
Osprey	556,500	0.4369	0.4612	690	18 × 0.1758	1 × 0.1758	0.879	0.1758	604.1	522.2	81.9	13,700	13,500	13,400	12,900
Parakeet	556,500	0.4372	0.4938	700	24 × 0.1523	7 × 0.1015	0.914	0.3045	716.9	525.1	191.8	19,800	19,300	18,700	18,500
Dove	556,500	0.4371	0.5083	710	26 × 0.1463	7 × 0.1138	0.927	0.3414	766.0	524.9	241.1	22,600	21,900	21,200	20,900
Eagle	556,500	0.4371	0.5391	710	30 × 0.1362	7 × 0.1362	0.953	0.4086	871.9	526.2	345.7	27,800	26,800	25,800	24,800
Peacock	605,000	0.4753	0.5370	740	24 × 0.1588	7 × 0.1059	0.953	0.318	779.7	570.9	208.8	21,600	21,000	20,400	20,100
Squab	605,000	0.4749	0.5522	745	26 × 0.1525	7 × 0.1186	0.966	0.356	832.3	570.4	261.9	24,300	23,600	22,800	22,500
Teal	605,000	0.4751	0.5834	750	30 × 0.1420	19 × 0.0852	0.994	0.426	939.5	572.0	367.5	30,000	29,000	28,000	28,000
Swift	636,000	0.4994	0.5133	745	36 × 0.1329	1 × 0.1329	0.930	0.1329	643.7	596.9	46.8	13,800	13,600	13,500	13,400
Kingbird	636,000	0.4997	0.5275	750	18 × 0.1880	1 × 0.1880	0.940	0.1880	690.8	597.2	93.6	15,700	15,400	15,300	14,800
Rook	636,000	0.4996	0.5643	765	24 × 0.1628	7 × 0.1085	0.977	0.326	819.2	600.0	219.2	22,600	22,000	21,400	21,100
Grosbeak	636,000	0.4995	0.5808	775	26 × 0.1564	7 × 0.1216	0.990	0.365	875.2	599.9	275.3	25,200	24,400	23,600	22,900
Egret	636,000	0.4995	0.6135	775	30 × 0.1456	19 × 0.0874	1.019	0.437	988.2	601.4	386.8	31,500	30,500	29,400	29,400
	653,900	0.5136	0.5321	760	18 × 0.1906	3 × 0.0885	0.953	0.1906	676.2	613.8	62.4	14,800	14,700	14,500	14,500
Flamingo	666,600	0.5238	0.5917	790	24 × 0.1667	7 × 0.1111	1.000	0.333	858.9	629.1	229.8	23,700	23,100	22,400	22,100
Gannet	666,600	0.5234	0.6086	795	26 × 0.1601	7 × 0.1245	1.014	0.373	917.3	628.7	288.6	26,400	25,600	24,800	24,000
Starling	715,500	0.5620	0.6535	835	26 × 0.1659	7 × 0.1290	1.051	0.387	984.8	675.0	309.8	28,400	27,500	26,600	25,700
Redwing	715,500	0.5617	0.6896	840	30 × 0.1544	19 × 0.0926	1.081	0.463	1110	676	434	34,600	33,300	32,700	31,600
Coot	795,000	0.6243	0.6416	860	36 × 0.1486	1 × 0.1486	1.040	0.1486	804.7	746.2	58.5	16,800	16,600	16,500	16,300
Tern	795,000	0.6242	0.6674	875	45 × 0.1329	7 × 0.0886	1.063	0.266	895.8	749.7	146.1	22,100	21,700	21,200	21,600
Cuckoo	795,000	0.6244	0.7053	885	18 × 0.1820	7 × 0.1213	1.092	0.364	1024	750	274	27,900	27,100	26,400	25,600
Condor	795,000	0.6240	0.7049	885	54 × 0.1213	7 × 0.1213	1.093	0.364	1024	750	274	28,200	27,400	26,600	25,800
Drake	795,000	0.6247	0.7264	890	26 × 0.1749	7 × 0.1360	1.108	0.408	1094	750	344	31,500	30,500	29,600	28,600
Mallard	795,000	0.6245	0.7669	900	30 × 0.1628	19 × 0.0977	1.140	0.489	1235	752	483	38,400	37,100	35,800	35,100
Ruddy	900,000	0.7066	0.7555	945	45 × 0.1414	7 × 0.0943	1.131	0.283	1015	849	166	24,400	24,000	23,500	23,300
Canary	900,000	0.7068	0.7984	955	54 × 0.1291	7 × 0.1291	1.162	0.387	1159	849	310	31,900	31,000	30,200	29,300

*Ampacity for conductor temperature rise of 40°C over 40°C ambient with a 2 ft/s crosswind and an emissivity factor of 0.5 without sun.

†Nominal conductor weights are based on ASTM standard stranding increments. Actual weights will vary within standard tolerances for wire diameters and lay lengths. Invoicing will be based on actual weights.

NOTE: 1 in² = 64.5 cm²; 1 in = 2.54 cm; 1 lb = 0.4536 kg; 1 ft = 0.3048 m.

TABLE 2-24 Aluminum-Clad Steel Wire and Cable—Weight, Breaking Strength, DC Resistance

(Based on ASTM Specifications B415 and B416)

Conductor stranding		Conductor diam., in	Conductor area		Conductor weight, lb		Breaking strength, min., lb	D-c resistance at 20°C (68°F) Ω/1,000 ft
No. of wires	Wire size, AWG		Cmils	In²	Per 1,000 ft	Per mi		
Solid (B415)								
1	4	0.2043	41,740	0.03278	93.63	494.3	5,081	1.222
1	5	0.1819	33,100	0.02600	74.25	392.0	4,290	1.541
1	6	0.1620	26,250	0.02062	58.88	310.9	3,608	1.943
1	7	0.1443	20,820	0.01635	46.69	246.6	3,025	2.450
1	8	0.1285	16,510	0.01297	37.03	195.6	2,529	3.089
1	9	0.1144	13,090	0.01028	29.37	155.1	2,005	3.896
1	10	0.1019	10,380	0.008155	23.29	123.0	1,590	4.912
1	11	0.09074	8,234	0.006467	18.47	97.52	1,261	6.194
1	12	0.08081	6,530	0.005129	14.65	77.33	1,000	7.811
Stranded (B416)								
19	5	0.910	628,900	0.4940	1,430	7,552	73,350	0.08224
19	6	0.810	498,800	0.3917	1,134	5,990	61,700	0.1037
19	7	0.721	395,500	0.3107	899.5	4,750	51,730	0.1308
19	8	0.642	313,700	0.2464	713.5	3,767	43,240	0.1649
19	9	0.572	248,800	0.1954	565.8	2,987	34,290	0.2079
19	10	0.509	197,300	0.1549	448.7	2,369	27,190	0.2622
7	5	0.546	231,700	0.1820	524.9	2,772	27,030	0.2264
7	6	0.486	183,800	0.1443	416.3	2,198	22,730	0.2803
7	7	0.433	145,700	0.1145	330.0	1,743	19,060	0.3535
7	8	0.385	115,600	0.09077	261.8	1,382	15,930	0.4458
7	9	0.343	91,650	0.07198	207.6	1,096	12,630	0.5621
7	10	0.306	72,680	0.05708	164.7	869.4	10,020	0.7088
7	11	0.272	57,640	0.04527	130.6	689.4	7,945	0.8938
7	12	0.242	45,710	0.03590	103.6	546.8	6,301	1.127
3	5	0.392	99,310	0.07800	224.5	1,186.0	12,230	0.5177
3	6	0.349	78,750	0.06185	178.1	940.2	10,280	0.6528
3	7	0.311	62,450	0.04905	141.2	745.6	8,621	0.8232
3	8	0.277	49,530	0.03890	112.0	591.3	7,206	1.038
3	9	0.247	39,280	0.03085	88.81	468.9	5,715	1.309
3	10	0.220	31,150	0.02446	70.43	371.8	4,532	1.651

NOTE: 1 in = 2.54 cm; 1 in² = 64.5 cm²; 1 lb = 0.4536 kg; 1 ft = 0.3048 m; 1 mi = 1.61 km.

TABLE 2-25 Galvanized-Steel Wire—Weight, Breaking Strength, DC Resistance

(ASTM Specifications A111 and A326)

Conductor size, BWG	Conductor diam., in	Conductor area, in²	Weight at 20°C* (68°F), lb/mi	Breaking strength, min., lb					D-c resistance at 20°C (68°F), max., Ω/mi				
				Grade EBB†	Grade BB†	Grade 85	Grade 135	Grade 195	Grade EBB	Grade BB	Grade 85	Grade 135	Grade 195
4	0.238	0.04449	797	2,028	2,270	.....	.....	.....	6.27	7.02			
6	0.203	0.03237	580	1,475	1,650	.....	.....	.....	8.62	9.65			
8	0.165	0.02138	383	975	1,090	.....	.....	.....	13.0	14.6			
9	0.148	0.01720	308	785	880	1,462	.....	.....	16.2	18.2	18.8		
10	0.134	0.01410	253	645	720	1,199	.....	.....	19.8	22.2	22.9		
11	0.120	0.01131	203	515	575	.....	.....	.....	24.7	27.6			
12	0.109	0.009331	167	425	475	793	1,213	1,800	29.9	33.5	34.7	38.9	38.9
14	0.083	0.005411	97.0	247	275	460	.....	.....	51.6	57.7	59.8		

*Density = 7.83 g per cu cm at 20°C.
†ASTM designation: Extra Best Best (EBB), Best Best (BB).
NOTE: 1 in = 2.54 cm; 1 in² = 64.5 cm²; 1 lb = 0.4536 kg; 1 mi = 1.61 km.

TABLE 2-26 Galvanized-Steel Strand—Dimensions, Weight, Breaking Strength (ASTM Specifications A363, A475)

Strand diameter, in.		Stranding		Strand area, sq in.	Strand weight, lb per 1000 ft	Breaking strength, minimum, lb							
						Utilities grade*				Com-mon	Siemens-Martin	High strength	Extra-high strength
Nominal	Actual	No. of wires	Diam-eter of coated wires, in.			1	2	3	4				
1¼	1.253	37	0.179	0.9311	3248	....	....	....	....	44,600	73,000	113,600	162,200
1⅛	1.127	37	0.161	0.7533	2691	....	....	....	....	36,000	58,900	91,600	130,800
1	1.001	37	0.143	0.5942	2057	....	....	....	....	28,300	46,200	71,900	102,700
1	1.000	19	0.200	0.5969	2073	....	....	....	....	28,700	47,000	73,200	104,500
⅞	0.885	19	0.177	0.4675	1581	....	....	....	....	21,900	35,900	55,800	79,700
¾	0.750	19	0.150	0.3358	1155	....	....	....	....	16,000	26,200	40,800	58,300
⅝	0.625	19	0.125	0.2332	796	....	....	....	....	11,000	18,100	28,100	40,200
⅝	0.621	7	0.207	0.2356	813	....	....	....	....	11,600	19,100	29,600	42,400
9⁄16	0.565	19	0.113	0.1905	637	....	....	....	....	9,640	16,100	24,100	33,700
9⁄16	0.564	7	0.188	0.1943	671	....	....	....	....	9,600	15,700	24,500	35,000
½	0.500	19	0.100	0.1492	504	....	....	....	....	7,620	12,700	19,100	26,700
½	0.495	7	0.165	0.1497	517	....	....	....	25,000	7,400	12,100	18,800	26,900
7⁄16	0.435	7	0.145	0.1156	399	....	....	....	18,000	5,700	9,350	14,500	20,800
⅜	0.360	7	0.120	0.07917	273	....	....	....	11,500	4,250	6,950	10,800	15,400
⅜	0.356	3	0.165	0.06415	220.3	....	....	8500					
5⁄16	0.327	7	0.109	0.06532	225	6000							
5⁄16	0.312	7	0.104	0.05946	205	....				3,200	5,350	8,000	11,200
5⁄16	0.312	3	0.145	0.04954	170.6			6500					
9⁄32	0.279	7	0.093	0.04755	164	4600	....			2,570	4,250	6,400	8,950
¼	0.240	7	0.080	0.03519	121	....				1,900	3,150	4,750	6,650
¼	0.259	3	0.120	0.03393	116.7	....	3150	4500					
7⁄32	0.216	7	0.072	0.02850	98.3	....				1,540	2,560	3,850	5,400
3⁄16	0.195	7	0.065	0.02323	80.3	2400	....						
3⁄16	0.186	7	0.062	0.02113	72.9	....				1,150	1,900	2,850	3,990
5⁄32	0.156	7	0.052	0.01487	51.3	....	....	....		870	1,470	2,140	2,940
⅛	0.123	7	0.041	0.00924	31.8	....				540	910	1,330	1,830
Elongation in 24 in.:						%							
....	.....	..		......	......	10	8	5	4	10	8	5	4

*Used principally by communication and power and light industries.

NOTE: Sizes and grades in bold-faced type are those most commonly used and readily available. 1 in = 2.54 cm; 1 in^2 = 64.5 cm^2; 1 lb = 0.4536 kg; 1 ft = 0.3048 m.

TABLE 2-27 Copper Wire and Cable—Electrical Characteristics

(Compiled from tables published by Westinghouse Electric Corp., "Electrical Transmission and Distribution Reference Book")

Conductor size, Awg or Mcm	Stranding		Conductor diameter, in	Breaking strength, lb	Conductor weight, lb per mile	Geometric mean radius at 60 cycles, ft	Resistance at 25°C (77°F)*				Resistance at 50°C (122°F)*				Inductive reactance (series) at 1 ft spacing (x_a)†			Capacitive reactance (shunt) at 1 ft spacing (x_a')‡			Current-carrying capacity at 60 cycles§ (approx), amp
	No. of wires	Wire diameter, in					D-c	25 cycles	50 cycles	60 cycles	D-c	25 cycles	50 cycles	60 cycles	25 cycles	50 cycles	60 cycles	25 cycles	50 cycles	60 cycles	
							Ohms per conductor per mile								Ohms per conductor per mile			Megohms per conductor per mile			
Solid conductors:																					
2	1	.....	0.258	3,003	1,061	0.00836	0.864	0.864	0.864	0.864	0.945	0.945	0.945	0.945	0.242	0.484	0.581	0.323	0.1614	0.1345	220
3	1	.....	0.229	2,439	841	0.00745	1.090	1.090	1.090	1.090	1.192	1.192	1.192	1.192	0.248	0.496	0.595	0.331	0.1656	0.1380	190
4	1	.....	0.204	1,970	667	0.00663	1.374	1.374	1.374	1.374	1.503	1.503	1.503	1.503	0.254	0.507	0.609	0.339	0.1697	0.1415	170
5	1	.....	0.1819	1,591	529	0.00590	1.733	1.733	1.733	1.733	1.895	1.895	1.895	1.895	0.260	0.519	0.623	0.348	0.1738	0.1449	140
6	1	.....	0.1620	1,280	420	0.00526	2.18	2.18	2.18	2.18	2.39	2.39	2.39	2.39	0.265	0.531	0.637	0.356	0.1779	0.1483	120
7	1	.....	0.1443	1,030	333	0.00468	2.75	2.75	2.75	2.75	3.01	3.01	3.01	3.01	0.271	0.542	0.651	0.364	0.1821	0.1517	110
8	1	.....	0.1285	826	264	0.00417	3.47	3.47	3.47	3.47	3.80	3.80	3.80	3.80	0.277	0.554	0.665	0.372	0.1862	0.1552	90
Stranded conductors:																					
1000	37	0.1644	1.151	43,830	16,300	0.0368	0.0585	0.0594	0.0620	0.0634	0.0640	0.0648	0.0672	0.0685	0.1666	0.333	0.400	0.216	0.1081	0.0901	1300
900	37	0.1560	1.092	39,510	14,670	0.0349	0.0650	0.0658	0.0682	0.0695	0.0711	0.0718	0.0740	0.0752	0.1693	0.339	0.406	0.220	0.1100	0.0916	1220
800	37	0.1470	1.029	35,120	13,040	0.0329	0.0731	0.0739	0.0760	0.0772	0.0800	0.0806	0.0826	0.0837	0.1722	0.344	0.413	0.224	0.1121	0.0934	1130
750	37	0.1424	0.997	33,400	12,230	0.0319	0.0780	0.0787	0.0807	0.0818	0.0853	0.0859	0.0878	0.0888	0.1739	0.348	0.417	0.226	0.1132	0.0943	1090
700	37	0.1375	0.963	31,170	11,410	0.0308	0.0836	0.0842	0.0861	0.0871	0.0914	0.0920	0.0937	0.0947	0.1759	0.352	0.422	0.229	0.1145	0.0954	1040

600	37	0.1273	0.891	27,020	9,781	0.0285	0.0975	0.0981	0.0997	0.1006	0.1066	0.1071	0.1086	0.1095	0.1799	0.360	0.432	0.235	0.1173	0.0977	940
500	37	0.1162	0.814	22,510	8,151	0.0260	0.1170	0.1175	0.1188	0.1196	0.1280	0.1283	0.1296	0.1303	0.1845	0.369	0.443	0.241	0.1205	0.1004	840
500	19	0.1622	0.811	21,590	8,151	0.0256	0.1170	0.1175	0.1188	0.1196	0.1280	0.1283	0.1296	0.1303	0.1853	0.371	0.445	0.241	0.1206	0.1005	840
450	19	0.1539	0.770	19,750	7,336	0.0243	0.1300	0.1304	0.1316	0.1323	0.1422	0.1426	0.1437	0.1443	0.1879	0.376	0.451	0.245	0.1224	0.1020	780
400	19	0.1451	0.726	17,560	6,521	0.0229	0.1462	0.1466	0.1477	0.1484	0.1600	0.1603	0.1613	0.1619	0.1909	0.382	0.458	0.249	0.1245	0.1038	730
350	19	0.1357	0.679	15,590	5,706	0.0214	0.1671	0.1675	0.1684	0.1690	0.1828	0.1831	0.1840	0.1845	0.1943	0.389	0.466	0.254	0.1269	0.1058	670
350	12	0.1708	0.710	15,140	5,706	0.0225	0.1671	0.1675	0.1684	0.1690	0.1828	0.1831	0.1840	0.1845	0.1918	0.384	0.460	0.251	0.1253	0.1044	670
300	19	0.1257	0.629	13,510	4,891	0.01987	0.1950	0.1953	0.1961	0.1966	0.213	0.214	0.214	0.215	0.1982	0.396	0.476	0.259	0.1296	0.1080	610
300	12	0.1581	0.657	13,170	4,891	0.0208	0.1950	0.1953	0.1961	0.1966	0.213	0.214	0.214	0.215	0.1957	0.392	0.470	0.256	0.1281	0.1068	610
250	19	0.1147	0.574	11,360	4,076	0.01813	0.234	0.234	0.235	0.235	0.256	0.256	0.257	0.257	0.203	0.406	0.487	0.266	0.1329	0.1108	540
250	12	0.1443	0.600	11,130	4,076	0.01902	0.234	0.234	0.235	0.235	0.256	0.256	0.257	0.257	0.200	0.401	0.481	0.263	0.1313	0.1094	540
4/0	19	0.1055	0.528	9,617	3,450	0.01668	0.276	0.277	0.277	0.278	0.302	0.303	0.303	0.303	0.207	0.414	0.497	0.272	0.1359	0.1132	480
4/0	12	0.1328	0.552	9,483	3,450	0.01750	0.276	0.277	0.277	0.278	0.302	0.303	0.303	0.303	0.205	0.409	0.491	0.269	0.1343	0.1119	490
4/0	7	0.1739	0.522	9,154	3,450	0.01579	0.276	0.277	0.277	0.278	0.302	0.303	0.303	0.303	0.210	0.420	0.503	0.273	0.1363	0.1136	480
3/0	12	0.1183	0.492	7,556	2,736	0.01559	0.349	0.349	0.349	0.350	0.381	0.381	0.382	0.382	0.210	0.421	0.505	0.277	0.1384	0.1153	420
3/0	7	0.1548	0.464	7,366	2,736	0.01404	0.349	0.349	0.349	0.350	0.381	0.381	0.382	0.382	0.216	0.431	0.518	0.281	0.1405	0.1171	420
2/0	7	0.1379	0.414	5,926	2,170	0.01252	0.440	0.440	0.440	0.440	0.481	0.481	0.481	0.481	0.222	0.443	0.532	0.289	0.1445	0.1205	360
1/0	7	0.1228	0.368	4,752	1,720	0.01113	0.555	0.555	0.555	0.555	0.606	0.607	0.607	0.607	0.227	0.455	0.546	0.298	0.1488	0.1240	310
1	7	0.1093	0.328	3,804	1,364	0.00992	0.699	0.699	0.699	0.699	0.765	0.765	0.765	0.765	0.233	0.467	0.560	0.306	0.1528	0.1274	270
1	3	0.1670	0.360	3,620	1,351	0.01016	0.692	0.692	0.692	0.692	0.757	0.757	0.757	0.757	0.232	0.464	0.557	0.299	0.1495	0.1246	270
2	7	0.0974	0.292	3,045	1,082	0.00883	0.881	0.882	0.882	0.882	0.964	0.964	0.964	0.964	0.239	0.478	0.574	0.314	0.1570	0.1308	230
2	3	0.1487	0.320	2,913	1,071	0.00903	0.873	0.873	0.873	0.873	0.955	0.955	0.955	0.955	0.238	0.476	0.571	0.307	0.1537	0.1281	240
3	7	0.0867	0.260	2,433	858	0.00787	1.112	1.112	1.112	1.112	1.216	1.216	1.216	1.216	0.245	0.490	0.588	0.322	0.1611	0.1343	200
3	3	0.1325	0.285	2,359	850	0.00805	1.101	1.101	1.101	1.101	1.204	1.204	1.204	1.204	0.244	0.488	0.585	0.316	0.1578	0.1315	200
4	3	0.1180	0.254	1,879	674	0.00717	1.388	1.388	1.388	1.388	1.518	1.518	1.518	1.518	0.250	0.499	0.599	0.324	0.1619	0.1349	180
5	3	0.1050	0.226	1,505	534	0.00638	1.750	1.750	1.750	1.750	1.914	1.914	1.914	1.914	0.256	0.511	0.613	0.332	0.1661	0.1384	150
6	3	0.0935	0.201	1,205	424	0.00568	2.21	2.21	2.21	2.21	2.41	2.41	2.41	2.41	0.262	0.523	0.628	0.341	0.1703	0.1419	130

*Resistance is based on conductivity = 97.3% IACS.

Resistance is increased to allow for stranding: 3-wire conductors = 1%; all others = 2%. For resistance temperance temperature conversion see Par. 2-23.

†See Table 2-32.

‡See Table 2-33.

§For conductor at 75°C, air at 25°C, wind 2 ft/s (1.4 mi/h), average tarnished surface.

NOTE: 1 in = 2.54 cm; 1 lb = 0.4536 kg; 1 mi = 1.61 km; 1 ft = 0.3048 m.

TABLE 2-28 Copper-Clad-Steel Cable—Electrical Characteristics

(Compiled from tables published by Westinghouse Electric Corp., "Electrical Transmission and Distribution Reference Book"; and Copperweld Steel Co.)

Nominal conductor size, in	Conductor stranding		Conductor diam., in	Conductor area, cir mils	Breaking strength, rated, lb		Conductor weight, lb per mile	Geometric mean radius at 60 cycles, average currents, ft	Resistance at 25°C (77°F),* small currents				Inductive reactance (series) at 1 ft spacing, average currents (x_a)†			Capacitive reactance (shunt) at 1 ft spacing (x_a')‡			Current-carrying capacity at 60 cycles§ (approx), amp
	No. of wires	Wire size, AWG			High strength	Extra high strength			D-c	25 cycles	50 cycles	60 cycles	25 cycles	50 cycles	60 cycles	25 cycles	50 cycles	60 cycles	
									Ohms per conductor per mile				Ohms per conductor per mile			Megohms per conductor per mile			
30% conductivity																			
7/8	19	5	0.910	628,900	56,570	66,910	9344	0.00758	0.306	0.316	0.326	0.331	0.261	0.493	0.592	0.233	0.1165	0.0971	620
13/16	19	6	0.810	498,800	45,830	55,530	7410	0.00675	0.386	0.396	0.406	0.411	0.267	0.505	0.606	0.241	0.1206	0.1005	540
23/32	19	7	0.721	395,500	37,740	45,850	5877	0.00601	0.486	0.496	0.506	0.511	0.273	0.517	0.621	0.250	0.1248	0.1040	470
21/32	19	8	0.642	313,700	31,040	37,690	4660	0.00535	0.613	0.623	0.633	0.638	0.279	0.529	0.635	0.258	0.1289	0.1074	410
9/16	19	9	0.572	248,800	25,500	30,610	3696	0.00477	0.773	0.783	0.793	0.798	0.285	0.541	0.649	0.266	0.1330	0.1109	360
5/8	7	4	0.613	292,200	24,780	29,430	4324	0.00511	0.656	0.664	0.672	0.676	0.281	0.533	0.640	0.261	0.1306	0.1088	410
9/16	7	5	0.546	231,700	20,470	24,650	3429	0.00455	0.827	0.835	0.843	0.847	0.287	0.548	0.654	0.269	0.1347	0.1122	360
1/2	7	6	0.486	183,800	16,890	20,460	2719	0.00405	1.043	1.050	1.058	1.062	0.293	0.557	0.668	0.278	0.1388	0.1157	310
7/16	7	7	0.433	145,700	13,910	16,890	2157	0.00361	1.315	1.323	1.331	1.335	0.299	0.569	0.683	0.286	0.1429	0.1191	270
3/8	7	8	0.385	115,600	11,440	13,890	1710	0.00321	1.658	1.666	1.674	1.678	0.305	0.581	0.697	0.294	0.1471	0.1226	230
11/32	7	9	0.343	91,650	9,393	11,280	1356	0.00286	2.09	2.10	2.11	2.11	0.311	0.592	0.711	0.303	0.1512	0.1260	200
5/16	7	10	0.306	72,680	7,758	9,196	1076	0.00255	2.64	2.64	2.65	2.66	0.316	0.604	0.725	0.311	0.1553	0.1294	170
..	3	5	0.392	99,310	9,262	11,860	1467	0.00457	1.926	1.931	1.936	1.938	0.289	0.545	0.654	0.293	0.1465	0.1221	220
..	3	6	0.349	78,750	7,639	9,754	1163	0.00407	2.43	2.43	2.44	2.44	0.295	0.556	0.668	0.301	0.1506	0.1255	190
..	3	7	0.311	62,450	6,291	7,922	922.4	0.00363	3.06	3.07	3.07	3.07	0.301	0.568	0.682	0.310	0.1547	0.1289	160

..	3	8	0.277	49,530	5,174	6,282	731.5	0.00323	3.86	3.87	3.87	3.87	0.307	0.580	0.696	0.318	0.1589	0.1324	140
..	3	9	0.247	39,280	4,250	5,129	580.1	0.00288	4.87	4.87	4.88	4.88	0.313	0.591	0.710	0.326	0.1629	0.1358	120
..	3	10	0.220	31,150	3,509	4,160	460.0	0.00257	6.14	6.14	6.15	6.15	0.319	0.603	0.724	0.334	0.1671	0.1392	100

40% conductivity

7/8	19	5	0.910	628,900	50,240	......	9344	0.01175	0.229	0.239	0.249	0.254	0.236	0.449	0.539	0.233	0.1165	0.0971	690
13/16	19	6	0.810	498,800	41,600	......	7410	0.01046	0.289	0.299	0.309	0.314	0.241	0.461	0.553	0.241	0.1206	0.1005	610
23/32	19	7	0.721	395,500	34,390	......	5877	0.00931	0.365	0.375	0.385	0.390	0.247	0.473	0.567	0.250	0.1248	0.1040	530
21/32	19	8	0.642	313,700	28,380	......	4660	0.00829	0.460	0.470	0.480	0.485	0.253	0.485	0.582	0.258	0.1289	0.1074	470
9/16	19	9	0.572	248,800	23,390	......	3696	0.00739	0.580	0.590	0.600	0.605	0.259	0.496	0.595	0.266	0.1330	0.1109	410
5/8	7	4	0.613	292,200	22,310	......	4324	0.00792	0.492	0.500	0.508	0.512	0.255	0.489	0.587	0.261	0.1306	0.1088	470
9/16	7	5	0.546	231,700	18,510	......	3429	0.00705	0.620	0.628	0.636	0.640	0.261	0.501	0.601	0.269	0.1347	0.1122	410
1/2	7	6	0.486	183,800	15,330	......	2719	0.00628	0.782	0.790	0.798	0.802	0.267	0.513	0.615	0.278	0.1388	0.1157	350
7/16	7	7	0.433	145,700	12,670	......	2157	0.00559	0.986	0.994	1.002	1.006	0.273	0.524	0.629	0.286	0.1429	0.1191	310
3/8	7	8	0.385	115,600	10,460	......	1710	0.00497	1.244	1.252	1.260	1.264	0.279	0.536	0.644	0.294	0.1471	0.1226	270
11/32	7	9	0.343	91,650	8,616	......	1356	0.00443	1.568	1.576	1.584	1.588	0.285	0.548	0.658	0.303	0.1512	0.1260	230
5/16	7	10	0.306	72,680	7,121	......	1076	0.00395	1.978	1.986	1.994	1.998	0.291	0.559	0.671	0.311	0.1553	0.1294	200
..	3	5	0.392	99,310	8,373	......	1467	0.00621	1.445	1.450	1.455	1.457	0.269	0.514	0.617	0.293	0.1465	0.1221	250
..	3	6	0.349	78,750	6,934	......	1163	0.00553	1.821	1.826	1.831	1.833	0.275	0.526	0.631	0.301	0.1506	0.1255	220
..	3	7	0.311	62,450	5,732	......	922.4	0.00492	2.30	2.30	2.31	2.31	0.281	0.537	0.645	0.310	0.1547	0.1289	190
..	3	8	0.277	49,530	4,730	......	731.5	0.00439	2.90	2.90	2.91	2.91	0.286	0.549	0.659	0.318	0.1589	0.1324	160
..	3	9	0.247	39,280	3,898	......	580.1	0.00391	3.65	3.66	3.66	3.66	0.292	0.561	0.673	0.326	0.1629	0.1358	140
..	3	10	0.220	31,150	3,221	......	460.0	0.00348	4.61	4.61	4.62	4.62	0.297	0.572	0.687	0.334	0.1671	0.1392	120
..	3	12	0.174	19,590	2,236	......	289.5	0.00276	7.32	7.33	7.33	7.34	0.310	0.596	0.715	0.351	0.1754	0.1462	90

*For resistance temperature conversion see Pars. 2-23 and 2-26.
†See Table 2-32.
‡See Table 2-33.
§For conductor at 125°C, air at 25°C, wind 2 ft/s (1.4 mi/h), average tarnished surface.
NOTE: 1 in = 2.54 cm; 1 lb = 0.4536 kg; 1 mi = 1.61 km; 1 ft = 0.3048 m.

TABLE 2-29 Copper-Clad-Steel Copper Cable—Electrical Characteristics

(Compiled from tables published by Westinghouse Electric Corp., "Electrical Transmission and Distribution Reference Book"; and Copperweld Steel Co.)

Conductor size / Type (Mcm or AWG)	Hard-drawn copper equivalent	EHS 30% Copper-clad steel wires No.	Diam., in	Hard-drawn copper wires No.	Diam., in	Conductor diam., in	Breaking strength, rated, lb	Conductor weight, lb/mi	Geometric mean radius at 60 cycles, ft	R25 D-c	R25 25 cyc	R25 50 cyc	R25 60 cyc	R50 D-c	R50 25 cyc	R50 50 cyc	R50 60 cyc	Ind 25 cyc	Ind 50 cyc	Ind 60 cyc	Cap 25 cyc	Cap 50 cyc	Cap 60 cyc	Current-carrying capacity at 60 cycles§ (approx), amp
350 E	350	7	0.1576	12	0.1576	0.788	32,420	7,409	0.0220	0.1658	0.1728	0.1789	0.1812	0.1812	0.1915	0.201	0.204	0.1929	0.386	0.463	0.243	0.1216	0.1014	660
350 EK	350	4	0.1470	15	0.1470	0.735	23,850	6,536	0.0245	0.1658	0.1682	0.1700	0.1705	0.1812	0.1845	0.1873	0.1882	0.1875	0.375	0.450	0.248	0.1241	0.1034	680
300 E	300	7	0.1459	12	0.1459	0.729	27,770	6,351	0.0204	0.1934	0.200	0.207	0.209	0.211	0.222	0.232	0.235	0.1969	0.394	0.473	0.249	0.1244	0.1037	600
300 EK	300	4	0.1361	15	0.1361	0.680	20,960	5,602	0.0227	0.1934	0.1958	0.1976	0.1981	0.211	0.215	0.218	0.219	0.1914	0.383	0.460	0.254	0.1269	0.1057	610
250 E	250	7	0.1332	12	0.1332	0.666	23,920	5,292	0.01859	0.232	0.239	0.245	0.248	0.254	0.265	0.275	0.279	0.202	0.403	0.484	0.255	0.1276	0.1064	540
250 EK	250	4	0.1242	15	C.1242	0.621	17,840	4,669	0.0207	0.232	0.235	0.236	0.237	0.254	0.258	0.261	0.261	0.1960	0.392	0.471	0.260	0.1301	0.1084	540
4/0 E	4/0	7	0.1225	12	0.1225	0.613	20,730	4,479	0.01711	0.274	0.281	0.287	0.290	0.300	0.312	0.323	0.326	0.206	0.411	0.493	0.261	0.1306	0.1088	480
4/0 EK	4/0	4	0.1143	15	0.1143	0.571	15,370	3,951	0.01903	0.274	0.277	0.278	0.279	0.300	0.304	0.307	0.308	0.200	0.401	0.481	0.266	0.1331	0.1109	490
4/0 F	4/0	1	0.1833	6	0.1833	0.550	12,290	3,750	0.01558	0.273	0.280	0.285	0.287	0.299	0.309	0.318	0.322	0.210	0.421	0.505	0.269	0.1344	0.1120	470
3/0 E	3/0	7	0.1091	12	0.1091	0.545	16,800	3,552	0.01521	0.346	0.353	0.359	0.361	0.378	0.391	0.402	0.407	0.212	0.423	0.508	0.270	0.1348	0.1123	420
3/0 EK	3/0	4	0.1018	4	0.1018	0.509	12,370	3,134	0.01697	0.346	0.348	0.350	0.351	0.378	0.382	0.386	0.386	0.206	0.412	0.495	0.274	0.1372	0.1143	420
3/0 F	3/0	1	0.1632	6	0.1632	0.490	9,980	2,974	0.01388	0.344	0.351	0.356	0.358	0.377	0.388	0.397	0.401	0.216	0.432	0.519	0.277	0.1385	0.1155	410
2/0 F	2/0	1	0.1454	6	0.1454	0.436	8,094	2,359	0.01235	0.434	0.441	0.446	0.448	0.475	0.487	0.497	0.501	0.222	0.444	0.533	0.285	0.1427	0.1189	350
1/0 F	1/0	1	0.1294	6	0.1294	0.388	6,536	1,870	0.01099	0.548	0.554	0.559	0.562	0.599	0.612	0.622	0.627	0.228	0.456	0.547	0.294	0.1469	0.1224	310
2 A	2	1	0.1699	2	0.1699	0.366	5,876	1,356	0.00763	0.869	0.875	0.880	0.882	0.950	0.962	0.973	0.979	0.247	0.493	0.592	0.298	0.1489	0.1241	240
2 F	2	1	0.1026	6	0.1026	0.308	4,233	1,176	0.00873	0.871	0.878	0.884	0.885	0.952	0.967	0.979	0.985	0.230	0.479	0.575	0.310	0.1551	0.1292	230
4 A	4	1	0.1347	2	0.1347	0.290	3,938	853	0.00604	1.382	1.388	1.393	1.395	1.511	1.525	1.540	1.545	0.258	0.517	0.620	0.314	0.1572	0.1310	180
6 A	6	1	0.1068	2	0.1068	0.230	2,585	536	0.00479	2.20	2.20	2.21	2.21	2.40	2.42	2.44	2.44	0.270	0.540	0.648	0.331	0.1655	0.1379	140
8 A	8	1	0.1127	2	0.0797	0.199	2,233	392	0.00394	3.49	3.50	3.51	3.51	3.82	3.84	3.86	3.87	0.280	0.560	0.672	0.341	0.1706	0.1422	100

Resistance at 25°C (77°F),* small currents; Resistance at 50°C (122°F),* current approximately 75% capacity — Ω/(conductor) (mi). Inductive reactance (series) at 1-ft spacing (x_a)† — Ω/(conductor) (mi). Capacitive reactance (shunt) at 1-ft spacing (x_a')‡ — $M\Omega$/(conductor) (mi).

*For resistance temperature conversion see pars. **2-23** and **2.26**.

Resistance at 50°C total temperature, based on ambient of 25°C plus 25°C rise due to heating effect of current. The approximate magnitude of current necessary to produce the 25°C rise is 75% of the "approximate current-carrying capacity at 60 cycles."

†See Table 2-32.

‡See Table 2-33.

§For conductor at 75°C, air at 25°C, wind 2 ft/s (1.4 mi/h), average tarnished surface.

1 in = 2.54 cm; 1 lb = 0.4536 kg; 1 mi = 1.61 km; 1 ft = 0.3048 m.

TABLE 2-30 ACSR Conductor—Electrical Characteristics

(Compiled from tables published by Westinghouse Electric Corp., "Electrical Transmission and Distribution Reference Book"; and Aluminum Co. of America)

Conductor size, kcmil or Awg	Hard-drawn copper equivalent,* kcmil or Awg	Geometric mean radius at 60 Hz, ft	Resistance at 25°C (77°F),† small currents				Resistance at 50°C (122°F)† current approx 75% capacity				Inductive reactance (series) at 1 ft spacing (x_a)‡			Capacitive reactance (shunt) at 1 ft spacing (x_a')§			Current-carrying capacity at 60 Hz¶ (approx), A
			DC	25 Hz	50 Hz	60 Hz	DC	25 Hz	50 Hz	60 Hz	25 Hz	50 Hz	60 Hz	25 Hz	50 Hz	60 Hz	
			Ω/conductor/mile								Ω/conductor/mile			MΩ/conductor/mile			
Multilayer conductors																	
1590	1000	0.0520	.0587	.0588	.0590	.0591	.0646	.0656	.0675	.0684	.1495	.299	.359	.1953	.0977	.0814	1380
1510.5	950	0.0507	.0618	.0619	.0621	.0622	.0680	.0690	.0710	.0720	.1508	.302	.362	.1971	.0986	.0821	1340
1431	900	0.0493	.0652	.0653	.0655	.0656	.0718	.0729	.0749	.0760	.1522	.304	.365	.1991	.0996	.0830	1300
1351	850	0.0479	.0691	.0692	.0694	.0695	.0761	.0771	.0792	.0803	.1536	.307	.369	.201	.1006	.0838	1250
1272	800	0.0465	.0734	.0735	.0737	.0738	.0808	.0819	.0840	.0851	.1551	.310	.372	.203	.1016	.0847	1200
1192.5	750	0.0450	.0783	.0784	.0786	.0788	.0862	.0872	.0894	.0906	.1568	.314	376	.206	.1028	.0857	1160
1113	700	0.0435	.0839	.0840	.0842	.0844	.0924	.0935	.0957	.0969	.1585	.317	.380	.208	.1040	.0867	1110
1033.5	650	0.0420	.0903	.0905	.0907	.0909	.0994	.1005	.1025	.1035	.1603	.321	.385	.211	.1053	.0878	1060
954	600	0.0403	.0979	.0980	.0981	.0982	.1078	.1088	.1118	.1128	.1624	.325	.390	.214	.1068	.0890	1010
900	566	0.0391	.104	.104	.104	.104	.1145	.1155	.1175	.1185	.1639	.328	.393	.216	.1078	.0898	970
874.5	550	0.0386	.107	.107	.107	.108	.1178	.1188	.1218	.1228	.1646	.329	.395	.217	.1083	.0903	950
795	500	0.0368	.117	.118	.118	.119	.1288	.1308	.1358	.1378	.1670	.334	.401	.220	.1100	.0917	900
795	500	0.0375	.117	.117	.117	.117	.1288	.1288	.1288	.1288	.1660	.332	.399	.219	.1095	.0912	900
795	500	0.0393	.117	.117	.117	.117	.1288	.1288	.1288	.1288	.1637	.327	.393	.217	.1085	.0904	910
715.5	450	0.0349	.131	.131	.131	.132	.1442	.1452	.1472	.1482	.1697	.339	.407	.224	.1119	.0932	830
715.5	450	0.0355	.131	.131	.131	.131	.1442	.1442	.1442	.1442	.1687	.337	.405	.223	.1114	.0928	840
715.5	450	0.0372	.131	.131	.131	.131	.1442	.1442	.1442	.1442	.1664	.333	.399	.221	.1104	.0920	840
666.6	419	0.0337	.140	.140	.141	.141	.1541	.1571	1591	.1601	.1715	.343	.412	.226	.1132	.0943	800
636	400	0.0329	.147	.147	.148	.148	.1618	.1638	.1678	.1688	.1726	.345	.414	.228	.1140	.0950	770
636	400	0.0335	.147	.147	.147	.147	.1618	.1618	.1618	.1618	.1718	.344	412	.227	.1135	.0946	780
636	400	0.0351	.147	.147	.147	.147	.1618	.1618	.1618	.1618	.1693	.339	.406	.225	.1125	.0937	780
605	380.5	0.0321	.154	.155	.155	.155	.1695	.1715	.1755	.1775	.1739	.348	.417	.230	.1149	.0957	750
605	380.5	0.0327	.154	.154	.154	.154	.1700	.1720	.1720	.1720	.1730	.346	.415	.229	.1144	.0953	760
556.5	350	0.0313	.168	.168	.168	.168	.1849	.1859	.1859	.1859	.1751	.350	.420	.232	.1159	.0965	730

TABLE 2-30 ACSR Conductor—Electrical Characteristics (*Continued*)

Conductor size, kcmil or Awg	Hard-drawn copper equivalent,* kcmil or Awg	Geometric mean radius at 60 Hz, ft	Resistance at 25°C (77°F),† small currents				Resistance at 50°C (122°F)† current approx 75% capacity				Inductive reactance (series) at 1 ft spacing (x_a)‡			Capacitive reactance (shunt) at 1 ft spacing (x_a')§			Current-carrying capacity at 60 Hz¶ (approx), A
			DC	25 Hz	50 Hz	60 Hz	DC	25 Hz	50 Hz	60 Hz	25 Hz	50 Hz	60 Hz	25 Hz	50 Hz	60 Hz	
			Ω/conductor/mile								Ω/conductor/mile			MΩ/conductor/mile			
Multilayer conductors																	
556.5	350	0.0328	.168	.168	.168	.168	.1849	.1859	.1859	.1859	.1728	.346	.415	.230	.1149	.0957	730
500	314.5	0.0311	.187	.187	.187	.187	.206	.206	.206	.206	.1754	.351	.421	.234	.1167	.0973	690
477	300	0.0290	.196	.196	.196	.196	.216	.216	.216	.216	.1790	.358	.430	.237	.1186	.0988	670
477	300	0.0304	.196	.196	.196	.196	.216	.216	.216	.216	.1766	.353	.424	.235	.1176	.0980	670
397.5	250	0.0265	.235	.235	.235	.235	.259	.259	.259	.259	.1836	.367	.441	.244	.1219	.1015	590
397.5	250	0.0278	.235	.235	.235	.235	.259	.259	.259	.259	.1812	.362	.435	.242	.1208	.1006	600
336.4	4/0	0.0244	.278	.278	.278	.278	.306	.306	.306	.306	.1872	.376	.451	.250	.1248	.1039	530
336.4	4/0	0.0255	.278	.278	.278	.278	.306	.306	.306	.306	.1855	.371	.445	.248	.1238	.1032	530
300	188.7	0.0230	.311	.311	.311	.311	.342	.342	.342	.342	.1908	.382	.458	.254	.1269	.1057	490
300	188.7	0.0241	.311	.311	.311	.311	.342	.342	.342	.342	.1883	.377	.452	.252	.1258	.1049	500
266.8	3/0	0.0217	.350	.350	.350	.350	.385	.385	.385	.385	.1936	.387	.465	.258	.1289	.1074	460

Conductor size, kcmil or Awg	Hard-drawn copper equivalent,* kcmil or Awg	Current approx 75% capacity	Resistance at 25°C (77°F),† small currents				Resistance at 50°C (122°F)† current approx 75% capacity				Small currents, Hz			Current approx 75% capacity, Hz			Capacitive reactance (shunt) at 1 ft spacing (x_a')§			Current-carrying capacity at 60 Hz¶ (approx), A
			DC	25 Hz	50 Hz	60 Hz	DC	25 Hz	50 Hz	60 Hz	25	50	60	25	50	60	25 Hz	50 Hz	60 Hz	
			Ω/conductor/mile														MΩ/conductor/mile			
Single-layer conductors																				
266.8	3/0	0.00684	0.351	0.351	0.351	0.352	0.386	0.430	0.510	0.552	.194	.388	.466	.252	.504	.605	.259	.1294	.1079	460
4/0	2/0	0.00814	0.441	0.442	0.444	0.445	0.485	0.514	0.567	0.592	.218	.437	.524	.242	.484	.581	.267	.1336	.1113	340
3/0	1/0	0.00600	0.556	0.557	0.559	0.560	0.612	0.642	0.697	0.723	.225	.450	.540	.259	.517	.621	.275	.1377	.1147	300
2/0	1	0.00510	0.702	0.702	0.704	0.706	0.773	0.806	0.866	0.895	.231	.462	.554	.267	.534	.641	.284	.1418	.1182	270
1/0	2	0.00446	0.885	0.885	0.887	0.888	0.974	1.01	1.08	1.12	.237	.473	.568	.273	.547	.656	.292	.1460	.1216	230
1	3	0.00418	1.12	1.12	1.12	1.12	1.23	1.27	1.34	1.38	.242	.483	.580	.277	.554	.665	.300	.1500	.1250	200
2	4	0.00418	1.41	1.41	1.41	1.41	1.55	1.59	1.66	1.69	.247	.493	.592	.277	.554	.665	.308	.1542	.1285	180
2	4	0.00504	1.41	1.41	1.41	1.41	1.55	1.59	1.62	1.65	.247	.493	.592	.267	.535	.642	.306	.1532	.1276	180
3	5	0.00430	1.78	1.78	1.78	1.78	1.95	1.98	2.04	2.07	.252	.503	.604	.275	.551	.661	.317	.1583	.1320	160
4	6	0.00437	2.24	2.24	2.24	2.24	2.47	2.50	2.54	2.57	.257	.514	.611	.274	.549	.659	.325	.1627	.1355	140
4	6	0.00452	2.24	2.24	2.24	2.24	2.47	2.50	2.53	2.55	.257	.515	.618	.273	.545	.655	.323	.1615	.1346	140
5	7	0.00416	2.82	2.82	2.82	2.82	3.10	3.12	3.16	3.18	.262	.525	.630	.279	.557	.665	.333	.1666	.1388	120
6	8	0.00394	3.56	3.56	3.56	3.56	3.92	3.94	3.97	3.98	.268	.536	.643	.281	.561	.673	.342	.1708	.1423	100

*Area of hard-drawn copper cable (conductivity = 97% IACS) having the same dc resistance as that of the ACSR aluminum (conductivity = 61% IACS).

†For resistance temperature conversion see Pars. **2-23** and **2-27**.
Resistances at 50°C total temperature, based on ambient of 25°C plus 25°C rise due to heating effect of current. The approximate magnitude of current necessary to produce the 25°C rise is 75% of the "approximate current-carrying capacity at 60 Hz."

‡See Table 2-32.
§See Table 2-33.
¶For conductor at 75°C, air at 25°C, wind 2 ft/s (1.4 mi/h), average tarnished surface.
NOTE: 1 ft = 0.3048 m; 1 mi = 1.61 km.

TABLE 2-31 Aluminum-Clad Steel Cable—Electrical Characteristics

(Compiled from tables published by Copperweld Steel Company)

Conductor stranding		Geometric mean radius at 60 cycles, average currents, ft	Resistance at 25°C (77°F),* small currents			Resistance at 75°C (167°F), current approx 75% of capacity			Reactance at 1-ft spacing				Current-carrying capacity at 60 cycles§ (approx), A
									Inductive (series) (x_a)†		Capacitive (shunt) (x_a')‡		
No. of wires	Wire size, AWG		D-c	50 cycles	60 cycles	D-c	50 cycles	60 cycles	50 cycles	60 cycles	50 cycles	60 cycles	
			Ω/(conductor) (mi)			Ω/(conductor) (mi)			Ω/(conductor) (mi)		MΩ/(conductor) (mi)		
19	5	0.004929	0.4420	0.4507	0.4507	0.5202	0.7203	0.7585	0.537	0.645	0.1165	0.0971	485
19	6	0.004387	0.5574	0.5683	0.5683	0.6559	0.8517	0.8886	0.548	0.658	0.1206	0.1005	425
19	7	0.003905	0.7030	0.7171	0.7171	0.8273	1.027	1.064	0.561	0.673	0.1248	0.1040	380
19	8	0.003478	0.8864	0.9038	0.9038	1.043	1.243	1.280	0.572	0.687	0.1289	0.1074	335
19	9	0.003098	1.118	1.140	1.140	1.315	1.518	1.554	0.584	0.701	0.1331	0.1109	295
10	10	0.002757	1.409	1.437	1.437	1.658	1.861	1.896	0.596	0.715	0.1360	0.1133	260
7	5	0.002958	1.217	1.240	1.240	1.432	1.634	1.669	0.589	0.707	0.1346	0.1122	280
7	6	0.002633	1.507	1.536	1.536	1.773	1.977	2.01	0.601	0.721	0.1388	0.1157	250
7	7	0.002345	1.900	1.937	1.937	2.24	2.44	2.47	0.612	0.735	0.1429	0.1191	220
7	8	0.002085	2.40	2.44	2.44	2.82	3.03	3.06	0.624	0.749	0.1471	0.1226	190
7	9	0.001858	3.02	3.08	3.08	3.56	3.77	3.80	0.636	0.763	0.1512	0.1260	160
7	10	0.001658	3.81	3.88	3.88	4.48	4.70	4.73	0.647	0.777	0.1552	0.1294	140
3	5	0.002940	2.78	2.78	2.78	3.27	3.52	3.56	0.589	0.707	0.1465	0.1221	170
3	6	0.002618	3.51	3.51	3.51	4.13	4.36	4.41	0.601	0.721	0.1506	0.1255	150
3	7	0.002333	4.42	4.42	4.42	5.21	5.43	5.47	0.612	0.735	0.1547	0.1289	130
3	8	0.002078	5.58	5.58	5.58	6.57	6.78	6.82	0.624	0.749	0.1589	0.1324	110

*For resistance temperature conversion see Pars. **2-23** and **2-29**.
†See Table **2-32**.
‡See Table **2-33**.
§For conductor at 125°C, air at 25°C, wind 2 ft/s (1.4 mi/h), average tarnished surface.
NOTE: 1 ft = 0.3048 m; 1 mi = 1.61 km.

TABLE 2-32 Inductive-Reactance Spacing Factors (x_d)

Frequency, cycles	Units / Tens	Equivalent conductor spacing, ft									
		0	1	2	3	4	5	6	7	8	9
		Ohms per conductor per mile									
25	0	......	0	0.0350	0.0555	0.0701	0.0814	0.0906	0.0984	0.1051	0.1111
	1	0.1164	0.1212	0.1256	0.1297	0.1334	0.1369	0.1402	0.1432	0.1461	0.1489
	2	0.1515	0.1539	0.1563	0.1585	0.1607	0.1627	0.1647	0.1666	0.1685	0.1702
	3	0.1720	0.1736	0.1752	0.1768	0.1783	0.1798	0.1812	0.1826	0.1839	0.1852
	4	0.1865	0.1878	0.1890	0.1902	0.1913	0.1925	0.1936	0.1947	0.1957	0.1968
50	0	......	0	0.0701	0.1111	0.1402	0.1627	0.1812	0.1968	0.2103	0.2222
	1	0.2328	0.2425	0.2513	0.2594	0.2669	0.2738	0.2804	0.2865	0.2923	0.2977
	2	0.3029	0.3079	0.3126	0.3170	0.3214	0.3255	0.3294	0.3333	0.3369	0.3405
	3	0.3439	0.3472	0.3504	0.3536	0.3566	0.3595	0.3624	0.3651	0.3678	0.3704
	4	0.3730	0.3755	0.3779	0.3803	0.3826	0.3849	0.3871	0.3893	0.3914	0.3935
60	0	......	0	0.0841	0.1333	0.1682	0.1953	0.2174	0.2361	0.2523	0.2666
	1	0.2794	0.2910	0.3015	0.3112	0.3202	0.3286	0.3364	0.3438	0.3507	0.3573
	2	0.3635	0.3694	0.3751	0.3805	0.3856	0.3906	0.3953	0.3999	0.4043	0.4086
	3	0.4127	0.4167	0.4205	0.4243	0.4279	0.4314	0.4348	0.4382	0.4414	0.4445
	4	0.4476	0.4506	0.4535	0.4564	0.4592	0.4619	0.4646	0.4672	0.4697	0.4722

Total inductive reactance = $x_a + x_d$.
See Par. **2-138**.
NOTE: 1 ft = 0.3048 m; 1 mi = 1.61 km.

CONDUCTOR MATERIALS

TABLE 2-33 Capacitive-Reactance Spacing Factors (x'_d)

Frequency, cycles	Units Tens	Equivalent conductor spacing, ft									
		0	1	2	3	4	5	6	7	8	9
		Megohms per conductor per mile									
25	0	......	0	0.0494	0.0782	0.0987	0.1146	0.1276	0.1386	0.1481	0.1565
	1	0.1640	0.1707	0.1769	0.1826	0.1879	0.1928	0.1974	0.2017	0.2058	0.2097
	2	0.2133	0.2168	0.2201	0.2233	0.2263	0.2292	0.2320	0.2347	0.2373	0.2398
	3	0.2422	0.2445	0.2468	0.2490	0.2511	0.2532	0.2552	0.2571	0.2590	0.2609
	4	0.2627	0.2644	0.2661	0.2678	0.2695	0.2711	0.2726	0.2742	0.2756	0.2771
50	0	......	0	0.0247	0.0391	0.0494	0.0573	0.0638	0.0693	0.0740	0.0782
	1	0.0820	0.0854	0.0885	0.0913	0.0940	0.0964	0.0987	0.1009	0.1029	0.1048
	2	0.1067	0.1084	0.1100	0.1116	0.1131	0.1146	0.1160	0.1173	0.1186	0.1199
	3	0.1211	0.1223	0.1234	0.1245	0.1255	0.1266	0.1276	0.1286	0.1295	0.1304
	4	0.1313	0.1322	0.1331	0.1339	0.1347	0.1355	0.1363	0.1371	0.1378	0.1386
60	0	......	0	0.0206	0.0326	0.0411	0.0478	0.0532	0.0577	0.0617	0.0652
	1	0.0683	0.0711	0.0737	0.0761	0.0783	0.0803	0.0823	0.0841	0.0858	0.0874
	2	0.0889	0.0903	0.0917	0.0930	0.0943	0.0955	0.0967	0.0978	0.0989	0.0999
	3	0.1009	0.1019	0.1028	0.1037	0.1046	0.1055	0.1063	0.1071	0.1079	0.1087
	4	0.1094	0.1102	0.1109	0.1116	0.1123	0.1129	0.1136	0.1142	0.1149	0.1155

Total capacitive reactance = $x'_a + x'_d$.
See Pars. **2-140** and **2-141**.
NOTE: 1 ft = 0.3048 m; 1 mi = 1.61 km.

2-172. *Fusing currents of different kinds of wire* were investigated by W. H. Preece, who developed the formula

$$I = ad^{3/2} \tag{2-34}$$

where I = fusing current in amperes, d = diameter of the wire in inches, a = a constant depending upon the material. He found the following values for a:

Copper	10,244	Iron	3,148
Aluminum	7,585	Tin	1,642
Platinum	5,172	Alloy (2Pb-1Sn)	1,318
German silver	5,230	Lead	1,379
Platinoid	4,750		

Although this formula has been used to a considerable extent in the past, it gives values that usually are erroneous in practice, because it is based on the assumption that all heat loss is due to radiation. A formula of the general type

$$I = kd^n \tag{2-35}$$

can be used with accuracy if k and n are known for the particular case (material, wire size, installation conditions, etc.).

2-173. *Fusing current-time for copper conductors and connections* may be determined by an equation developed by I. M. Onderdonk:

$$33 \left(\frac{I}{A} \right)^2 S = \log \left(\frac{T_m - T_a}{234 + T_a} + 1 \right) \tag{2-36}$$

$$I = A \sqrt{\frac{\log \left(\dfrac{T_m - T_a}{234 + T_a} + 1 \right)}{33S}} \tag{2-37}$$

TABLE 2-34 Properties of Resistance Metals and Alloys

Material	Chemical composition	Resistivity at 20°C (68°F) Ω·cmils/ft	Resistance Temperature coefficient, per °C	Resistance Temperature range, °C	Linear expansion Temperature coefficient, per °C	Linear expansion Temperature range, °C	Melting point, approx, °C	Tensile strength at 20°C (68°F), min, psi	Specific gravity	Weight, lb/in³
Driver-Harris Co., Harrison, N. J.										
Karma*.............	Ni 73%-Cr 20% + Al + Fe	800	....	−50–105	0.00001	20–100	1400	130,000	8.105	0.292
Nichrome*.............	Ni 60%-Cr 16%-balance Fe	675	0.00015	20–500	0.000017	20–1000	1350	95,000	8.247	0.2979
Nichrome V*..........	Ni 80%-Cr 20%	650	0.00011	20–500	0.000017	10–1000	1400	100,000	8.412	0.3039
Chromax*.............	Ni 35%-Cr 20%-balance Fe	600	0.00036	20–500	0.0000158	20–500	1380	70,000	7.950	0.2872
Nilvar*.............	Ni 36%-balance Fe	484	0.00135	20–100	0.000001	20–100	1425	70,000	8.08	0.292
Stainless type 304......	Cr 18%-Ni 8%-balance Fe	438	0.00094	20–500	0.000020	0–1000	1399	100,000	7.93	0.286
142 alloy..............	Ni 42%-balance Fe	400	0.0012	20–500	0.0000053	20–400	1425	70,000	8.12	0.293
Advance*.............	Ni 43%-balance Cu	294	±0.00002	20–100	0.0000149	20–100	1210	60,000	8.9	0.321
Therlo*.............	Ni 29%-Co 17%-balance Fe	294	0.0038	0–100	0.000006	30–500	1450	75,000	8.36	0.302
Manganin.............	Mn 13%-balance Cu	290	±0.000015	15–35	0.0000187	15–35	1020	40,000	8.192	0.296
146 alloy.............	Ni 46%-balance Fe	275	0.0027	20–500	0.000008	25–425	1425	70,000	8.17	0.295
152 alloy (52)..........	Ni 51%-balance Fe	260	0.0029	20–500	0.0000095	20–500	1425	70,000	8.247	0.2979
Duranickel.............	Nickel plus additions	260	0.001	20–500	0.000014	20–500	1435	90,000	8.75	0.316
Midohm*.............	Ni 23%-balance Cu	180	0.00018	−50–150	0.0000175	20–500	1100	50,000	8.9	0.321
R-63 alloy.............	Mn 4%-Si 1%-balance Ni	130	0.003	20–250	0.0000152	20–500	1425	70,000	8.72	0.315
Hytemco*.............	Ni 72%-balance Fe	120	0.0042	20–100	0.000015	20–1000	1425	70,000	8.46	0.305
Permanickel.............	Nickel plus additions	100	0.0036	30–500	0.000014	30–1000	1450	90,000	8.75	0.316
90 alloy.............	Ni 11%-balance Cu	90	0.00049	−50–150	0.0000175	20–500	1100	35,000	8.9	0.321
Gr. A nickel.............	Ni 99%	60	0.0050	0–100	0.000015	20–500	1450	60,000	8.9	0.321
Lohm*.............	Ni 6%-balance Cu	60	0.0008	−50–150	0.000018	20–500	1100	50,000	8.9	0.321
99 alloy.............	Ni 99.8%	48	0.0060	0–100						
30 Alloy.............	Ni 2.25%-balance Cu	30	0.0015	−50–150	0.0000175	20–500	1100	30,000	8.9	0.321

Hoskins Manufacturing Co., Detroit, Mich.

Chromel AA* Ni 68%-Cr 20%-Fe 8%	700	0.00011	20-500	0.0000135	20-1000	1390	120,000	8.33	0.301
Chromel A* Ni 80%-Cr 20%	650	0.00011	20-500	0.000017	10-1000	1400	100,000	8.412	0.3039
Chromel C* Ni 60%-Cr 16%-balance Fe	675	0.00015	20-500	0.000017	20-1000	1350	95,000	8.247	0.2979
Chromel D* Ni 35%-Cr 20%-balance Fe	600	0.00036	20-500	0.0000158	20-500	1380	70,000	7.950	0.2872
Copel* Ni 43%-balance Cu	294	±0.00002	20-100	0.0000149	20-100	1210	60,000	8.9	0.321
Alloy 875 Cr 22.5%-Al 5.5%-balance Fe	875	0.00002	20-500	0.0000174	20-1000	1520	110,000	7.10	0.256
Alloy 815 Cr 22.5%-Al 4.6%-balance Fe	815	0.00008	20-500	0.0000159	20-1000	1520	110,000	7.25	0.262
Alloy 750 Cr 15%-Al 4%-balance Fe	750	0.00015	20-500	0.0000150	20-1000	1520	110,000	7.43	0.268

The Kanthal Corp, Bethel, Conn.

Kanthal DR* Fe 75%-Cr 20%-Al 4.5%-Co 0.5%	812	0.00007	20-150	0.0000119	20-100	1505	100,000	7.2	0.262
Nikrothal L* Ni 75%-Cr 17%-balance Si + Mn	800	0.000003	20-150	0.0000126	20-100	1410	150,000	8.1	0.292
Nikrothal 6* Ni 60%-Cr 16%-balance Fe	675	0.000140	20-100	0.000013	20-100	1350	90,000	8.25	0.298
Nikrothal 8* Ni 80%-Cr 20%	650	0.000080	20-100	0.000014	20-100	1400	95,000	8.41	0.304
Cuprothal 294* Ni 45%-balance Cu	294	0.00002	20-100				60,000	8.9	0.321
Cuprothal 180* Ni 22%-balance Cu	180	0.00018	20-100				50,000	8.9	0.321
Cuprothal 90* Ni 11%-balance Cu	90	0.00045	20-100				35,000	8.9	0.321
Cuprothal 60* Ni 6%-balance Cu	60	0.0008	20-100				35,000	8.9	0.321
Cuprothal 30* Ni 2%-balance Cu	30	0.0014	20-100				30,000	8.9	0.321

Pure Metals

Platinum	63.80	0.00300	20	0.0000089	20	1773		21.45	0.7750
Iron	60.14	0.0050	20	0.0000117	20	1535	50,000	7.86	0.2840
Molybdenum	34.27	0.0033	18	0.000005	20	2625	100,000	10.2	0.3685
Tungsten	33.22	0.0045		0.000024	18	3410 ± 20	490,000	19.3	0.6973
Aluminum	16.06	0.00446				660	35,000	2.7	0.0975
Gold	14.55	0.0034	20	0.0000142	20	1063		19.3	0.6973
Copper	10.37	0.00393	20	0.0000166	20	1083	35,000	8.92	0.3223
Silver	9.796	0.0038	20	0.0000189	20	960		10.5	0.3793

*Trademark.

NOTE: 1 ft = 0.3048 m; 1 lb/in² = 6.895 kPa; 1 lb/in³ = 27,680 kg/m³.

ELECTRICAL ENGINEERING MATERIALS REFERENCE GUIDE

TABLE 2-35 Compositions and Melting Points of Fusible Alloys

("International Critical Tables," vol. 2. p. 391)

Chemical composition, %					Melting point, deg C	Chemical composition, %					Melting point, deg C
Bi	Pb	Sn	Cd	Hg		Bi	Pb	Sn	Cd	Hg	
20	20	..	..	60	20	..	32	50	18	..	145
50	27	13	10	..	72	50	50	..	..	..	160
52	40	..	8	..	92	15	41	44	..	..	164
53	32	15	..	..	96	33	..	67	..	..	166
54	26	..	20	..	103	20	..	80	..	..	200
29	43	28	..	..	132	..	..	..	..	..	

Compositions and melting points are approximate.

where I = current in amperes, A = conductor area in circular mils, S = time current applied in seconds, T_m = melting point of copper in degrees Celsius, T_a = ambient temperature in degrees Celsius.

2-174. Copper Conductors. E. R. Stauffacher has prepared a chart of the fusing current for sizes from 30 AWG to 500,000 cmils from 0.1 to 10 s (see Fig. 2-15). This chart is based on the assumptions that (1) radiation may be neglected owing to the short time involved, that is, 10 s; (2) resistance of 1 cm cube of copper at 0°C is 1.589 $\mu\Omega$; (3) temperature-resistance coefficient of copper at 0°C is ½₃₄; (4) melting point of copper is 1083°C; and (5) ambient temperature is 40°C.

For most practical purposes, Eq. (2-37) also may be applied where the melting temperature T_m of solder or other materials used in making connections is the determining factor; for example:

Soldered Connections. Select a value of T_m corresponding to the melting temperature for the composition of tin-lead alloy used. This may be determined by test or approximated from Fig. 2-16 prepared from the "Smithsonian Physical Tables" (for example, T_m = 183°C for 70:30 solder).

Brazed Connections. A reasonable value of T_m is 450°C.

Bolted Connections. Generally accepted value of T_m is 250°C.

Miscellaneous Metals and Alloys

2-175. Contact metals may be grouped into three general classifications:

Hard metals, which have melting points, for example, tungsten and molybdenum. Contacts of these metals are employed usually where operations are continuous or very frequent and current has nominal value of 5 to 10 A. Hardness to withstand mechanical wear and high melting point to resist arc erosion and welding are their outstanding advantages. Tendency to form high-resistance oxides is a disadvantage, but this can be overcome by several methods, such as using high-contact force, a hammering or wiping action, and a properly balanced electric circuit.

Highly conductive metals, of which silver is the best for both electric current and heat. Its disadvantages are softness and a tendency to pit and transfer. In sulfurous atmosphere, a resistant sulfide surface will form on silver, which results in high contact-surface resistance. These disadvantages are overcome usually by alloying.

Noncorroding metals, which for the most part consist of the noble metals, such as gold and the platinum group. Contacts of these metals are used on sensitive devices, employing extremely light pressures or low currents in which clean contact surfaces are essential. Because most of these metals are soft, they are usually alloyed.

The metals commonly used are tungsten, molybdenum, platinum, palladium, gold, silver, and their alloys. Alloying materials are copper, nickel, cadmium, iron, and the rarer metals such as iridium and ruthenium. Some are prepared by powder metallurgy.

Commercial grades are available under trade names or alloy numbers from Baker & Co.,

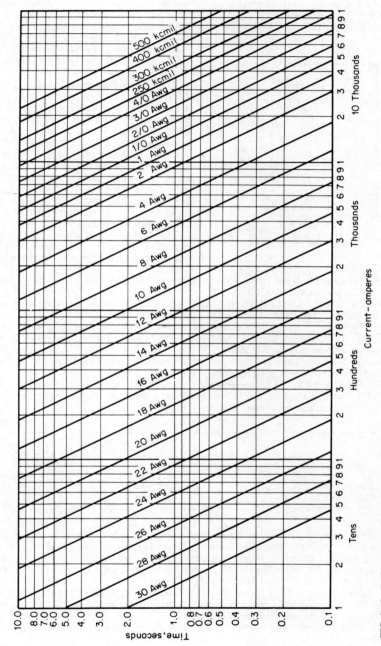

FIG. 2-15 Fusing current time for copper conductors.

Newark, N.J.; Fansteel Metallurgical Corp., North Chicago (Fasaloy, Fastell); North American Phillips Co., New York (Elmet); P. R. Mallory & Co., Indianapolis; and others.

2-176. Tungsten (W) is a hard, dense, slow-wearing metal, a good thermal and electrical conductor, characterized by its high melting point and freedom from sticking or welding. It is manufactured in several grades having various grain sizes.

2-177. Molybdenum (Mo) has contact characteristics about midway between tungsten and fine silver. It often replaces either metal where greater wear resistance than that of silver or lower contact-surface resistance than that of tungsten is desired.

2-178. Platinum (Pt) is one of the most stable of all metals under the combined action of corrosion and electrical erosion. It has a high melting point and does not corrode, and surfaces remain clean and low in resistance under most adverse atmospheric and electrical conditions.

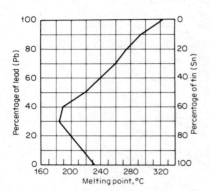

FIG. 2-16 Melting temperatures (T_m) for sodder.

Platinum alloys of iridium (Ir), ruthenium (Ru), silver (Ag), or other metals are used to increase hardness and resistance to wear.

2-179. Palladium (Pd) has many of the properties of platinum and frequently is used as an alternate for platinum and its alloys.

Palladium alloys of silver (Ag), ruthenium (Ru), nickel (Ni), and other metals are used to increase hardness and resistance to wear.

2-180. Gold (Au) is similar to platinum in corrosion resistance but has a much lower melting point. Gold and its alloys are ductile and easily formed into a variety of shapes. Because of its softness it is usually alloyed.

Gold alloys of silver (Ag) and other metals are used to impart hardness and improve resistance to mechanical wear and electrical erosion.

2-181. Silver (Ag) has the highest thermal and electrical conductivity (110%, IACS) of any metal. It has low contact-surface resistance, since its oxide decomposes at approximately 300°F. It is available commercially in three grades:

| | Typical composition, % | |
Grade	Silver	Copper
Fine silver	99.95+	
Sterling silver	92.5	7.5
Coin silver	90	10

Fine silver is used extensively under low contact pressure where sensitivity and low contact-surface resistance are essential or where the circuit is operated infrequently.

Sterling and coin silvers are harder than fine silver and resist transfer at low voltage (6 to 8 V) better than fine silver. Since their contact-surface resistance is greater than that of fine silver, higher contact-closing forces should be used.

Silver alloys of copper (Cu), nickel (Ni), cadmium (Cd), iron (Fe), carbon (C), tungsten (W), molybdenum (Mo), and other metals are used to improve hardness, resistance to wear and arc erosion, and for special applications.

2-182. High-Melting-Point Metals. Table 2-36 shows the melting-point range of all the metals.

Table 2-37 contains the physical properties of four of the metals in the highest melting-point range.

2-183. Selenium is a nonmetallic element chemically resembling sulfur and tellurium

CONDUCTOR MATERIALS

TABLE 2-36 Melting Points of Metals

(Compiled from tables published by Fansteel Metallurgical Corp.)

Melting point, deg C	Metal	Melting point, deg C	Metal
3500–3400		1500–1400	Silicon, nickel, cobalt, yttrium
3400–3300	**Tungsten**	1400–1300	Beryllium
3300–3200		1300–1200	Manganese
3200–3100	Rhenium	1200–1100	
3100–3000		1100–1000	Gold, copper
3000–2900	**Tantalum**	1000– 900	Praseodymium, germanium, radium, **silver**
2900–2800		900– 800	Cerium, arsenic, neodymium, calcium, lanthanum
2800–2700		800– 700	Barium, strontium
2700–2600	**Molybdenum**	700– 600	Antimony, magnesium, aluminum
2600–2500	Osmium, iridium	600– 500	
2500–2400	**Columbium**	500– 400	Zinc, tellurium
2400–2300	Boron	400– 300	Cadmium, thallium, lead
2300–2200	Hafnium	300– 200	Selenium, tin, bismuth
2200–2100	Zirconium	200– 100	Indium, lithium
2100–2000		100– 0	Cesium, gallium, rubidium, **potassium, sodium**
2000–1900	Chromium, ruthenium, rhodium	0 to −100	Mercury
1900–1800	Thorium		
1800–1700	Vanadium, titanium, platinum		
1700–1600	Uranium		
1600–1500	Iron, palladium		

TABLE 2-37 Properties of High-Melting-Point Metals

(Fansteel Metallurgical Corp.)

Property	Tungsten	Tantalum	Molybdenum	Columbium
Specific gravity at 20 C..........................	19.3	16.6	10.2	8.57
Electrical resistivity at 20 C, microhm-cm.......	5.5	12.4	5.17	13.1
Electrical resistance at 20 C, ohm (mil, ft)......	33.1	74.6	31.1	79.0
Temperature coefficient of resistance at 20 C....	0.0051	0.0036	0.0047	0.00395
Tensile strength, unannealed wire, lb per sq in...	200,000	130,000	105,000	96,000
Coefficient of linear expansion, per deg C.......	4.3×10^{-6}	6.5×10^{-6}	4.9×10^{-6}	7.1×10^{-6}
Specific heat at 20 C, cal per g atom per deg C..	6.18	6.512	6.24	6.012
Melting point, deg C..........................	3410	2996	2625	2415

NOTE: 1 ft = 0.3048 in; 1 lb/in^2 = 6.895 kPa.

and occurs in several allotropic forms varying in specific gravity from 4.3 to 4.8. It melts at 217°C and boils at 690°C. At 0°C it has a resistivity of approximately 60,000 $\Omega \cdot$ cm. The dielectric constant ranges from 6.1 to 7.4. It has the peculiar property that its resistivity decreases upon exposure to light; the resistivity in darkness may be anywhere from 5 to 200 times the resistivity under exposure to light (see paper by W. J. Hammer, *Trans. AIEE,* 1903, vol. 21, pp. 372–393).

2-184. Bibliography

Electric Utility Engineering Reference Book, vol. 3; Westinghouse Electric Corporation, East Pittsburgh, Pa., 1959.

Underground Systems Reference Book; Edison Electric Institute, Transmission and Distribution Committee, 1957.

Aluminum Electrical Conductor Handbook, 2nd ed; The Aluminum Association, 818 Connecticut Avenue, N.W., Washington, D.C. 20006, 1982.

Metals Handbook, 9th ed.; American Society for Metals, 1979, vol. 1, Properties and Selection of Nonferrous Alloys, p. 1007.

Smith, C. S.: Thermal Conductivity of Copper Alloys, I. Copper-Zinc Alloys; *Trans. AIMME, Inst. Metals Div.,* 1930, vol. 89, p. 84.

ELECTRICAL ENGINEERING MATERIALS REFERENCE GUIDE

Johnson, J. B.: *Materials of Construction;* New York, John Wiley & Sons, Inc., 1925.

Preece, W. H.: On the Heating Effects of Electric Currents; *Proc. Roy. Soc. (London),* April 1884; December 1887; April 1888.

Stauffacher, E. R.: Short-Time Current-Carrying Capacity of Copper Wires; *Gen. Elec. Rev.,* June 1928.

Resistance and Reactance of Aluminum Conductors, Alcoa Conductor Products Company, 510 One Allegheny Square, Pittsburgh, Pa. 15212.

Current Temperature Characteristics of Aluminum Conductors, Alcoa Conductor Products Company, Pittsburgh, Pa.

Overload and Fault Current Limitations of Bare Aluminum Conductors, Alcoa Conductor Products Company, Pittsburgh, Pa.

Chapter 3

CARBON AND GRAPHITE

Prepared by THE CARBON AND MANUFACTURED GRAPHITE SECTIONS, NATIONAL ELECTRICAL MANUFACTURERS ASSOCIATION

3-1 Forms of Carbon   3-1
3-2 Temperature Coefficient of Resistance   3-1

3-3 Carbon-Brush Applications   3-1
3-4 Arc-Lamp Carbons   3-5

3-1. Forms of Carbon. Carbon occurs in two forms, amorphous and crystalline. The crystalline forms include diamond and graphite. The amorphous forms include charcoal, coke, and carbon black; coal is an impure variety of amorphous carbon. Some of the typical properties of carbon and electrographites manufactured from these carbons are listed in Tables 3-1 and 3-2.

Most carbon used for electrical purposes is made from a mixture of powdered carbon and/or graphite (such as lampblack and petroleum coke) and binders (such as pitch and resins), which are mixed into a homogeneous mass, extruded or molded, and then baked. When the mixture is baked to approximately 900°C with the air excluded, the volatile part of the binding material is driven off and the remaining binder is carbonized. The resulting product can be converted into electrographite by furnacing it in the absence of oxygen to a temperature of not less than 2200°C, usually higher.

3-2. Temperature Coefficient of Resistance. Carbon exhibits an increasing electrical and thermal conductivity with rising temperature. Graphite can exhibit a complicated change in electrical conductivity with rising temperature (see Fig. 3-1), and its thermal conductivity decreases markedly with rising temperature.

3-3. Carbon-Brush Applications. The term *carbon brush* is used to designate all types of sliding electrical contacts that contain any appreciable percentage of carbon or graphite in their composition. Other ingredients may be metals and suitable binders.

Carbon and graphite brushes for commutator-type machines and collector rings of ac machines are made in various grades with appropriate characteristics for the types of service, including atmospheric conditions and load cycles.

The functions to be performed by carbon brushes on electrical machines are these:

Carbon brushes on a slip-ring machine have only to provide a suitable sliding electrical connection between the line and the rotor with reasonable life.

Carbon brushes on a dc machine have three entirely distinct functions to perform: (1) They must carry the current into and out of the commutator. (2) They participate in the reversal of all or part of the current in the armature coils during the time they are short-circuited. Performance of both functions simultaneously complicates the problem of design. In order that the current in the short-circuited armature coils may be reversed without arcing or sparking, there must be an appreciable resistance in the contact between the brush and commutator. The amount of resistance depends upon the magnitude of the reactance voltage. (3) And they must have reasonable life.

The carbon brush in its various types and compositions provides the characteristics needed for an ideal sliding electrical connection, namely, good conductivity, low coefficient of friction, and high durability.

The resistances of carbon contacts vary with pressure, current, and time (see *Bibliography and Abstracts on Electrical Contacts,* ASTM, Holm Conference, IEEE, 1934 to date, and R. Holm, *Electric Contacts,* Springer, 1967). The property of variable contact resistance with varying contact pressure is also very useful in carbon-pile resistors, which can be varied over

TABLE 3-1 Typical Properties of Carbon

Material	True specific gravity*	Specific heat at temperatures, g·cal/g·°C						Threshold temp. of oxidation in air, °C
		26–76 °C	26–282 °C	26–538 °C	36–902 °C	47–1193 °C	56–1450 °C	
Carbon, coke base, gas calcined	1.98–2.10†	0.168	0.200	0.199	0.315	0.352	0.387	350
Carbon, coke base, graphitized	2.20–2.24	0.168	0.200	0.199	0.315	0.352	0.387	400
Carbon, lampblack base, gas calcined	1.80–1.85†							350
Carbon, lampblack base, graphitized	1.98–2.08							400
Anthracite, gas calcined	1.79†							—
Anthracite, electric calcined	1.90–1.97†							—
Graphite, pure	2.25	0.165	0.195	0.234	0.324	0.350	0.390	—
Diamond	3.51	0.160	0.315	0.415				—

*As measured by pycnometer.
†Dependent on source and degree of calcination.

TABLE 3-2 Properties of Manufactured Carbon and Graphite*

Material	Form	Apparent density, g/cm^3	Strength, lb/in^2			Elastic modulus, lb/in^2 × 10^6	Electrical resistivity at 20°C (68°F), Ω·in	Thermal expansion per °C × 10^{-7} (RT-100°C)[†]
			Tensile	Compressive	Flexural			
Carbon rounds (coal based)	Up to 55 in diam.	1.6	200	1800/2000	400/700	0.6/0.8	0.0012/0.0020	30
Carbon blocks (coal based)	Up to 30 × 36 in	1.6	200	2500/2700	600/750	0.8/0.9	0.0014/0.0020	35
Graphite rounds (mold stock & electrodes)	Up to 3 in diam.	1.6/1.8	800/2500	4000/8000	2000/5500	1.5/2.5	0.00025/0.00040	10/15
	4 to 10 in diam.	1.6/1.8	500/2500	3000/6500	1500/3500	1.0/2.0	0.00025/0.00040	10/20
	12 to 28 in diam.	1.6/1.8	450/2500	2000/6500	1000/3500	1.0/2.0	0.00025/0.00040	5/20
	30 in diam. & larger	1.6/1.8	450/1500	3000/5000	1400/2000	1.0/2.0	0.00030/0.00040	20/30
Graphite blocks	Up to 6 in thick	1.6/1.8	700/2500	3000/8000	1700/4500	1.5/2.0	0.00025/0.00040	10/15
	6 to 20 in thick	1.6/1.8	700/2500	3000/6000	1500/3500	1.0/2.0	0.00025/0.00040	10/20
	Over 20 in thick	1.6/1.8	550/2000	3000/5000	1500/2500	1.0/2.0	0.00030/0.00040	10/20

*Characteristics shown are typical values and will differ with individual grades and manufacturers.
[†]RT = room temperature.
NOTE: With grain, properties are listed as minimum/maximum values. 1 in = 2.54 cm; 1 lb/in^2 = 6.895 kPa.

TABLE 3-3 Range of Properties and Characteristics of Typical Brush Materials*

Physical properties and characteristics	Types of brush materials				
	Carbon	Carbon graphite	Graphite (resin bonded)	Electrographitic	Metal graphite
Resistivity, $\Omega\cdot$in	0.0010–0.0035	0.0005–0.0025	0.0003–0.050	0.0004–0.0035	0.000002–0.0001
Scleroscope hardness	45–85	40–75	10–35	12–75	7–35
Flexural strength, lb/in^2	3500–10,000	3000–7500	1000–5000	1500–10,000	2500–10,000
Current-carrying capacity, A/in^2	35–45	45–50	10–60	35–70	75–150
Contact drop†	Medium–high	Low–medium	Low–very high	Medium–high	Very low–low
Coefficient of friction‡	Medium–high	Low–medium	Low–medium	Low–medium	Very low–medium
Abrasiveness (cleaning action)	Pronounced	Medium–pronounced	Slight	Slight	Slight–medium
Peripheral speed, ft/min	2000–4000	3000–5000	4000–12,000	3000–9000	3000–6000

*See NEMA Standard CB1—1984 for test procedures.
†These terms have the following meanings: high, over 2.5 V; medium, 1.8 to 2.5 V; low, 1.0 to 1.8 V; very low, below 1.0 V.
‡These terms have the following meanings: high, over 0.26; medium, 0.20 to 0.26; low, 0.15 to 0.20; very low, 0.15 and below.
NOTE: 1 in = 2.54 cm; 1 lb/in^2 = 6.895 kPa; 1 in^2 = 64.5 cm^2; 1 ft/min = 0.00508 m/s.

a wide range, from practically open circuit to very low resistance with manipulation of the pressure on the pile.

Carbon contacts for relays have the same characteristics as low-resistance carbon brushes (Table 3-3).

For references on electrical brushes and their design, see NEMA Standard CB1.

3-4. Arc-Lamp Carbons. Carbons are produced in forms especially adapted for arc-lamp projectors of various types for the motion picture industry, for searchlights, and for irradiation applications. They contain various salts or compounds of metals such as cerium, calcium, cobalt, and strontium to control the wavelength of the arc output from ultraviolet through infrared. They can also be made to simulate the properties of sunlight.

The resistance of a ½-in-diameter by 12-in-long enclosed-arc carbon varies from 0.012 to 0.015 Ω/lin in. The resistances of other diameters vary according to their cross-sectional areas. For many applications, these carbons are copper-coated.

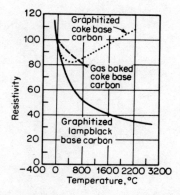

FIG. 3-1 Temperature-resistance relationship of carbon. 100 = resistivity at 20°C.

Chapter 4

MAGNETIC MATERIALS

By ANTHONY L. VON HOLLE and KENNETH L. LATIMER

4-1 Definitions 4-1
4-2 Magnetic Properties and Their
Application 4-8
4-9 Carbon Steels 4-11

4-14 Materials for Special Purposes 4-19
4-15 Alloys 4-20
4-24 High-Frequency Materials Applications
4-25

4-1. Definitions. The following definitions of terms relating to magnetic materials and to the properties and testing of these materials have been selected from ASTM Standard A340.[1] Terms primarily related to magnetostatics are indicated by the symbol * and those related to magnetodynamics are indicated by the symbol **. General (nonrestricted) terms are not marked.

AC Excitation $N_1 I/l_1$. The ratio of the rms ampere-turns of exciting current in the primary winding of an inductor to the effective length of the magnetic path.

Active (Real) Power P. The product of the rms current I in an electric circuit, the rms voltage E across the circuit, and the cosine of the angular phase difference θ between the current and the voltage.

$$P = EI \cos \theta \qquad (4-1)$$

NOTE: The portion of the active power that is expended in a magnetic core is the total core loss P_c.

Aging, Magnetic. The change in the magnetic properties of a material resulting from metallurgical change. This term applies whether the change results from a continued normal or a specified accelerated aging condition.

NOTE: This term implies a deterioration of the magnetic properties of magnetic materials for electronic and electrical applications, unless otherwise specified.

Ampere-turn. Unit of magnetomotive force in the rationalized mksa system. One ampere-turn equals $4\pi/10$ or 1.257 gilberts.

Ampere-turn per Meter. Unit of magnetizing force (magnetic field strength) in the rationalized mksa system. One ampere-turn per meter is $4\pi \times 10^{-3}$ or 0.01257 oersted.

Anisotropic Material. A material in which the magnetic properties differ in various directions.

---

[1] *1983 Annual Book of ASTM Standards,* sec3. vol. 03.04; ASTM, Philadelphia., 1983.

4-2　　ELECTRICAL ENGINEERING MATERIALS REFERENCE GUIDE

Antiferromagnetic Material. A feebly magnetic material in which almost equal magnetic moments are lined up antiparallel to each other. Its susceptibility increases as the temperature is raised until a critical (Neél) temperature is reached; above this temperature the material becomes paramagnetic.

**Apparent Power P_a. The product (volt-amperes) of the rms exciting current and the applied rms *terminal* voltage in an *electric* circuit containing inductive impedance. The components of this impedance due to the winding will be linear, while the components due to the magnetic core will be nonlinear.

**Apparent Power; Specific, $P_{a(B,f)}$. The value of the apparent power divided by the active mass of the specimen (volt-amperes per unit mass) taken at a specified maximum value of cyclically varying induction B and at a specified frequency f.

*Coercive Force H_c. The (dc) magnetizing force at which the magnetic induction is zero when the material is in a symmetrically cyclically magnetized condition.

*Coercive Force, Intrinsic, H_{ci}. The (dc) magnetizing force at which the intrinsic induction is zero when the material is in a symmetrically cyclically magnetized condition.

*Coercivity H_{cs}. The maximum value of coercive force.

**Core Loss, Specific, $P_{c(B,f)}$. The active power (watts) expended per unit mass of magnetic material in which there is a cyclically varying induction of a specified maximum value B at a specified frequency f.

**Core Loss (Total) P_c. The active power (watts) expended in a magnetic circuit in which there is a cyclically alternating induction.

NOTE: Measurements of core loss are normally made with sinusoidally alternating induction, or the results are corrected for deviations from the sinusoidal condition.

Curie Temperature T_c. The temperature above which a ferromagnetic material becomes paramagnetic.

*Demagnetization Curve. That portion of a normal (dc) hysteresis loop which lies in the second or fourth quadrant, that is, between the residual induction point B_r and the coercive force point H_c. Points on this curve are designated by the coordinates B_d and H_d.

Diamagnetic Material. A material whose relative permeability is less than unity.

NOTE: The intrinsic induction B_i, is oppositely directed to the applied magnetizing force H.

Domains, Ferromagnetic. Magnetized regions, either macroscopic or microscopic in size, within ferromagnetic materials. Each domain, per se, is magnetized to intrinsic saturation at all times, and this saturation induction is unidirectional within the domain.

**Eddy-Current Loss, Normal, P_e. That portion of the core loss which is due to induced currents circulating in the magnetic material subject to an *SCM* excitation.

*Energy Product $B_d H_d$. The product of the coordinate values of any point on a demagnetization curve.

*Energy-Product Curve, Magnetic. The curve obtained by plotting the product of the corresponding coordinates B_d and H_d of points on the demagnetization curve as abscissa against the induction B_d as ordinates.

NOTE 1: The maximum value of the energy product $(B_d H_d)_m$ corresponds to the maximum value of the external energy.

NOTE 2: The demagnetization curve is plotted to the left of the vertical axis and usually the energy-product curve to the right.

**Exciting Power, rms, P_z. The product of the rms exciting current and the rms voltage induced in the exciting (primary) winding on a magnetic core.

NOTE: This is the apparent volt-amperes required for the excitation of the magnetic core only. When the core has a secondary winding, the induced primary voltage is obtained from the measured open-circuit secondary voltage multiplied by the appropriate turns ratio.

**Exciting Power, Specific $P_{z(B,f)}$. The value of the rms exciting power divided by the active mass of the specimen (volt-amperes/unit mass) taken at a specified maximum value of cyclically varying induction B and at a specified frequency f.

Ferrimagnetic Material. A material in which unequal magnetic moments are lined up antiparallel to each other. Permeabilities are of the same order of magnitude as those of ferromagnetic materials, but are lower than they would be if all atomic moments were parallel and in the same direction. Under ordinary conditions the magnetic characteristics of ferrimagnetic materials are quite similar to those of ferromagnetic materials.

MAGNETIC MATERIALS

Ferromagnetic Material. A material that, in general, exhibits the phenomena of hysteresis and saturation, and whose permeability is dependent on the magnetizing force.

Gauss (Plural Gausses). The unit of magnetic induction in the cgs electromagnetic system. The gauss is equal to 1 maxwell per square centimeter or 10^{-4} tesla. See *magnetic induction (flux density).*

Gilbert. The unit of magnetomotive force in the cgs electromagnetic system. The gilbert is a magnetomotive force of $10/4\pi$ ampere-turns. See *magnetomotive force.*

*Hysteresis Loop, Intrinsic. A hysteresis loop obtained with a ferromagnetic material by plotting (usually to rectangular coordinates) corresponding dc values of intrinsic induction B_i for ordinates and magnetizing force H for abscissas.

*Hysteresis Loop, Normal. A closed curve obtained with a ferromagnetic material by plotting (usually to rectangular coordinates) corresponding dc values of magnetic induction B for ordinates and magnetizing force H for abscissas when the material is passing through a complete cycle between equal definite limits of either magnetizing force $\pm H_m$ or magnetic induction $\pm B_m$. In general the normal hysteresis loop has mirror symmetry with respect to the origin of the B and H axes, but this may not be true for special materials.

*Hysteresis-Loop Loss W_h. The energy expended in a single slow excursion around a normal hysteresis loop is given by the following equation:

$$W_h = \int HdB/4\pi \qquad \text{ergs} \qquad (4\text{-}2)$$

where the integrated area enclosed by the loop is measured in gauss-oersteds.

**Hysteresis Loss, Normal, P_h. 1. The power expended in a ferromagnetic material, as a result of hysteresis, when the material is subjected to an *SCM* excitation.

2. The energy loss/cycle in a magnetic material as a result of magnetic hysteresis when the induction is cyclic (but not necessarily periodic).

Hysteresis, Magnetic. The property of a ferromagnetic material exhibited by the lack of correspondence between the changes in induction resulting from increasing magnetizing force and from decreasing magnetizing force.

Induction B. See *magnetic induction (flux density).*

*Induction, Intrinsic, B_i. The vector difference between the magnetic induction in a magnetic material and the magnetic induction that would exist in a vacuum under the influence of the same magnetizing force. This is expressed by the equation

$$B_i = B - \Gamma_m H \qquad (4\text{-}3)$$

NOTE: In the cgs-em system $B_i/4\pi$ is often called magnetic polarization.

Induction, Maximum:

*1. B_m—The maximum value of B in a hysteresis loop. The tip of this loop has the magnestostatic coordinates H_m, B_m, which exist simultaneously.

**2. B_{max}—the maximum value of induction in a flux-current loop.

NOTE: In a flux-current loop, the magnetodynamic values B_{max} and H_{max} do not exist simultaneously; B_{max} occurs later than H_{max}.

*Induction, Normal, B. The maximum induction, in a magnetic material that is in a symmetrically cyclically magnetized condition.

NOTE: Normal induction is a magnetostatic parameter usually measured by ballistic methods.

*Induction, Remanent, B_d. The magnetic induction that remains in a magnetic circuit after the removal of an applied magnetomotive force.

NOTE: If there are no air gaps or other inhomogeneities in the magnetic circuit the remanent induction B_r will equal the residual induction B_r; if air gaps or other inhomogeneities are present, B_d will be less than B_r.

*Induction, Residual, B_r. The magnetic induction corresponding to zero magnetizing force in a magnetic material that is in a symmetrically cyclically magnetized condition.

*Induction, Saturation, B_s. The maximum intrinsic induction possible in a material.

*Induction Curve, Intrinsic (Ferric). A curve of a previously demagnetized specimen depicting the relation between intrinsic induction and corresponding ascending values of magnetizing force. This curve starts at the origin of the B_i and H axes.

ELECTRICAL ENGINEERING MATERIALS REFERENCE GUIDE

Induction Curve, Normal.   A curve of a previously demagnetized specimen depicting the relation between normal induction and corresponding ascending values of magnetizing force. This curve starts at the origin of the B and H axes.

Isotropic Material.   Material in which the magnetic properties are the same for all directions.

Magnetic Circuit.   A region at whose surface the magnetic induction is tangential.

NOTE: A practical magnetic circuit is the region containing the flux of practical interest, such as the core of a transformer. It may consist of ferromagnetic material with or without air gaps or other feebly magnetic materials such as procelain and brass.

Magnetic Constant (Permeability of Space) Γ_m.   The dimensional scalar factor that relates the mechanical force between two currents to their intensities and geometrical configurations. That is,

$$dF = \Gamma_m I_1 I_2 \, dl_1 \times (dl_2 \times r_1)/nr^2 \tag{4-4}$$

where Γ_m = magnetic constant when the element of force dF of a current element $I_1 \, dl_1$ on another current element $I_2 \, dl_2$ is at a distance r

r_1 = unit vector in the direction from dl_1 to dl_2

n = dimensionless factor; the symbol n is unity in unrationalized systems and 4π in rationalized systems

NOTE 1: The numerical values of Γ_m depend upon the system of units employed. In the cgs-em system $\Gamma_m = 1$, in the rationalized mksa system $\Gamma_m = 4\pi \times 10^{-7} \, h/m$.

NOTE 2: The magnetic constant expresses the ratio of magnetic induction to the corresponding magnetizing force at any point in a vacuum and therefore is sometimes called the permeability of space μ_r.

NOTE 3: The magnetic constant times the relative permeability is equal to the absolute permeability.

$$\mu_{abs} = \Gamma_m \mu_r \tag{4-5}$$

Magnetic Field Strength H.   See *magnetizing force.*

Magnetic Flux ϕ.   The product of the magnetic induction B and the area of a surface (or cross section) A when the magnetic induction B is uniformly distributed and normal to the plane of the surface.

$$\phi = BA \tag{4-6}$$

where ϕ = magnetic flux

B = magnetic induction

A = area of the surface

NOTE 1: If the magnetic induction is not uniformly distributed over the surface, the flux ϕ is the surface integral of the normal component of B over the area.

$$\phi = \int \int_s B \, dA \tag{4-7}$$

NOTE 2: Magnetic flux is scalar and has no direction.

Magnetic Flux Density B.   See *magnetic induction (flux density).*

Magnetic Induction (Flux Density) B.   That magnetic vector quantity which at any point in a magnetic field is measured either by the mechanical force experienced by an element of electric current at the point, or by the electromotive force induced in an elementary loop during any change in flux linkages with the loop at the point.

NOTE 1: If the magnetic induction B is uniformly distributed and normal to a surface or cross section, then the magnetic induction is

$$B = \phi/A \tag{4-8}$$

where B = magnetic induction

ϕ = total flux

A = area

MAGNETIC MATERIALS

4-5

NOTE 2: B_{in} is the instantaneous value of the magnetic induction and B_m is the maximum value of the magnetic induction.

Magnetizing Force (Magnetic Field Strength) H. That magnetic vector quantity at a point in a magnetic field which measures the ability of electric currents or magnetized bodies to produce magnetic induction at the given point.

NOTE 1: The magnetizing force H may be calculated from the current and the geometry of certain magnetizing circuits. For example, in the center of a uniformly wound long solenoid

$$H = C(NI/l) \tag{4-9}$$

where H = magnetizing force
 C = constant whose value depends on the system of units
 N = number of turns
 I = current
 l = axial length of the coil

If I is expressed in amperes and l is expressed in centimeters, then $C = 4\pi/10$ in order to obtain H in the cgs = em unit, the oersted.

If I is expressed in amperes and l is expressed in meters, then $C = 1$ in order to obtain H in the mksa unit, ampere-turn per meter.

NOTE 2: The magnetizing force H at a point in air may be calculated from the measured value of induction at the point by dividing this value by the magnetic constant Γ_m.

Magnetizing Force, AC Three different values of dynamic magnetizing force parameters are in common use:

a. H_L—an assumed peak value computed in terms of peak magnetizing current (considered to be sinusoidal).

b. H_x—an assumed peak value computed in terms of measured rms exciting current (considered to be sinusoidal).

c. H_p—computed in terms of a measured peak value of exciting current, and thus equal to the value H'_{max}.

Magnetodynamic. The magnetic condition when the values of magnetizing force and induction vary, usually periodically and repetitively, between two extreme limits.

Magnetomotive Force $\mathcal{F}$. The line integral of the magnetizing force around any flux loop in space.

$$\mathcal{F} = \oint H \, dl \tag{4-10}$$

where $\mathcal{F}$ = magnetomotive force
 H = magnetizing force
 dl = unit length along the loop

NOTE: The magnetomotive force is proportional to the net current linked with any closed loop of flux or closed path.

$$\mathcal{F} = CNI \tag{4-11}$$

where $\mathcal{F}$ = magnetomotive force
 N = number of turns linked with the loop
 I = current in amperes
 C = constant whose value depends on the system of units. In the cgs system $C = 4\pi/10$. In the mksa system $C = 1$

Magnetostatic. The magnetic condition when the values of magnetizing force and induction are considered to remain invariant with time during the period of measurement. This is often referred to as a dc (direct-current) condition.

Magnetostriction. Changes in dimensions of a body resulting from magnetization.

Maxwell. The unit of magnetic flux in the cgs electromagnetic system. One maxwell equals 10^{-8} weber. See *magnetic flux*.

4-6   ELECTRICAL ENGINEERING MATERIALS REFERENCE GUIDE

NOTE:

$$e = -N \, d\phi/dt \times 10^{-8} \tag{4-12}$$

where e = induced instantaneous emf volts

$d\phi/dt$ = time rate of change of flux, maxwells per second

N = number of turns surrounding the flux, assuming each turn is linked with all the flux

Oersted. The unit of magnetizing force (magnetic field strength) in the cgs electromagnetic system. One oersted equals a magnetomotive force of 1 gilbert/cm of flux path. One oersted equals $100/4\pi$ or 79.58 ampere-turns per meter. See *magnetizing force (magnetic field strength)*.

Paramagnetic Material. A material having a relative permeability which is slightly greater than unity, and which is practically independent of the magnetizing force.

***Permeability, AC.** A generic term used to express various dynamic relationships between magnetic induction B and magnetizing force H for magnetic material subjected to a cyclic excitation by alternating or pulsating current. The values of ac permeability obtained for a given material depend fundamentally upon the excursion limits of dynamic excitation and induction, the method and conditions of measurement, and also upon such factors as resistivity, thickness of laminations, frequency of excitation, etc.

NOTE: The numerical value for any permeability is meaningless unless the corresponding B or H excitation level is specified. For incremental permeabilities not only the corresponding dc B or H excitation level must be specified, but also the dynamic excursion limits of dynamic excitation range (ΔB or ΔH).

AC permeabilities in common use for magnetic testing are

a. ***Impedance (rms) Permeability** μ_z.* The ratio of the measured peak value of magnetic induction to the value of the apparent magnetizing force H_z calculated from the measured rms value of the exciting current, for a material in the *SCM* condition.

NOTE: The value of the current used to compute H_z is obtained by multiplying the measured value of rms exciting current by 1.414. This assumes that the total exciting current is magnetizing current and is sinusoidal.

b. ***Inductance Permeability** μ_L.* For a material in an *SCM* condition, the permeability is evaluated from the measured inductive component of the electric circuit representing the magnetic specimen. This circuit is assumed to be composed of paralleled linear inductive and resistive elements ωL_1 and R_1.

c. ***Peak Permeability** μ_p.* The ratio of the measured peak value of magnetic induction to the peak value of the magnetizing force H_p, calculated from the measured peak value of the exciting current, for a material in the *SCM* condition.

Other ac permeabilities are

d. ***Ideal Permeability** μ_a.* The ratio of the magnetic induction to the corresponding magnetizing force after the material has been simultaneously subjected to a value of ac magnetizing force approaching saturation (of approximate sine waveform) superimposed on a given dc magnetizing force, and the ac magnetizing force has thereafter been gradually reduced to zero. The resulting ideal permeability is thus a function of the dc magnetizing force used.

NOTE: Ideal permeability, sometimes called anhysteretic permeability, is principally significant to feebly magnetic material and to the Rayleigh range of soft magnetic material.

e. ***Impedance, Permeability, Incremental,** $\mu_{\Delta z}$.* Impedance permeability μ_z obtained when an ac excitation is superimposed on a dc excitation, *CM* condition.

f. ***Inductance Permeability, Incremental,** $\mu_{\Delta L}$.* Inductance permeability μ_L obtained when an ac excitation is superimposed on a dc excitation, *CM* condition.

g. ***Initial Dynamic Permeability** μ_{0d}.* The limiting value of inductance permeability μ_L reached in a ferromagnetic core when, under *SCM* excitation, the magnetizing current has been progressively and gradually reduced from a comparatively high value to zero value.

NOTE: This same value, μ_{0d}, is also equal to the initial values of both impedance permeability μ_x and peak permeability μ_p.

h. ***Instantaneous Permeability (Coincident with B_{max})** μ_t.* With *SCM* excitation, the ratio of the maximum induction B_{max} to the instantaneous magnetizing force H_t, which is

MAGNETIC MATERIALS

4-7

the value of apparent magnetizing force H' determined at the instant when B reaches a maximum.

i. **Peak Permeability, Incremental, $\mu_{\Delta p}$. Peak permeability μ_p obtained when an ac excitation is superimposed on dc excitation, *CM* condition.

*Permeability, DC. Permeability is a general term used to express relationships between magnetic induction B and magnetizing force H under various conditions of magnetic excitation. These relationships are either (*1*) absolute permeability, which in general is the quotient of a change in magnetic induction divided by the corresponding change in magnetizing force, or (*2*) relative permeability, which is the ratio of the absolute permeability to the magnetic constant Γ_m.

NOTE 1: The magnetic constant Γ_m is a scalar quantity differing in value and uniquely determined by each electromagnetic system of units. In the unrationalized cgs system Γ_m is 1 gauss/oersted and in the mksa rationalized system $\Gamma_m = 4\pi \times 10^{-7}$ H/m.

NOTE 2: Relative permeability is a pure number which is the same in all unit systems. The value and dimension of absolute permeability depend on the system of units employed.

NOTE 3: For any ferromagnetic material permeability is a function of the degree of magnetization. However, initial permeability μ_0 and maximum permeability μ_m are unique values for a given specimen under specified conditions.

NOTE 4: Except for initial permeability μ_0, a numerical value for any of the dc permeabilities is meaningless unless the corresponding B or H excitation level is specified.

NOTE 5: For the incremental permeabilities μ_Δ and $\mu_{\Delta i}$, a numerical value is meaningless unless both the corresponding values of mean excitation level (B or H) and the excursion range (ΔB or ΔH) are specified.

The following dc permeabilities are frequently used in magnetostatic measurements primarily concerned with the testing of materials destined for use with permanent or dc excited magnets.

a. *Absolute Permeability μ_{abs}. The sum of the magnetic constant and the intrinsic permeability. It is also equal to the product of the magnetic constant and the relative permeability:

$$\mu_{abs} = \Gamma_m + \mu_i = \Gamma_m \mu_r \qquad (4\text{-}13)$$

b. *Differential Permeability μ_d. The absolute value of the slope of the hysteresis loop at any point, or the slope of the normal magnetizing curve at any point.

c. *Effective Circuit Permeability μ_{eff}. When a magnetic circuit consists of two or more components, each individually homogeneous throughout but having different permeability values, the effective (overall) permeability of the circuit is that value computed in terms of the total magnetomotive force, the total resulting flux, and the geometry of the circuit.

NOTE: For a symmetrical series circuit in which each component has the same cross-sectional area, reluctance values add directly, giving

$$\mu_{eff} = \frac{l_1 + l_2 + l_3 + \cdots}{l_1/\mu_1 + l_2/\mu_2 + l_3/\mu_3 + \cdots} \qquad (4\text{-}14)$$

For a symmetrical parallel circuit in which each component has the same flux path length, permeance values add directly, giving

$$\mu_{eff} = \frac{\mu_1 A_1 + \mu_2 A_2 + \mu_3 A_3 + \cdots}{A_1 + A_2 + A_3 + \cdots} \qquad (4\text{-}15)$$

d. *Incremental Intrinsic Permeability $\mu_{\Delta i}$. The ratio of the change in intrinsic induction to the corresponding change in magnetizing force when the mean induction differs from zero.

e. *Incremental Permeability μ_Δ. The ratio of a change in magnetic induction to the corresponding change in magnetizing force when the mean induction differs from zero. It equals the slope of a straight line joining the excursion limits of an incremental hysteresis loop.

NOTE: When the change in H is reduced to zero, the incremental permeability μ_Δ becomes the reversible permeability μ_{rev}.

4-8 ELECTRICAL ENGINEERING MATERIALS REFERENCE GUIDE

*f. *Initial Permeability* μ_0.* The limiting value approached by the normal permeability as the applied magnetizing force H is reduced to zero. The permeability is equal to the slope of the normal induction curve at the origin of linear B and H axes.

*g. *Intrinsic Permeability* μ_i.* The ratio of intrinsic induction to the corresponding magnetizing force.

*h. *Maximum Permeability* μ_m.* The value of normal permeability for a given material where a straight line from the origin of linear B and H axes becomes tangent to the normal induction curve.

*i. *Normal Permeability* μ *(without subscript).* The ratio of the normal induction to the corresponding magnetizing force. It is equal to the slope of a straight line joining the extrusion limits of a normal hysteresis loop, or the slope of a straight line joining any point (H_m, B_m) on the normal induction curve to the origin of the linear B and H axes.

*j. *Relative Permeability* μ_r.* The ratio of the absolute permeability of a material to the magnetic constant Γ_m, giving a pure numeric parameter.

NOTE: In the cgs-em system of units the relative permeability is numerically the same as the absolute permeability.

*k. *Reversible Permeability* μ_{rev}.* The limit of the incremental permeability as the change in magnetizing force approaches zero.

*l. *Space Permeability* μ_0.* The permeability of space (vacuum), identical with the magnetic constant Γ_m.

****Reactive Power (Quadrature Power)* P_q.* The product of the rms current in an electric circuit, the rms voltage across the circuit, and the sine of the angular phase difference between the current and the voltage.

$$P_q = EI \sin \theta \qquad (4\text{-}16)$$

where P_q = reactive power, vars
E = voltage, volts
I = current, amperes
θ = angular phase by which E leads I

NOTE: The reactive power supplied to a magnetic core having an *SCM* excitation is the product of the magnetizing current and the voltage induced in the exciting winding.

**Remanence* B_{dm}.* The maximum value of the remanent induction for a given geometry of the magnetic circuit.

NOTE: If there are no air gaps or other inhomogeneities in the magnetic circuit, the remanence B_{dm} is equal to the retentivity B_{rs}; if air gaps or other inhomogeneities are present, B_{dm} will be less than B_{rs}.

**Retentivity* B_{rs}.* That property of a magnetic material which is measured by its maximum value of the residual induction.

NOTE: Retentivity is usually associated with saturation induction.

Symmetrically Cyclically Magnetized Condition, SCM. A magnetic material is in an *SCM* condition when, under the influence of a magnetizing force that varies cyclically between two equal positive and negative limits, its successive hysteresis loops or flux-current loops are both identical and symmetrical with respect to the origin of the axes.

Tesla. The unit of magnetic induction in the mksa (Giorgi) system. The tesla is equal to 1 Wb/m^2 or 10^4 gausses.

Var. The unit of reactive (quadrature) power in the mksa (Giorgi) and the practical systems.

Volt-Ampere. The unit of apparent power in the mksa (Giorgi) and the practical systems.

Watt. The unit of active power in the mksa (Giorgi) and the practical systems. One watt is a power of one joule/second.

Weber. The unit of magnetic flux in the mksa and in the practical system. The weber is the magnetic flux whose decrease to zero when linked with a single turn induces in the turn a voltage whose time integral is one volt-second. One weber equals 10^8 maxwells. See *magnetic flux.*

4-2. Magnetic Properties and Their Application. The relative importance of the various magnetic properties of a magnetic material varies from one application to another. In gen-

MAGNETIC MATERIALS 4-9

eral, properties of interest may include normal induction, hysteresis, dc permeability, ac permeability, core loss, and exciting power. It should be noted that there are various means of expressing ac permeability. The choice depends primarily on the ultimate use.

Techniques for the magnetic testing of many magnetic materials are described in the ASTM standards.[1] The magnetic and electric circuits employed in magnetic testing of a specimen are as free as possible from any unfavorable design factors which would prevent the measured magnetic data from being representative of the inherent magnetic properties of the specimen. The flux "direction" in the specimen is normally specified, since most magnetic materials are magnetically anisotropic. In most ac magnetic tests, the waveform of the flux is required to be sinusoidal.

As a result of the existence of unfavorable conditions, such as those listed and described below, the performance of a magnetic material in a magnetic device can be greatly deteriorated from that which would be expected from magnetic testing of the material. Allowances for these conditions, if present, must be made during the design of the device if the performance of the device is to be correctly predicted.

Leakage. A principal difficulty in the design of many magnetic circuits is due to the lack of a practicable material which will act as an insulator with respect to magnetic flux. This results in magnetic flux seldom being completely confined to the desired magnetic circuit. Estimates of leakage flux for a particular design may be made based on experience and/or experimentation.

Flux Direction. Some magnetic materials have a very pronounced directionality in their magnetic properties. Failure to utilize these materials in their preferred directions results in impaired magnetic properties.

Fabrication. Stresses introduced into magnetic materials by the various fabricating techniques often adversely affect the magnetic properties of the materials. This occurs particularly in materials having high permeability. Stresses may be eliminated by a suitable stress-relief anneal after fabrication of the material to final shape.

Joints. Joints in an electromagnetic core may cause a large increase in total excitation requirements. In some cores operated on ac, core loss may also be increased.

Waveform. When a sinusoidal voltage is applied to an electromagnetic core, the resulting magnetic flux is not necessarily sinusoidal in waveform, especially at high inductions. Any harmonics in the flux waveform cause increases in core loss and required excitation power.

Flux Distribution. If the maximum and minimum lengths of the magnetic path in an electromagnetic core differ too much, the flux density may be appreciably greater at the inside of the core structure than at the outside. For cores operated on ac, this can cause the waveform of the flux at the extremes of the core structure to be distorted even when the total flux waveform is sinusoidal.

4-3. Types of Magnetism. Any substance may be classified into one of the following categories according to the type of magnetic behavior it exhibits:

1. Diamagnetic.

2. Paramagnetic.

3. Antiferromagnetic.

4. Ferromagnetic.

5. Ferrimagnetic.

Substances which fall into the first three categories are so weakly magnetic that they are commonly thought of as "nonmagnetic." In contrast, ferromagnetic and ferrimagnetic substances are strongly magnetic and are thereby of interest as "magnetic materials." The magnetic behavior of any ferromagnetic or ferrimagnetic material is a result of its spontaneously magnetized magnetic domain structure and is characterized by a nonlinear normal induction curve, hysteresis, and saturation.

The pure elements which are ferromagnetic are iron, nickel, cobalt, and some of the rare

---

[1] *1983 Annual Book of ASTM Standards,* sec3. vol. 03.04; ASTM, Philadelphia., 1983.

earths. Typical normal induction curves of annealed samples of iron, nickel, and cobalt of comparatively high purity are shown in Fig. 4-1 for the purpose of general comparison. Ferromagnetic materials of value to industry for their magnetic properties are almost invariably alloys of the metallic ferromagnetic elements with one another and/or with other elements.

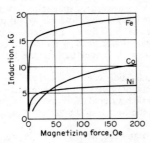

FIG. 4-1 Typical normal-induction curves of annealed samples of iron, nickel, and cobalt.

Ferrimagnetism occurs mainly in the ferrites, which are chemical compounds having ferric oxide (Fe_2O_3) as a component. In recent years, some of the magnetic ferrites have become very important in certain magnetic applications. The magnetic ferrites saturate magnetically at lower inductions than do the great majority of metallic ferromagnetic materials. However, the electrical resistivities of ferrites are at least several orders of magnitude greater than those of metals.

4-4. Commercial magnetic materials are generally divided into two main groups, each composed of ferromagnetic and ferrimagnetic substances:

1. Magnetically "soft" materials.
2. Magnetically "hard" materials.

The distinguishing characteristic of "soft" magnetic materials is high permeability. These materials are employed as core materials in the magnetic circuits of electromagnetic equipment.

"Hard" magnetic materials are characterized by a high maximum magnetic energy product (BH)$_{max}$. These materials are employed as permanent magnets to provide a constant magnetic field when it is inconvenient or uneconomical to produce the field by electromagnetic means.

Typical normal-induction curves for a wide range of commercial magnetic materials are shown in Fig. 4-2.

4.5. "Soft" Magnetic Materials. A wide variety of "soft" magnetic materials have been developed to meet the many different requirements imposed on magnetic cores for modern electrical apparatus and electronic devices. The various soft magnetic materials will be considered under three classifications:

1. Materials for solid cores.
2. Materials for laminated cores.
3. Materials for special purposes.

4-6. Materials for Solid Cores. These materials are used in dc applications such as yokes of dc dynamos, rotors of synchronous dynamos, and cores of dc electromagnets and relays. Proper annealing of these materials improves their magnetic properties. The principal magnetic requirements for the solid-core materials are high saturation, high permeability at relatively high inductions, and at times, low coercive force.

4-7. Wrought iron is a ferrous material, aggregated from a solidifying mass of pasty particles of highly refined metallic iron, into which is incorporated, without subsequent fusion, a minutely and uniformly distributed quantity of slag. The better types of wrought iron are known as Norway iron and Swedish iron and are widely used in relays after being annealed to reduce coercive force and to minimize magnetic aging. A normal-induction curve for wrought iron is shown in Fig. 4-3.

4-8. Cast irons are irons which contain carbon in excess of the amount which can be retained in solid solution in austenite at the eutectic temperature. The minimum carbon content is about 2%, while the practical maximum carbon content is about 4.5%. Cast iron was used in the yokes of dc dynamos in the early days of such machines.

MAGNETIC MATERIALS

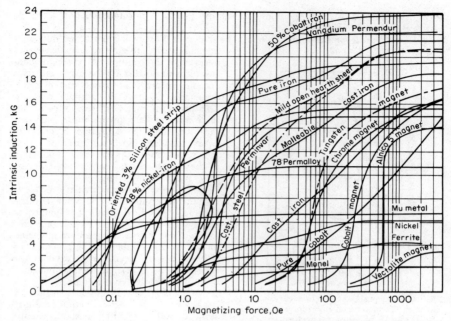

FIG. 4-2 Typical normal-induction curves of a wide range of commercial magnetic materials.

Gray cast iron is a cast iron in which graphite is present in the form of flakes. It has very poor magnetic properties, inferior mechanical properties, and practically no ductility. It does lend itself well to the casting of complex shapes and is readily machinable. A normal-induction curve and a permeability curve for gray cast iron are shown in Figs. 4-3 and 4-4, respectively.

Malleable cast iron is a cast iron in which the graphite is present as temper carbon nodules. It is magnetically better than gray cast iron. A permeability curve for malleable cast iron is shown in Fig. 4-4.

Ductile (nodular) cast iron is a cast iron with the graphite essentially spheroidal in shape. It is magnetically better than gray cast iron. Ductile cast iron has the good castability and machinability of gray cast iron together with much greater strength, ductility, and shock resistance. Typical normal-induction curves and permeability curves for ductile cast iron are shown in Figs. 4-3 and 4-4, respectively.

4-9. Carbon Steels. Carbon steels may contain from less than 0.1% carbon to more than 1% carbon. The magnetic properties of a carbon steel are greatly influenced by the carbon content and the disposition of the carbon. Low-carbon steels (less than 0.2% carbon) have magnetic properties which are similar to those of wrought iron and far superior to those of any of the cast irons.

Wrought carbon steels are widely used as solid-core materials. The low-carbon types are preferred in most applications.

Cast carbon steels replaced cast iron many years ago as the material used in the yokes of dc machines but have since been largely supplanted in this application by

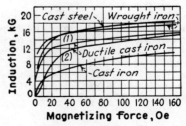

FIG. 4-3 Normal-induction curves of typical wrought iron, cast iron, ductile cast iron (3% Si) as (1) cast and (2) annealed, and cast steel.

wrought (hot-rolled) carbon-steel plates of welding quality. A normal-induction curve for cast steel is shown in Fig. 4-3.

4-10. Materials for Laminated Cores. The materials most widely employed in wound or stacked cores in electromagnetic devices operated at the commercial power frequencies (50 and 60 Hz) are the electrical steels and the specially processed carbon steels designated as magnetic lamination steels. The principal magnetic requirements for these materials are low core loss, high permeability, and high saturation. ASTM publishes standard specifications for these materials.[1] On a tonnage basis, production of these materials far exceeds that of any other magnetic material.

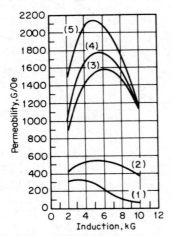

FIG. 4-4 Permeability-induction curves of (1) cast iron, (2) ductile cast iron (normal composition) as cast, (3) annealed, (4) ductile iron (3% Si) annealed, and (5) malleable cast iron.

4-11. Electrical steels are flat-rolled low-carbon silicon-iron alloys. Since applications for electrical steels lie mainly in energy-loss-limited equipment, the core losses of electrical steels are normally guaranteed by the producers. The general category of electrical steels may be divided into classifications of (1) nonoriented materials and (2) grain-oriented materials.

4-12. Grading. Electrical steels are usually graded by high-induction core loss. Both ASTM and AISI have established and published designation systems for electrical steels based on core loss.[1,2]

The ASTM core loss type designation (ASTM Standard A664) consists of six or seven characters. The first two characters are 100 times the nominal thickness of the material in millimeters. The third character is a code letter which designates the class of the material and specifies the sampling and testing practices. The last three or four characters are 100 times the maximum permissible core loss in watts per pound at a specified test frequency and induction.

The AISI designation system has been discontinued but is still widely used. The AISI type designation for a grade consisted of the letter M followed by a number. The letter M stood for magnetic material, and the number was approximately equal to 10 times the maximum permissible core loss in watts per pound for 0.014-in material at 15 kG, 60 Hz in 1947.

Nonoriented electrical steels have approximately the same magnetic properties in all directions in the plane of the material (see Figs. 4-5 and 4-6). The common application is in punched laminations for large and small rotating machines and for small transformers. Today, nonoriented materials are always cold-rolled to final thickness. Hot rolling to final thickness is no longer practiced. Nonoriented materials are available in both fully processed and semiprocessed conditions.

Fully processed nonoriented materials (ASTM Standard A677) have their magnetic properties completely developed by the producer. Stresses introduced into these materials during fabrication of magnetic cores must be relieved by annealing to achieve optimum magnetic properties in the cores. In many applications, however, the degradation of the magnetic properties during fabrication is slight and/or can be tolerated and the stress-relief anneal is omitted. Fully processed nonoriented materials contain up to about 3.5% silicon. Additionally a small amount (about 0.5%) of aluminum is usually present. The common thicknesses are 0.014, 0.0185, and 0.025 in. Grade designations and maximum core-loss

---

[1] *1983 Annual Book of ASTM Standards,* sec3. vol. 03.04; ASTM, Philadelphia., 1983.
[2] *Steel Products Manual, Electrical Steel;* AISI, Washington, D.C., January 1983.

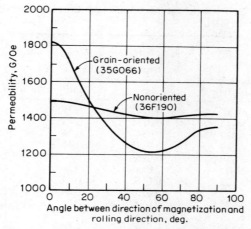

FIG. 4-5 Effect of direction of magnetization on normal permeability at 10 Oe of fully processed electrical steels.

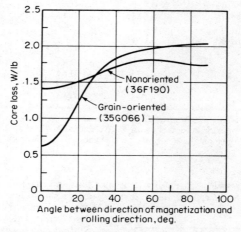

FIG. 4-6 Effect of direction of magnetization on core loss at 15 kG, 60 Hz of fully processed electrical steel.

limits are given in Table 4-1. Some typical magnetic properties are shown in Figs. 4-7, 4-8, 4-9, and 4-10. Some general properties are presented in Table 4-4. Some characteristics and applications are described in Table 4-5.

Semiprocessed nonoriented materials (ASTM Standard A683) do not have their inherent magnetic properties completely developed by the producer and must be annealed properly to achieve both decarburization and grain growth. These materials are used primarily in high-volume production of small laminations and cores which would require stress-relief annealing if made from fully processed material. Semiprocessed nonoriented materials contain up to about 3% silicon. Additionally a small amount (about 0.5%) of aluminum is usually present. The carbon content may be as high as 0.05% but should be reduced to 0.005% or less by the required anneal. The common thicknesses of semiprocessed nonoriented

TABLE 4-1 Grade Designations and Maximum Core-Loss Limits for Fully Processed Nonoriented Electrical Steels*

ASTM type	Former AISI type	Nominal thickness		Max. core loss at 15 kG†	
		in	mm	W/lb, 60 Hz	W/kg, 60 Hz
36F145	M-15	0.014	0.36	1.45	2.53
47F168	M-15	0.0185	0.47	1.68	2.93
36F158	M-19	0.014	0.36	1.58	2.75
47F174	M-19	0.0185	0.47	1.74	3.03
64F208	M-19	0.025	0.64	2.08	3.62
36F168	M-22	0.014	0.36	1.68	2.93
47F185	M-22	0.0185	0.47	1.85	3.22
64F218	M-22	0.025	0.64	2.18	3.80
36F180	M-27	0.014	0.36	1.80	3.13
47F190	M-27	0.0185	0.47	1.90	3.31
64F225	M-27	0.025	0.64	2.25	3.92
36F190	M-36	0.014	0.36	1.90	3.31
47F205	M-36	0.0185	0.47	2.05	3.57
64F240	M-36	0.025	0.64	2.40	4.18
47F230	M-43	0.0185	0.47	2.30	4.01
64F270	M-43	0.025	0.64	2.70	4.70
47F290	M-45	0.0185	0.47	2.90	5.05
64F340	M-45	0.025	0.64	3.40	5.92
47F380	M-47	0.0185	0.47	3.80	6.62
64F470	M-47	0.025	0.64	4.70	8.19
47F450	—	0.0185	0.47	4.50	7.84
64F550	—	0.025	0.64	5.50	9.58
47F490	—	0.0185	0.47	4.90	8.53
64F600	—	0.025	0.64	6.00	10.45

*Adapted from *Steel Products Manual, Electrical Steels;* AISI, Washington, D.C., January 1983.
†Core loss is determined in accordance with ASTM Method A343 on as-sheared Epstein specimens consisting of strips of which half have been cut parallel and half have been cut transverse to the rolling direction.

materials are 0.0185 and 0.025 in. Grade designations and maximum core-loss limits are given in Table 4-2. Some typical magnetic properties are shown in Figs. 4-11, 4-12, and 4-13. Some characteristics and applications are described in Table 4-5.

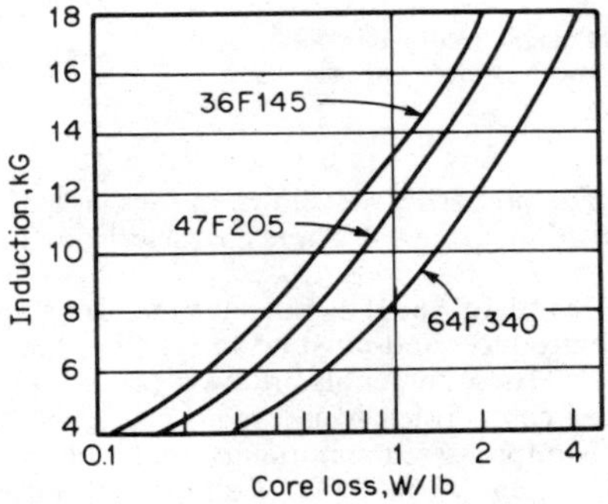

Grain-oriented electrical steels (ASTM Standards A665 and A725) have a pronounced directionality in their magnetic properties (Figs. 4-5 and 4-6). This directionality is a result of the "cube-on-edge" crystal structure achieved by proper composition and processing. Grain-oriented materials are employed most effectively in magnetic cores in which the flux path lies entirely or predominately in the rolling direction of the material. The common application is in cores of power and distribution transformers for electric utilities.

Grain-oriented materials are produced in a fully processed condition, either unflattened or thermally flattened, in thicknesses of 0.0090, 0.0106, 0.0118, and 0.0138 in.

FIG. 4-7 Typical 60-Hz core loss of as-sheared 50/50 Epstein specimens of fully processed nonoriented electrical steels.

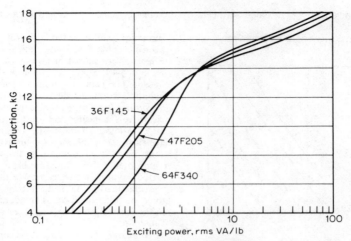

FIG. 4-8 Typical 60-Hz exciting power of as-sheared 50/50 Epstein specimens of fully processed nonoriented electrical steels.

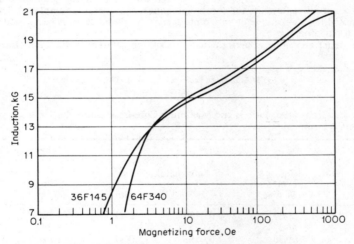

FIG. 4-9 Typical normal-induction curves of as-sheared 50/50 Epstein specimens of fully processed nonoriented electrical steels.

Unflattened material has appreciable coil set or curvature. It is used principally in making spirally wound or formed cores. These cores must be stress-relief annealed to relieve fabrication stresses. Thermally flattened material is employed principally in making sheared or stamped laminations. Annealing of the laminations to remove both residual stresses from the thermal-flattening and fabrication stresses is usually recommended. However, special thermally flattened materials are available which do not require annealing when used in the form of wide flat laminations.

Two types of grain-oriented electrical steels are currently being produced commercially. The regular type, which was introduced many years ago, contains about 3.15% silicon and has grains about 3 mm in diameter. The high-permeability type, which was introduced more recently, contains about 2.9% silicon and has grains about 8 mm in diameter. In comparison with the regular type, the high-permeability type has better core loss and permeability at high inductions.

Grade designations and maximum core-loss limits for grain-oriented electrical steels are given in Table 4-3. Some typical magnetic properties are shown in Figs. 4-14, 4-15, 4-16, and 4-17. Some general properties are presented in Table 4-4. Some characteristics and applications are described in Table 4-5.

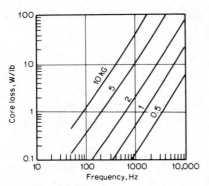

FIG. 4-10 Effect of frequency on core loss of as-sheared 50/50 Epstein specimens of 36F145 fully processed nonoriented electrical steels.

Surface insulation of the surfaces of electrical steels is needed to limit the interlaminar core losses of magnetic cores made of electrical steels. Numerous surface insulations have been developed to meet the requirements of various applications. The various types of surface insulations have been classified by AISI.[1] Each class has been given an identification. The identification and a description of each class are given in Table 4-6.

Annealing of laminations or cores made from electrical steels is performed to accomplish either stress relief in fully processed material or decarburization and grain growth in semiprocessed material. Both batch-type annealing furnaces and continuous annealing furnaces are employed. The former is best suited for low-volume or varied production, while the latter is best suited for high-volume production.

Stress-relief annealing is performed at a soak temperature in the range from 730 to

---

[1]*1983 Annual Book of ASTM Standards,* sec3. vol. 03.04; ASTM, Philadelphia., 1983.

TABLE 4-2 Grade Designations and Maximum Core-Loss Limits for Semiprocessed Nonoriented Electrical Steels*

ASTM type	Former AISI type	Nominal thickness in	mm	Max. core loss at 15 kG† W/lb, 60 Hz	W/kg, 60 Hz
47S178	M-27	0.0185	0.47	1.78	3.10
64S194	M-27	0.025	0.64	1.94	3.38
47S188	M-36	0.0185	0.47	1.88	3.27
64S213	M-36	0.025	0.64	2.13	3.71
47S200	M-43	0.0185	0.47	2.00	3.48
64S230	M-43	0.025	0.64	2.30	4.01
47S230	M-45	0.0185	0.47	2.30	4.01
64S260	M-45	0.025	0.64	2.60	4.53
47S250	—	0.0185	0.47	2.50	4.35
64S280	—	0.025	0.64	2.80	4.88
47S300	—	0.0185	0.47	3.00	5.22
64S350	—	0.025	0.64	3.50	6.10

*Adapted from *Steel Products Manual, Electrical Steels;* AISI, Washington, D.C., January 1983.
†Core loss is determined in accordance with ASTM Method A343 on quality-evaluation-annealed Epstein specimens consisting of strips of which half have been cut parallel and half have been cut transverse to the rolling direction. The quality-evaluation-annealing is customarily done at a soak temperature of 845°C for approximately 1 h in a suitable decarburizing atmosphere.

MAGNETIC MATERIALS

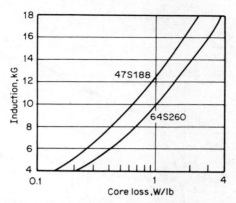

FIG. 4-11 Typical 60-Hz core loss of annealed 50/50 Epstein specimens of semiprocessed nonoriented electrical steels.

845°C. The soak time need be no longer than that required for the charge to reach soak temperature. The heating and cooling rates must be slow enough that excessive thermal gradients in the material are avoided. The annealing atmosphere and other annealing conditions must be such that chemical contamination of the material is avoided.

Annealing for decarburization and grain growth is performed at a soak temperature in the range from 760 to 870°C. Atmospheres of hydrogen or partially combusted natural gas and containing water vapor are often used. The soak time required for decarburization depends not only on the temperature and atmosphere but also on the dimensions of the laminations or cores being annealed. If the dimensions are large, long soak times may be required.

4-13. Magnetic lamination steels (ASTM Standard A726) are cold-rolled low-carbon steels intended for magnetic applications, primarily at power frequencies. The magnetic properties of magnetic lamination steels are not normally guaranteed and are generally inferior to those of electric steels. However, magnetic lamination steels are frequently used as

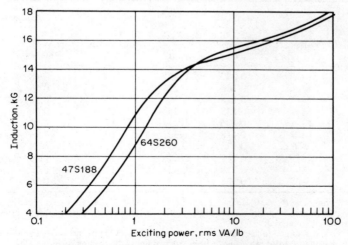

FIG. 4-12 Typical 60-Hz exciting power of annealed 50/50 Epstein specimens of semiprocessed nonoriented electrical steels.

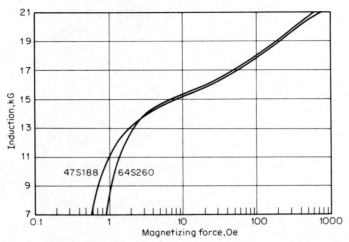

FIG. 4-13 Typical normal-induction curves of annealed 50/50 Epstein specimens of semiprocessed nonoriented electrical steels.

core materials in small electrical devices especially when the cost of the core material is a more important consideration than the magnetic performance.

Usually, but not always, stamped laminations or assembled core structures made from magnetic lamination steels are given a decarburizing anneal to enhance the magnetic properties. Optimum magnetic properties are obtained when the carbon content is reduced to 0.005% or less from its initial value, which may approach 0.1%. The soak temperature of the anneal is in the range from 730 to 790°C. The atmosphere most often used at the present

TABLE 4-3 Grade Designations and Maximum Core-Loss Limits for Fully Processed Grain-Oriented Electrical Steels*

ASTM type†	Former AISI type	Nominal thickness in	Nominal thickness mm	Induction, kG	Max. core loss‡ W/lb, 60 Hz	Max. core loss‡ W/kg, 60 Hz
23G048	—	0.087	0.22	15	0.48	0.80
27G053	M-4	0.106	0.27	15	0.53	0.89
30G058	M-5	0.118	0.30	15	0.58	0.97
35G066	M-6	0.138	0.35	15	0.66	1.11
27H076	M-4	0.106	0.27	17	0.76	1.27
30H083	M-5	0.118	0.30	17	0.83	1.39
35H094	M-6	0.138	0.35	17	0.94	1.57
27P066	—	0.106	0.27	17	0.66	1.11
30P070	—	0.118	0.30	17	0.70	1.17
35P076	—	0.138	0.35	17	0.76	1.27

*Adapted from *Steel Products Manual, Electrical Steels;* AISI, Washington, D.C., January 1983.
†The first seven entries in this column apply to the regular material. The last three entries apply to the high-permeability material.
‡Core loss is determined in accordance with ASTM Method A343 on stress-relief-annealed Epstein specimens consisting of strips which have been cut parallel to the rolling direction. The stress-relief annealing is customarily done at a soak temperature in the range of 790 to 845°C for approximately 1 h in an atmosphere comprised of a mixture of pure nitrogen and pure hydrogen with a dew point not greater than −20°C.

MAGNETIC MATERIALS 4-19

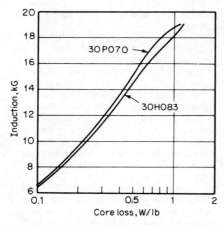

FIG. 4-14 Typical 60-Hz core loss of annealed parallel-grain Epstein specimens of fully processed grain-oriented electrical steels.

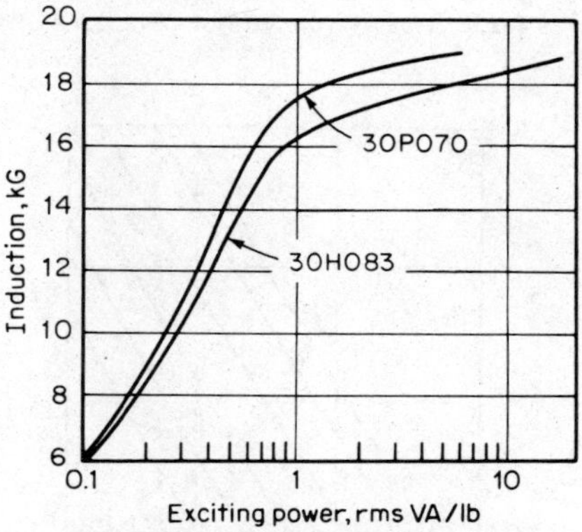

FIG. 4-15 Typical 60-Hz exciting power of annealed parallel-grain Epstein specimens of fully processed grain-oriented electrical steels.

time is partially combusted natural gas with a suitable dew point. Soak time is dependent to a considerable degree upon the dimensions of the laminations or core structures being annealed.

Three types of magnetic lamination steels are produced. Type 1 is usually made to a controlled chemical composition and is furnished in the full-hard or annealed condition without guaranteed magnetic properties. Type 2 is made to a controlled chemical composition, given special processing, and furnished in the annealed condition without guaranteed magnetic properties. After a suitable anneal, the magnetic properties of Type 2 are superior to those of Type 1. Type 2S is similar to Type 2, but the core loss is guaranteed. The maximum core-loss limits for Type 2S are shown in Table 4-7.

4-14. Materials for Special Purposes. For certain applications of soft, or nonretentive, materials, special alloys and other materials have been developed which, after proper fab-

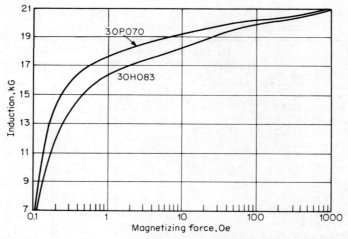

FIG. 4-16 Typical normal-induction curves of annealed parallel-grain Epstein specimens of fully processed grain-oriented electrical steels.

rication and heat-treatment, have superior properties in certain ranges of magnetization. Several of these alloys and materials will be described.

4-15. Nickel-Iron Alloys. Nickel alloyed with iron in various proportions produces a series of alloys with a wide range of magnetic properties. With 30% nickel, the alloy is practically nonmagnetic and has a resistivity of 86 $\mu\Omega$/cm. With 78% nickel the alloy, properly heat-treated, has very high permeability. These effects are shown in Figs. 4-18 and 4-19. Many variations of this series have been developed for special purposes. Table 4-8 lists some of the more important commercial types of nickel-iron alloys, with their approximate properties. These alloys are all very sensitive to heat-treatment; so their properties are largely influenced thereby. A comparison of their normal-induction curves is given in the curves of Fig. 4-20 and Fig. 3-2.

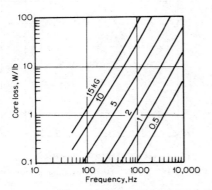

FIG. 4-17 Effect of frequency on core loss of annealed parallel-grain Epstein specimens of 30H083 fully processed grain-oriented electrical steels.

4-16. Permalloy[1] is a term applied to a number of nickel-iron alloys developed by the Bell Laboratories, each specified by a prefix number indicating the nickel content. The term is usually associated with the 78.5% nickel-iron alloys, the important properties of which are high permeability and low hysteresis loss in relatively low magnetizing fields. These properties are obtained by a unique heat-treatment consisting of a high-temperature anneal, preferably in hydrogen, with slow cooling followed by rapid cooling from about 625°C. The alloy is very sensitive to mechanical strain; so it is desirable to heat-treat the alloy in its final form. The addition of 3.8% chromium or molybdenum increases the resistivity from 16 to 65 and 55 $\mu\Omega \cdot$ cm, respectively, without seriously impairing the magnetic quality. In fact, low-density permeabilities are better with these additions. These alloys have found their principal application as a material for the continuous loading of submarine cables and in loading coils for land lines (see Figs. 4-20 and 4-21 and Table 4-8).

---

[1] *Elem. Bell Syst. Tech.*, 1936, vol. 15, p. 113.

TABLE 4-4 General Properties of Fully Processed Electrical Steels

ASTM type	Nominal alloy content (Si + Al), %	Assumed density, g/cm^3	Volume resistivity, $\Omega \cdot$cm $\times 10^6$	Saturation induction, kG
36F145 and 47F168	3.5	7.65	52	19.8
36F158 through 64F208	3.3	7.65	50	19.9
36F168 through 64F218	3.2	7.65	48	20.0
36F180 through 64F225	2.8	7.70	45	20.2
36F190 through 64F240	2.65	7.70	44	20.2
47F230 and 64F270	2.35	7.70	40	20.4
47F290 and 64F340	1.85	7.75	34	20.7
47F380 and 64F470	1.05	7.80	24	21.1
47F450 and 64F550	0.80	7.80	21	21.2
47F490 and 64F600	0.50	7.85	18	21.3
23G048 through 35G066	3.15	7.65	48	20.0
27H076 through 35H094	3.15	7.65	48	20.0
27P066 through 35P076	2.90	7.65	45	20.1

MAGNETIC MATERIALS

TABLE 4-5 Some Characteristics and Typical Applications for Specific Types of Electrical Steels*

Oriented types		
ASTM type	Some characteristics	Typical applications
23G048 through 35G066 or 27H076 through 35H094 or 27P066 through 35P076	Highly directional magnetic properties due to grain orientation. Very low core loss and high permeability in rolling direction.	Highest-efficiency power and distribution transformers with lower weight per kVA. Large generators and power transformers.

Nonoriented types		
ASTM type	Some characteristics	Typical applications
36F145 and 47F168	Lowest core loss, conventional grades. Excellent permeability at low inductions.	Small power transformers and rotating machines of high efficiency.
36F158 through 64F225 or 47S178 and 64S194	Low core loss, good permeability at low and intermediate inductions.	High-reactance cores, generators, stators of high-efficiency rotating equipment.
36F190 through 64F270 or 47S188 through 64S260	Good core loss, good permeability at all inductions, and low exciting current. Good stamping properties.	Small generators, high-efficiency, continuous duty rotating ac and dc machines.
47F290 through 64F600 or 47S250 through 64S350	Ductile, good stamping properties, good permeability at high inductions.	Small motors, ballasts, and relays.

*Adapted from *Steel Products Manual, Electrical Steels;* AISI, Washington, D.C., January 1983.

By special long-time high-temperature treatments, maximum permeability values greater than 1 million have been obtained. The double treatment required by the 78% Permalloy is most effective when the strip is thin, say under 10 mils. For greater thicknesses, the quick cooling from 625°C is not uniform throughout the section, and loss of quality results.

A 48% nickel-iron was developed for applications requiring a moderately high-permeability alloy with higher saturation density than 78 Permalloy. The same general composition is marketed under many names, such as Hyperm 50, Hipernik, Audiolloy, Allegheny Electric Metal, 4750, Carpenter 49 alloy. Annealing after all mechanical operations are completed is recommended. These alloys have found extensive use in radio, radar, instrument, and magnetic-amplifier components (see Fig. 4-20).

Deltamax. By the use of special techniques of cold reduction and annealing, the 48% nickel-iron alloy develops directional properties resulting in high permeability and a square hysteresis loop in the rolling direction (see Fig. 4-21). A similar product is sold under the name of Orthonic. For optimum properties, these materials are rapidly cooled after a 2-h anneal in pure hydrogen at 1100°C. They are generally used in wound cores of thin tape for applications such as pulse transformers and magnetic amplifiers.

4-17. Iron-Nickel-Copper-Chromium. The addition of copper and chromium to high nickel-iron alloys has the effect of raising the permeability at low flux density. Alloys of this type are marketed under the names of Mumetal, 1040 alloy, Hymu 80. A typical induction characteristic is curve *A* in Fig. 4-20; for optimum properties they are annealed after cutting

ELECTRICAL ENGINEERING MATERIALS REFERENCE GUIDE

TABLE 4-6 Descriptions of Flat-Rolled Electrical Steel Insulations or Core Plates*

Identification	Description
C-0	This identification is merely for the purpose of describing the natural oxide surface which occurs on flat-rolled silicon steel which gives a slight but effective insulating layer sufficient for most small cores and will withstand normal stress-relief annealing temperatures. This oxidized surface condition may be enhanced in the stress-relief anneal of finished cores by controlling the atmosphere to be more or less oxidizing to the surface.
C-2†	This identification is for the purpose of describing an inorganic insulation which consists of a glasslike film which forms during high-temperature hydrogen anneal of grain-oriented silicon steel as the result of the reaction of an applied coating of MgO and silicates in the surface of the steel. This insulation is intended for air-cooled or oil-immersed cores. It will withstand stress-relief annealing temperatures and has sufficient interlamination resistance for wound cores of narrow width strip such as used in distribution transformer cores. It is not intended for stamped laminations because of the abrasive nature of the coating.
C-3	This insulation consists of an enamel or varnish coating intended for air-cooled or oil-immersed cores. The interlamination resistance provided by this coating is superior to the C-1 type coating which is primarily utilized as a die lubricant. The C-3 coating will also enhance punchability, is resistant to normal operating temperatures, but will not withstand stress-relief annealing (see Note).
C-4	This insulation consists of a chemically treated or phosphated surface intended for air-cooled or oil-immersed cores requiring moderate levels of insulation resistance. It will withstand stress-relief annealing and serves to promote punchability.
C-5	This is an inorganic insulation similar to C-4 but with ceramic fillers added to enhance the interlamination resistance. It is typically applied over the C-2 coating on grain-oriented silicon steel. It is principally intended for air-cooled or oil-immersed cores which utilize sheared laminations and operate at high volts per turn, but finds application in all apparatus requiring high levels of interlaminar resistance. Like C-2, it will withstand stress-relief annealing in a neutral or slightly reducing atmosphere.

*From *Steel Products Manual, Electrical Steels;* AISI, Washington, D.C., January 1983.
†C-1 has been deleted from this table and is generally superseded by C-3.
NOTE: In fabricating operations involving the application of heat, such as welding and die casting, it may be desirable that a thinner than normal coating be used to leave as little residue as possible. These coatings can enhance punchability, and the producers should be consulted to obtain a correct weight of coating. To identify these coatings, various letter suffixes have been adopted, and the producer should be consulted for the proper suffix.

and forming for 4 h at 1100°C in pure hydrogen and cooled slowly. Important applications are as magnetic shielding for instruments and electronic equipment and as cores in magnetic amplifiers.

4-18. Constant-permeability alloys having a moderate permeability which is quite constant over a considerable range of flux densities are desirable for use in circuits in which waveform distortion must be kept at a minimum. Isoperm and Conpernik are two alloys of this type. They are nickel-iron alloys containing 40 to 55% nickel which have been severely cold-worked. Perminvar is the name given to a series of cobalt-nickel-iron alloys (for exam-

TABLE 4-7 Maximum Core-Loss Limits for Type 2S Magnetic Lamination Steel*

Actual thickness by micrometer†		Max. core loss at 15 kG, 60 Hz‡	
in	mm	W/lb	W/kg
0.016	0.41	3.3	7.3
0.018	0.46	3.6	7.9
0.021	0.53	4.1	9.0
0.025	0.64	4.9	10.8
0.028	0.71	5.6	12.3

*Adapted from *Steel Products Manual, Electrical Steels;* AISI, Washington, D.C., January 1983.

†Other thicknesses can be specified, and the assigned core loss values will be commensurate with these stated values.

‡Core loss is determined in accordance with ASTM Method A343 on quality-evaluation-annealed Epstein specimens consisting of strips of which half have been cut parallel and half have been cut transverse to the rolling direction. The quality-evaluation annealing is customarily done at a soak temperature of 790°C for approximately 1 h in a suitable decarburizing atmosphere.

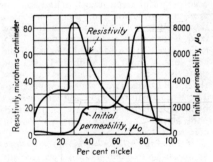

FIG. 4-18 Electrical resistivity and initial permeability of iron-nickel alloys with various nickel contents.

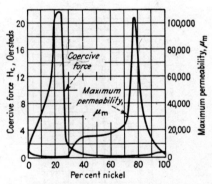

FIG. 4-19 Maximum permeability and coercive force of iron-nickel alloys with various nickel contents.

TABLE 4-8 Special-Purpose Materials

Name	Approximate composition, %	Saturation, G	Maximum permeability	Coercivity (from saturation), Oe	Initial permeability	Resistivity, microhm-cm
78 Permalloy	78.5 Ni	10,500	70,000	−0.05	8,000	16
MoPermalloy	79 Ni, 4.0 Mo	8,000	90,000	−0.05	20,000	55
Supermalloy	79 Ni, 5 Mo	7,900	900,000	−0.002	100,000	60
48% nickel-iron	48 Ni	16,000	60,000	−0.06	5,000	45
Monimax	47 Ni, 3 Mo	14,500	35,000	−0.10	2,000	80
Sinimax	43 Ni, 3 Si	11,000	35,000	−0.10	3,000	85
Mumetal	77 Ni, 5 Cu, 2 Cr	6,500	85,000	−0.05	20,000	60
Deltamax	50 Ni	15,500	85,000	−0.10	......	45

ple, 50% nickel, 25% cobalt, 25% iron) which also exhibit this characteristic of constant permeability over a low ($\cong 800$ G) density range. When magnetized to higher flux densities, they give a double loop constricted at the origin so as to give no measurable remanence or coercive force. The characteristics of the alloys in this group vary greatly with the chemical content and the heat-treatment. A sample containing approximately 45 Ni, 25 Co, and 30 Fe, baked for 24 h at 425°C and slowly cooled, had hysteresis losses as follows: At 100 G, 214 × 10^{-4} erg/(cm^3)(cycle); at 1003 G, 15.27 ergs; at 1604 G, 163 ergs; at 4950 G, 1736 ergs; and at 13,810 G, 4430 ergs. Over the range of flux densities in which the permeability is constant (from 0 to 600 G), the hysteresis loss is very small, or on the order of the foregoing figure for 100 G. The resistivity of the sample was 19.63 $\mu\Omega \cdot$cm.

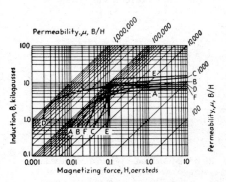

FIG. 4-20 Induction-permeability curves of some high-nickel alloy strip (0.014 in). A. Mumetal. B. Permalloy. C. 48% nickel-iron. D. Supermalloy. E. Deltamax. F. MoPermalloy.

4-19. Monel metal is an alloy of 67% nickel, 28% copper, and 5% other metals. It is slightly magnetic below 95°C (see Fig. 4-2).

4-20. Iron-Cobalt Alloys. The addition of cobalt to iron has the effect of raising the saturation intensity of iron up to about 36% cobalt (Fe$_2$Co). This alloy is useful for pole pieces of electromagnets and for any application where high magnetic intensity is desired. It is workable hot but quite brittle cold. *Hyperco* contains approximately ⅓ Co, ⅔ Fe, plus 1 to 2% "added element." Total core loss is about 2.5 W/lb at 15 kG and 0.010 in thick. It is available as hot-rolled sheet, cold-rolled strip, plates, and forgings. The 50% cobalt-iron alloy Permendur has a high permeability in fields up to 50 Oe and, with about 2% vanadium added, can be cold-rolled (see Fig. 4-2).

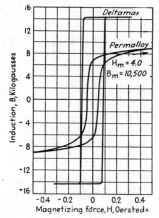

Fig. 4-21 Hysteresis curves of 78 Permalloy and Deltamax from saturation.

4-21. Iron-Silicon Aluminum Alloys. Aluminum in small percentages, usually under 0.5%, is a valuable addition to the iron-silicon alloy. Its principal function appears to be as a deoxidizer. Masumoto[1] has investigated soft magnetic alloys containing much higher percentages of aluminum and found several that have high permeabilities and low hysteresis losses. Certain compositions have very low magnetostriction and anisotropy, high initial permeability, and high electrical resistivity. An alloy of 9.6% silicon and 6% aluminum with iron has better low-flux-density properties than the Permalloys. However, poor ductility has limited these alloys to dc applications in cast configurations or in insulated pressed-powder cores for high-frequency uses. These alloys are commonly known as Sendust. The material has been prepared in sheet form[2] by special processes.

---

[1]Masumoto: On a New Alloy "Sendust" and Its Magnetic and Electrical Properties; *Tohoku Imp. Univ. Rept.*, 1936, Anniversary Vol., Ser. I, p. 388.

[2]Helms and Adams: Sendust Sheet-processing Techniques and Magnetic Propertiesl *J. Appl. Phys.*, March 1964, vol. 35.

MAGNETIC MATERIALS   4-25

4-22. Temperature-Sensitive Alloys.[3] Inasmuch as the Curie point of metal may be moved up or down the temperature scale by the addition of other elements, it is possible to select alloys which lose their ferromagnetism at almost any desired temperature up to 1115°C, the change point in cobalt. Iron-base alloys are ordinarily used to obtain the highest possible permeability at points below the Curie temperature. Nickel, manganese, chromium, and silicon are the most effective alloy elements for this purpose;[2,4] and most alloys made for temperature-control application, such as instruments, reactors, and transformers, use one or more of these. Figure 4-22 shows the magnetization-temperature characteristics of a group of these alloys. The Carpenter Temperature Compensator 30 is a nickel-copper-iron alloy which loses its magnetism at 55°C and is used for temperature compensation in meters.[5]

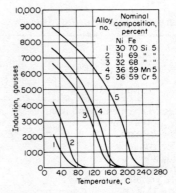

FIG. 4-22 Thermosensitive alloys. Temperature-induction for $H = 20$.

4-23. Heusler's alloys are ferromagnetic alloys composed of "nonmagnetic" elements. Copper, manganese, and aluminum are frequently used as the alloying elements. The saturation induction is about one-third that of pure iron.

4-24. High-Frequency Materials Applications. Magnetic materials used in reactors, transformers, inductors, and switch-mode devices are selected upon the basis of magnetic induction, permeability, and associated material power losses at the design frequency. Control of eddy currents becomes of primary importance to reduce losses and minimize skin effect produced by eddy-current shielding. This is accomplished by the use of high-permeability alloys in the form of wound cores of thin tape, or compressed, insulated powder iron-alloy cores, or sintered ferrite cores.

Typically, the thin magnetic strip material is used in applications where operating frequencies range from 400 Hz to 20 kHz. Power conditioning equipment frequently operates at 10 kHz and up, and the magnetic materials used are compressed, powdered iron-alloy cores or sintered ferrite cores. Power losses in magnetic materials are of great concern, especially so when operated at high frequencies. Table 4-9 lists representative values of losses that may be developed by magnetic core materials operating in the range of 400 Hz to 100 kHz.

---

[3]Jackson and Russell: Temperature Sensitive Magnetic Alloys and Their Uses; *Instruments,* November 1938, p. 280
[4]Shaw, J. L.: Cureie Temperature Alloys; *Prod. Eng.,* June 1948.
[5]Eberly, W. S.: Temperature-Compensator Alloys; *Mach. Des.,* May 1954.

TABLE 4-9 Comparative Power Losses of Magnetic Materials
(Core Losses in W/cm^3 at 1.0 kG)

	Test frequency		
Magnetic material	400 Hz	10.0 kHz	100.0 kHz
3% oriented silicon-iron, 0.004 in thick	0.00065	0.100	—
4-79 MoPermalloy, 0.004 in thick	0.0001	0.020	1.20
Iron powder (75 permeability)	0.015	0.30	—
2-81 Permalloy powder (60 permeability)	0.002	0.030	0.50
Ferrite (2000 permeability)	0.004	0.008	0.18

Other considerations are the ambient operating temperature of the magnetic component and the physical configurations in which the magnetic component can be supplied. A comparison of the dc properties of the more common commercially available cast and powdered alloys is shown in Table 4-10.

TABLE 4-10 Properties of Magnetic Core Materials

	dc properties			
Material	Permeability maximum, G/Oe	Magnetic induction, kG	Curie temp., C°	Electrical resistivity, $\mu\Omega \cdot$ cm
3% grain-oriented silicon iron	30,000	20.0	740	47
4-79 Moly Permalloy	250,000	7.8	454	58
5-79 MoPermalloy (Supermalloy)	600,000	8.2	400	60
48 nickel-iron alloy	100,000	15.5	482	47
Iron powder	75*	12.0	770	10^7
2-81 Permalloy powder	125*	7.0	400	10^6
Ferrite (MgZn)	2,500*	4.0	180	$10^2 - 10^8$*

*Varies by composition.

4-25. 3% Silicon-iron alloys for high-frequency use are available in an insulated 0.001- to 0.006-in-thick strip that exhibits high effective permeability and low losses at relatively high flux densities. This alloy, as well as other rolled to strip soft magnetic alloys, is used to make laminated magnetic cores by various methods, including (1) the wound-core approach for winding toroids and C and E cores; (2) stamped or sheared-to-length laminations for laid-up transformers; and (3) stamped laminations of various configurations (rings E, I, F, L, DU, etc.) for assembly into transformer cores. Laminated core materials usually are annealed after all fabricating and stamping operations have been completed in order to develop the desired magnetic properties of the material. Subsequent forming, bending, or machining may impair the magnetic characteristics developed by the anneal.

4-26. Nickel-alloy tape of high permeability is used in thickness of 0.000125 to 0.006 in for tape-wound cores designed for the frequency range of 0.1 to 100 kHz. Tapes less than 0.001 in are usually wound on nonmagnetic stainless steel or ceramic bobbins for support of the tape and to ensure stability of the magnetic properties. Commonly used nickel-iron alloys are 4-79 MoPermalloy, Mumetal, 48 nickel-iron, and Supermalloy with thickness and construction chosen to provide the desired permeability at the application frequency. Figure 4-23 shows the effect of tape thickness on the initial permeability of some of these materials.

The decrease in permeability with increasing frequency is a characteristic of all magnetic materials. Applications include uses as current transformers, transformers for inverter-converters, high-power pulse transformers, and magnetic pick-up heads. Magnetic core materials less than 0.001 in thick are used in timer circuits, high-frequency inverters, digital memory devices, pulse transformers, and magnetometer sensing.

Other tape materials of commercial significance are permendur (35 to 50% Co), vanadium permendur (49% Co, 2% V), and the amorphous metal alloys. Permendurs are high-induction (23.0 kG) materials which can be rolled to a 0.001-in-thick strip for use as recording head laminations.

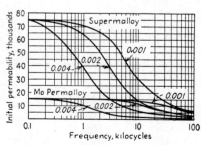

FIG. 4-23 Effect of tape thickness on the initial permeability of Supermalloy and MoPermalloy at various frequencies.

MAGNETIC MATERIALS

Vanadium permendur is used in tape-wound transformer cores where size and weight reduction is important and operating frequencies are below 3.0 kHz. A high degree of magnetostriction may be developed by annealing, thus enabling these materials to be used in transducer applications.

4-27. Amorphous metal alloys are made using a new technology which produces a thin (0.001 to 0.003 in) ribbon from rapidly quenched molten metal. The alloy solidifies before the atoms have a chance to segregate or crystallize, resulting in a glasslike atomic structure material of high electrical resistivity, 125 to 130 $\mu\Omega \cdot$ cm. A range of magnetic properties may be developed in these materials by using different alloying elements. Amorphous metal alloys may be used in the same high-frequency applications as the cast, rolled to strip, silicon-iron and nickel-iron alloys.

4-28. Nickel-iron powder cores are made of insulated alloy powder which is compressed to shape and heat-treated. The alloy composition most widely used is 2-81 Permalloy powder composed of 2% molybdenum, 81% nickel, and balance iron. Another less widely used powder, Sendust, is made of 7 to 13% silicon, 4 to 7% aluminum, and balance iron. Prior to pressing, the powder particles are thinly coated with an inorganic, high-temperature insulation which can withstand the high compacting pressures and the high-temperature (650°C) hydrogen atmosphere anneal. The insulation of the particles lowers eddy-current loss and provides a distributed air gap which can be controlled to provide cores in a range of permeabilities. The 2-81 Permalloy cores are commercially available in permeability ranges of 14 to 300, and Sendust cores have permeabilities ranging from 10 to 140.

These types of nickel-iron powder cores find use in applications where inductance must remain relatively constant when the magnetic component experiences changes in dc current or temperature. Additional stability over temperature can also be achieved by the addition of low Curie temperature powder materials to neutralize the naturally positive permeability-temperature coefficient of the alloy powder. Some applications are in telephone loading coils or filter chokes for power conditioning equipment where output voltage ripple must be minimized. Other uses are for pulse transformers and switch-mode power supplies where low power losses are desired. Operating frequencies can range from 1.0 kHz for 300 permeability materials to 500 kHz for the 14 permeability materials.

4-29. Powdered-iron cores are manufactured from various types of iron powders whose particle size range from 2 to 100 μm. The particles are electrically insulated from one another using special insulating materials. The insulated powder is blended with phenolic or epoxy binders and a mold release agent. The powder is then dry-pressed in a variety of shapes including toroids, E cores, threaded tuning cores, cups, sleeves, slugs, bobbins, and other special shapes. A low-temperature bake of the pressed product produces a solid component in which the insulated particles provide a built-in air gap, reducing eddy-current losses, increasing electrical Q, and thus allowing higher operating frequencies. The use of different iron powder blends and insulation systems provides a range of permeability, from 4 to 90, for use over the frequency spectrum of 50 Hz to 250 MHz. Applications include high-frequency transformers, tuning coils, variable inductors, rf chokes, and noise suppressors for power supply and power control circuits.

4-30. Ferrite cores are molded from a mixture of metallic oxide powders such that certain iron atoms in the cubic crystal of magnetite (ferrous ferrite) are replaced by other metal atoms, such as Mn and Zn, to form manganese zinc ferrite, or by Ni and Zn to form nickel zinc ferrite. Manganese zinc ferrite is the material most commercially available[1] and is used in devices operating below 1.5 MHz. Nickel zinc ferrites are used mainly for filter applications above that frequency. They resemble ceramic materials in production processes and physical properties. The electrical resistivities correspond to those of semiconductors, being at least 1 million times those of metals. Magnetic permeability μ_0 may be as high as 10,000. The Curie point is quite low, however, in the range 100 to 300°C. Saturation flux density is generally below 5000 G (see Fig. 4-24.) Ferrite materials are available in several compositions which, through processing, can improve one or two magnetic parameters (magnetic induction, permeability, low hysteresis loss, Curie temperature) at the expense of the other

---

[1]Roess, E.: Soft Magnetic Ferrites and Application in Telecommunications and Power Converters; IEEE Trans. Magnetics, vol. MAG-18, no. 6, November 1982.

parameters. The materials are fabricated into shapes such as toroids; E, U, and I cores; beads; and self-shielding pot cores. Ferrite cores find use in filter applications up to 1.0 MHz, high-frequency power transformers operating at 10 to 100 kHz, pulse transformer delay lines, adjustable air-gap inductors, recording heads, and filters used in high-frequency electronic circuits.

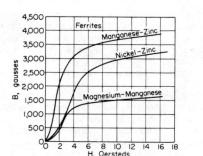

FIG. 4-24 Typical normal-induction curves for "soft" ferrites.

4-31. Permanent-magnet materials that are commercially available may be grouped into five classes:

a. Quench-hardened alloys.
b. Precipitation-hardened cast alloys.
c. Ceramic materials.
d. Powder compacts and elongated single-domain materials.
e. Ductile alloys.

4-32. Quench-Hardened Alloys. Early permanent magnets were made of low carbon steel (1% C) that was hardened by heat-treatment. Later developments saw improvements in the magnetic properties through the use of alloying elements of tungsten, chromium, and cobalt. The magnetic properties of these alloys and others discussed in this section are listed in Table 4-11. The chrome steels are less expensive than the cobalt steels, and both find use in hysteresis clutch and motor applications.

4-33. Precipitation-hardened cast alloys for magnetic applications are available in a wide range of magnetic properties. See the Alnicos listed in Table 4-11. Formed by conventional casting techniques, Alnico magnets are structurally weak and brittle and are not readily machined except by grinding. Alnico 5 DG and 5 Columnar grades are directionally solidified during casting. The Alnico 5 DG utilizes a chill plate in the mold to obtain a partially columnar crystallization in the casting, and the Alnico 5 Columnar uses a hot mold and chill plate procedure to develop a nearly 100% columnar structure throughout the magnet. All the Alnico grades listed, except grades 1, 2, 3, and 4, are heat-treated in a shape-oriented magnetic field which attains magnetic precipitate alignment and develops the excellent magnet anisotropic properties. Due to process limitations, Alnico 5 DG, 5 Columnar, 8, and 9 should be used only in straight flux line magnet designs. Alnico magnets find use in applications in loud speakers, microwave devices, motors, generators, meters, magnetos, separators, communication devices, and vending machines.

4-34. Ceramic magnet material usage is increasing yearly because of improved magnetic properties and the high cost of cobalt used in metallic alloy magnets. The basic raw material used in these magnets is iron oxide in combination with either strontium carbonate or barium carbonate. The iron oxide and carbonate mixture is calcined, and then the aggregate is ball-milled to a particle size of about 1.0 μm. The material is compacted in dies using the dry powder or a water-based slurry of the powder. High pressures are needed to press the parts to shape. In some ceramic grades, a magnetic field is applied during pressing to orient the material in order to obtain a preferred magnetic orientation. Parts are sintered at high temperatures and ground to finished size using diamond grinding wheels with suitable coolants. Ceramic magnets are hard and brittle, exhibit high electrical resistivities, and have lower densities than cast magnet alloys.

Made in the form of rings, blocks, and arcs, ceramic magnets find use in applications for loudspeakers, dc motors, microwave oven magnetron tubes, traveling wave tubes, holding magnets, chip collectors, and magnetic separator units. Ceramic magnet arcs find wide use in the auto industry in engine coolant pumps, heating-cooling fan motors, and window lift motors. As with other magnets, they are normally supplied nonmagnetized and are magnetized in the end-use structure using magnetizing fields of the order of 10,000 Oe to saturate the magnet. The brittleness of the material necessitates proper design of the magnet support structure so as not to impart mechanical stress to the magnet.

TABLE 4-11 Magnetic Properties and Chemical Compositions

Magnet material	Nominal chemical composition	Residual flux density B_r, G, nominal*	Coercive force H_c, Oe, nominal*	Max. energy prod (BH) max, MG·Oe, nominal*
3½% Cr steel	3.5 Cr, 1 C, bal. Fe	10,300	60	0.3
3% Co steel	3.25 Co, 4 Cr, 1 C, bal. Fe	9700	80	0.38
17% Co steel	18.5 Co, 3.75 Cr, 5 W, .75 C, bal. Fe	10,700	160	0.69
36% Co steel	38 Co, 3.8 Cr, 5 W, .75 C, bal. Fe	10,400	230	0.98
Alnico 1	12 Al, 21 Ni, 5 Co, 3 Cu, bal. Fe	7200	470	1.4
Alnico 2	10 Al, 19 Ni, 13 Co, 3 Cu, bal. Fe	7500	560	1.7
Alnico 3	12 Al, 25 Ni, 3 Cu, bal. Fe	7000	480	1.35
Alnico 4	12 Al, 27 Ni, 5 Co, bal. Fe	5600	720	1.35
†Alnico 5	8 Al, 14 Ni, 24 Co, 3 Cu, bal. Fe	12,800	640	5.5
†Alnico 5 DG	8 Al, 14 Ni, 24 Co, 3 Cu, bal. Fe	13,300	670	6.5
†Alnico 5 Col.	8 Al, 14 Ni, 24 Co, 3 Cu, bal. Fe	13,500	740	7.55
†Alnico 6	8 Al, 16 Ni, 24 Co, 3 Cu, 1 Ti, bal. Fe	10,500	780	3.9
†Alnico 8	7 Al, 15 Ni, 35 Co, 4 Cu, 5 Ti, bal. Fe	8200	1650	5.3
†Alnico 8 HC	8 Al, 14 Ni, 38 Co, 3 Cu, 8 Ti, bal. Fe	7200	1900	5.0
†Alnico 9	7 Al, 15 Ni, 35 Co, 4 Cu, 5 Ti, bal. Fe	10,500	1500	9.0
Ceramic 1	$MO \cdot 6\ Fe_2O_3$	2300	1860/3250‡	1.05
†Ceramic 2	$MO \cdot 6\ Fe_2O_3$ M represents one or more of	2900	2400/3000‡	1.8
†Ceramic 3	$MO \cdot 6\ Fe_2O_3$ the metals chosen from the	3300	2200/2400‡	2.6
†Ceramic 4	$MO \cdot 6\ Fe_2O_3$ group barium, strontium,	2500	2300/3800‡	1.45
†Ceramic 5	$MO \cdot 6\ Fe_2O_3$ lead.	3800	2400	3.4
†Ceramic 6	$MO \cdot 6\ Fe_2O_3$	3200	2820/3300‡	2.45
†Ceramic 7	$MO \cdot 6\ Fe_2O_3$	3400	3250/4000‡	2.75
†Ceramic 8	$MO \cdot 6\ Fe_2O_3$	3850	2950/3050‡	3.5

TABLE 4-11 Magnetic Properties and Chemical Compositions (*Continued*)

Magnet material	Nominal chemical composition		Residual flux density B_r, G, nominal*	Coercive force H_c, Oe, nominal*	Max. energy prod (BH) max, MG·Oe, nominal*
Sint. Alnico 2	10 Al, 19 Ni, 13 Co, 3 Cu, bal. Fe		7100	550	1.5
†Sint. Alnico 5	8 Al, 14 Ni, 24 Co, 3 Cu, bal. Fe		10,900	620	3.95
†Sint. Alnico 6	8 Al, 16 Ni, 24 Co, 3 Cu, 1 Ti, bal. Fe		9400	790	2.95
†Sint. Alnico 8	7 Al, 15 Ni, 35 Co, 4 Cu, 5 Ti, bal. Fe		7400	1500	4.0
†Sint. Alnico 8 HC	7 Al, 14 Ni, 38 Co, 3 Cu, 8 Ti, bal. Fe		6700	1800	4.5
†Rare earth cobalt 12	RE·Co	RE represents one or more of the	7200	6500/10,000‡	12.0
†Rare earth cobalt 15	RE·Co	metals chosen from the rare earth	8000	7000/14,000‡	15.0
†Rare earth cobalt 16	RE·Co	group: samarium, praseodymium,	8300	7500/18,000‡	16.0
†Rare earth cobalt 18	RE·Co	cerium, yttrium, misch metal, etc.	8700	8000/20,000‡	18.0
†ESD 31	20.7 Fe, 11.6 Co, 67.7 Pb		5000	1000	2.3
†ESD 32	18.3 Fe, 10.3 Co, 72.4 Pb		6800	960	3.0
ESD 41	20.7 Fe, 11.6 Co, 67.7 Pb		3600	970	1.1
ESD 42	18.3 Fe, 10.3 Co, 72.4 Pb		4800	830	1.25
†Cunife 1	60 Cu, 20 Ni, 20 Fe		5500	530	1.4
Vicalloy 1	10 V, 52 Co, bal. Fe		7500	250	0.8
Remalloy	12 Co, 15 Mo, bal. Fe		9700	250	1.0

*Values derived from major hysteresis loop.
†Anisotropic.
‡Intrinsic coercive force, H_{ci}.
NOTE: MG·Oe = megagauss · oersteds; ESD = elongated single domain.
SOURCE: Adapted from *Standard Specification for Permanent Magnet Materials;* Magnetic Materials Producers Association, Evanston, Ill.

MAGNETIC MATERIALS

4-35. *Powder compacts* for magnets are represented by the sintered Alnico and the rare earth cobalt magnets. These are listed in Table 4-11.

4-36. *Sintered Alnico magnets* are made of fine powders of the alloy which are compressed to the desired shape and size, then sintered at about 1200°C in a hydrogen atmosphere. During sintering, there is 2 to 10% shrinkage in part dimensions which must be considered in the die design. All of the sintered Alnico magnets listed, except sintered Alnico 2, receive a postsintering heat-treatment in a magnetic field. This treatment produces a preferred magnetic orientation in the part which should be considered in the magnet design.

Sintered magnets are structurally stronger than their cast equivalent, but they are limited to small sizes of 15 g or less. Larger sizes are generally more economically made by casting. Also, the sintered magnets usually exhibit slightly lower magnetic properties than equivalent grades of cast Alnico alloys.

4-37. *Rare earth cobalt magnets* have the highest energy product and coercivity of any commercially available magnetic material. Magnets are produced by powder metallurgy techniques from alloys of cobalt (65 to 77%), rare earth metals (23 to 35%), and sometimes copper and iron. The rare earth metal used is usually samarium, but other metals used are praseodymium, cerium, yttrium, neodymium, lathanum, and a rare earth metal mixture called misch metal. The rare earth alloy is ground to a fine particle size (1 to 10 μm), and the powder is then die-compacted in a strong magnetic field. The part is then sintered and abrasive-ground to finish tolerances.

Although this material uses comparatively expensive raw materials, the high value of coercive force (5500 to 9500 Oe) leads to small magnet size and good temperature stability. These magnets find use in miniature electronic devices such as motors, printers, electron beam focusing assemblies, magnetic bearings, and traveling wave tubes. Plastic-bonded rare earth magnets are also being made, but the magnetic value of the energy product is only a fraction of the sintered product.

4-38. *Elongated single-domain (ESD) materials* are available as alloys of iron and cobalt in the form of anisotropic particles (needles, or plates). A special process develops the magnetic particles in a lead (Pb) matrix which is able to flow and fill a compacting die in the form of the magnet. The particle magnetization is field-aligned in the long axis of the particles. When frozen in the lead matrix, the particles resist demagnetization. This magnetic material finds limited use due to its low operating temperature (300°F) and the high mass density due to the lead content.

4-39. *Ductile alloys* include the materials Cunife, Vicalloy, Remalloy, chromium-cobalt-iron (Cr-Co-Fe), and, in a limited sense, manganese-aluminum-carbon (Mn-Al-C). They are sufficiently ductile and malleable to be drawn, forged, or rolled into wire or strip forms. A final heat-treatment after forming develops the magnetic properties. Cunife has a directional magnetism developed as a result of cold working and finds wide use in meters and automotive speedometers. Vicalloy has been used as a high-quality and high-performance magnetic recording tape and in hysteresis clutch applications. Remalloy has been used extensively in telephone receivers but is now being replaced by a newer, less costly magnetic material.

New permanent-magnet materials that are now being produced are the Cr-Co-Fe alloy and the Mn-Al-C alloy. The Cr-Co-Fe alloy family contains 20 to 35% chromium and from 5 to 25% cobalt. This alloy is unique among permanent-magnet alloys due to its good hot and cold ductility, machinability, and excellent magnetic properties. The heat-treatment of the alloy involves a rapid cooling from approximately 1200°C to a spinoidal decomposition phase occurring at about 600°C. The magnetic phase developed in the spinoidal decomposition process may be oriented by a heat-treatment in a magnetic field, or the material may be magnetically oriented by "deformation aging" as would be accomplished in a wire drawing operation. The magnetic properties that can be developed are comparable to those of Alnico 5 and are superior to those of the other ductile alloys, Cunife, Vicalloy, and Remalloy. Western Electric has introduced a Cr-Co-Fe alloy which replaces Remalloy in the production of telephone receiver magnets and at a lower cost due to reduced cobalt.

4-40. *The Mn-Al-C alloy* achieves permanent-magnetic properties (B_r, 5500 G; H_c, 2300 Oe; Mg·Oe energy product, 5 Mg·Oe) when mechanical deformation of the alloy takes place at a temperature of about 720°C. Mechanical deformation may be performed by warm extrusion. Magnet size is limited by the amount of deformation needed to develop and orient the magnetic phase in the alloy. The alloying elements are inexpensive, but the tooling and equipment needed in the deformation process is expensive and may be a factor in the economical production of this magnet alloy. Magnets of this alloy would find use in loudspeakers, motor applications, and microwave oven magnetron tubes. The low density, 5.1 g/cm³, is desirable for motors where reduced inertia and weight savings are important. The low Curie temperature, 320°C, limits the use of this alloy to applications where the ambient temperature is less than 125°C.

4-41. Permanent-magnet design involves the calculation of magnet area and magnet length to produce a specific magnetic flux density across a known gap, usually with the magnet having the smallest possible volume. Designs are developed from magnet material hysteresis loop data of the second quadrant, commonly called *demagnetization curves*. An example of demagnetization curve data is shown in Fig. 4-25.

Other considerations are the operating temperature of the magnetic assembly, magnet weight, and cost. Also, care should be exercised in the calculation of any steel return path cross section to ensure that it is adequate to carry the flux output of the magnet. Table 4-12 illustrates the range of magnetic characteristics that may be considered in the design. Detailed magnetic and material specifications may be obtained from the magnet manufacturer.

4-42. Bibliography. Information on the properties of commercial magnetic materials may be obtained from technical bulletins and catalogs issued by the various producers of magnetic materials. Other sources of information include the following:

Bozorth, R. M.: *Ferromagnetism;* Princeton, N.J., D. Van Nostrand Company, Inc., 1951.

Cullity, B. D.: *Introduction to Magnetic Materials;* Reading, Mass., Addison-Wesley Publishing Company, 1972.

Heck, C.: *Magnetic Materials and Their Applications;* New York, Crane, Russack, and Company, Inc., 1974.

Parker, R. J. and Studders, R. J.: *Permanent Magnets and Their Applications;* New York, John Wiley and Sons, Inc., 1962.

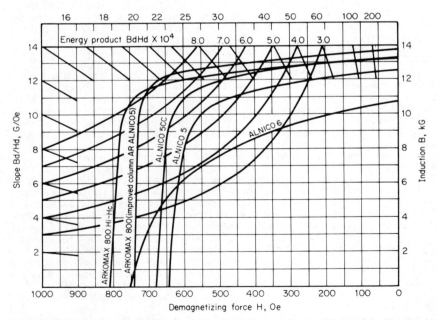

FIG. 4-25 Typical demagnetization curves for Alnico permanent-magnetic materials. (Reprinted with permission from *PM-121C Alnico Permanent Magnets;* The Arnold Engineering Company, Marengo, Illinois.)

MAGNETIC MATERIALS

TABLE 4-12 Comparison of Magnetic and Physical Properties of Selected Commercial Materials

	Alnico 5	Alnico 9	Ferrite	Co5R
B_r, G	12800	10500	4100	9500
H_c, Oe	640	1500	2900	6500
$B_d H_d$, Mg·Oe	5.5	9.0	4.0	22.0
Curie point, °C	850	815	470	740
Temperature coefficient, %/°C	0.02	0.02	0.19	0.03
Density, g/cm³	7.3	7.3	4.9	8.6
Energy/unit weight	0.8	1.2	0.8	2.6

Steel Products Manual, Electrical Steel; AISI, Washington, D.C., January 1983.

1983 Annual Book of ASTM Standards, sec. 3, vol. 03.04; ASTM, Philadelphia, Pa., 1983.

Standard Specifications for Permanent Magnet Materials; Magnetic Materials Producers Association, Evanston, Ill., 1983.

Permanent Magnet Guidelines; Magnetic Materials Producers Association, Evanston, Ill., 1983.

Chapter 5

GENERAL PROPERTIES OF INSULATING MATERIALS

By T. W. DAKIN

5-1	Electrical Insulation and Dielectric Defined 5-1	5-7	Composite Dielectrics 5-7
5-2	Circuit Analogy of a Dielectric or Insulation 5-1	5-8	Potential Distribution in Dielectrics 5-8
5-3	Capacitance and Permitivity or Dielectric Constant 5-2	5-9	Dielectric Strength 5-9
		5-10	Water Penetration 5-12
5-4	Resistance and Resistivity of Dielectrics and Insulation 5-3	5-11	Ionizing Radiation 5-12
		5-12	Arc Tracking of Insulation 5-12
5-5	Variation of Dielectric Properties with Frequency 5-6	5-13	Thermal Aging 5-12
		5-14	Application of Electrical Insulation 5-14
5-6	Variation of Dielectric Properties with Temperature 5-7	5-15	References on Insulating Materials 5-14

5-1. Electrical Insulation and Dielectric Defined. Electrical insulation is a medium or a material which, when placed between conductors at different potentials, permits only a small or negligible current in phase with the applied voltage to flow through it. The term dielectric is almost synonymous with electrical insulation, which can be considered the applied dielectric. A perfect dielectric passes no conduction current and only capacitive charging current between conductors. Only a vacuum at low stresses between uncontaminated metal surfaces satisfies this condition.

The range of resistivities of substances which can be considered insulators is from greater than 10^{20} Ω·cm downward to the vicinity of 10^6 Ω·cm, depending on the application and voltage stress. There is no sharp boundary defined between low-resistance insulators and semiconductors. If the voltage stress is low and there is little concern about the level of current flow (other than that which would heat and destroy the insulation), relatively low-resistance insulation can be tolerated.

5-2. Circuit Analogy of a Dielectric or Insulation. Any dielectric or electrical insulation can be considered as equivalent to a combination of capacitors and resistors which will duplicate the current-voltage behavior at a particular frequency or time of voltage application. In the case of some dielectrics, simple linear capacitors and resistors do not adequately represent the behavior. Rather, resistors and capacitors with particular nonlinear voltage-current or voltage-charge relations must be postulated to duplicate the dielectric current-voltage characteristic.

The simplest circuit representation of a dielectric is a parallel capacitor and resistor, as shown in Fig 5-1 for $R_S = 0$. The perfect dielectric would be simply a capacitor. Another representation of a dielectric is a series-connected capacitor and resistor as in Fig 5-1 for $R_p = \infty$, while still another involves both R_S and R_p.

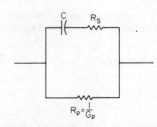

FIG. 5-1 Equivalent circuit of a dielectric.

The ac dielectric behavior is indicated by the phase diagram (Fig. 5-2). The perfect dielectric capacitor has a current which leads the voltage by 90°, but the imperfect dielectric has a current which leads the voltage by less than 90°. The dielectric phase angle is θ, and the difference, $90° - \theta = \delta$, is the loss angle. Most measurements of dielectrics give directly the tangent of the loss angle $\tan \delta$ (known as the *dissipation factor*) and the capacitance C. In Fig. 5-1, if $R_p = \infty$, the series $R_s - C$ has a $\tan \delta = 2\pi f C_s R_s$, and if $R_s = 0$, the parallel $R_p - C$ has a $\tan \delta = 1/2\pi f C_p R_p$.

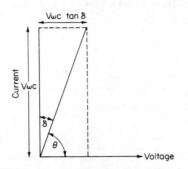

FIG. 5-2 Current-voltage phase relation in a dielectric.

The ac power or heat loss in the dielectric is $V^2 2\pi f C \tan \delta$ watts, or $VI \sin \delta$ watts, where $\sin \delta$ is known as the power factor, V is the applied voltage, I is the total current through the dielectric, and f is the frequency. From this it can be seen that the equivalent parallel conductance of the dielectric σ (the inverse of the equivalent parallel resistance ρ) is $2\pi f C \tan \delta$. The ac conductivity is

$$\sigma = (5/9) f \varepsilon' \tan \delta \times 10^{-12} \, \Omega^{-1} \, \text{cm}^{-1} = 1/\rho \qquad (5\text{-}1)$$

where ε' is the permittivity (or relative dielectric constant) and f is the frequency. [The IEEE now recommends the symbol ε' for the dielectric constant relative to a vacuum. The literature on dielectrics and insulation has also used κ (kappa) for this dimensionless quantity or ε'_r. In some places, ε' has been used to indicate the absolute dielectric constant, which is the product of the relative dielectric constant and the dielectric constant of a vacuum ε_0, which is equal to 8.85×10^{-12} F/m.] κ_0 has also been used to represent the dielectric constant of a vacuum. While the ac conductivity theoretically increases in proportion to the frequency, in practice, it will depart from this proportionality insofar as ε' and $\tan \delta$ change with frequency.

5-3. Capacitance and Permittivity or Dielectric Constant. The capacitance between plane electrodes in a vacuum (with fringing neglected) is

$$C = \varepsilon' \varepsilon_0 A/t = 0.0884 \times 10^{-12} A/t \quad \text{farads} \qquad (5\text{-}2)$$

where ε_0 is the dielectric constant of a vacuum, A the area in square centimeters, and t the spacing of the plates in centimeters. ε_0 is 0.225×10^{-12} F/in when A and t are expressed in inch units.

When a dielectric material fills the volume between the electrodes, the capacitance is higher by virtue of the charges within the molecules and atoms of the material, which attract more charge to the capacitor plates for the same applied voltage. The capacitance with the dielectric between the electrodes is

$$C = \varepsilon' \varepsilon_0 A/t \qquad (5\text{-}3)$$

where ε' is the relative dielectric constant of the material. The capacitance relations for several other commonly occurring situations are

Coaxial conductors: $\quad C = \dfrac{2\pi \varepsilon' \varepsilon_0 L}{\ln(r_2/r_1)} \quad$ farads $\qquad (5\text{-}4)$

Concentric spheres: $\quad C = \dfrac{4\pi \varepsilon' \varepsilon_0 r_1 r_2}{r_2 - r_1} \quad$ farads $\qquad (5\text{-}5)$

Parallel cylindrical conductors: $\quad C = \dfrac{\pi \varepsilon' \varepsilon_0 L}{\cosh^{-1}(D/2r)} \quad$ farads $\qquad (5\text{-}6)$

GENERAL PROPERTIES OF INSULATING MATERIALS

5-3

In these equations, L is the length of the conductors, r_2 and r_1 are the outer and inner radii and D is the separation between centers of the parallel conductors with radii r. For dimensions in centimeters ε_0 is 0.0884 F/cm.

The value of ε' depends on the number of atoms or molecules per unit volume and the ability of each to be polarized (that is, to have a net displacement of their charge in the direction of the applied voltage stress). Values of ε' range from unity for vacuum to slightly greater than unity for gases at atmospheric pressure, 2 to 8 for common insulating solids and liquids, 35 for ethyl alcohol and 91 for pure water, and 1000 to 10,000 for titanate ceramics (see Table 5-1 for typical values).

The relative dielectric constant of materials is not constant with temperature, frequency, and many other conditions and is more appropriately called the dielectric permittivity. Refer to the volume by Smyth[1]* for a discussion of the relation of ε' to molecular structure and to von Hippel[2,3] and other tables of dielectric materials[4] from the MIT Laboratory for Insulation Research. The permittivity of many liquids has been tabulated in *NBS Circ.* 514. The "Handbook of Chemistry and Physics" (Chemical Rubber Publishing Co.) also lists values for a number of plastics and other materials.

The permittivity of many plastics, ceramics, and glasses varies with the composition, which is frequently variable in nominally identical materials. In the case of some plastics, it varies with degree of cure and in the case of ceramics with the firing conditions. Plasticizers often have a profound effect in raising the permittivity of plastic compositions.

There is a force of attraction between the plates of a capacitor having an applied voltage. The stored energy is $1/2CV^2$J. The force equals the derivative of this energy with respect to the plate separation: $(1/2)\varepsilon'\varepsilon_0 E^2 \times 10^2 N/cm^2$ or $(1/2)\varepsilon'\varepsilon_0 E^2 \times 10$ bar, where E is the electric field in volts per centimeter. The force increases proportionally to the capacitance or permittivity. This leads to a force of attraction of dielectrics into an electric field, that is, a net force which tends to move them toward a region of high field. If two dielectrics are present, the one with higher permittivity will displace the one with lower permittivity in the higher-field region. For example, air bubbles in a liquid are repelled from high-field regions. Correspondingly, elongated dielectric bodies are rotated into the direction of the electric field. In general, if the voltage on a dielectric system is maintained constant, the dielectrics move (if they are able) to create a higher capacitance.

5-4. Resistance and Resistivity of Dielectrics and Insulation. The measured resistance

---

*Superscripts refer to bibliography for this subsection, Par. **5-15**.

TABLE 5-1 Dielectric Permittivity (Relative Dielectric Constant), ε^1

	k		k
Inorganic crystalline:		**Polymer resins:**	
NaCl, dry crystal	5.5	Nonpolar resins:	
CaCO₃ (av)	9.15	Polyethylene	2.3
Al₂O₃	10.0	Polystyrene	2.5–2.6
MgO	8.2	Polypropylene	2.2
BN	4.15	Polytetrafluoroethylene	2.0
TiO₂ (av)	100	Polar resins:	
BaTiO₃ crystal	4,100	Polyvinyl chloride (rigid)	3.2–3.6
Muscovite mica	7.0–7.3	Polyvinyl acetate	3.2
Fluorophlogopite (synthetic		Polyvinyl fluoride	8.5
mica)	6.3	Nylon	4.0–4.6
		Polyethylene terephthalate	3.25
Ceramics:		Cellulose cotton fiber (dry)	5.4
Alumina	8.1–9.5	Cellulose Kraft fiber (dry)	5.9
Steatite	5.5–7.0	Cellulose cellophane (dry)	6.6
Forsterite	6.2–6.3	Cellulose triacetate	4.7
Aluminum silicate	4.8	Tricyanoethyl cellulose	15.2
Typical high-tension porcelain	6.0–8.0	Epoxy resins unfilled	3.0–4.5
Titanates	50–10,000	Methylmethacrylate	3.6
Beryl	4.5	Polyvinyl acetate	3.7–3.8
Zirconia	8.0–10.5	Polycarbonate	2.9–3.0
Magnesia	8.2	Phenolics (cellulose-filled)	4–15
Glass-bonded mica	6.4–9.2	Phenolica (glass-filled)	5–7
		Phenolics (mica-filled)	4.7–7.5
Glasses:		Silicones (glass-filled)	3.1–4.5
Fused silica	3.8		
Corning 7740 (common laboratory Pyrex)	5.1		

R of insulation depends upon the geometry of the specimen or system measured, which for a parallel-plate arrangement is

$$R = \rho t/A \quad \text{ohms} \tag{5-7}$$

where t is the insulation thickness in centimeters, A is the area in square centimeters, and ρ is the dielectric resistivity in ohm-centimeters. If t and A vary from place to place, the effective "insulation resistance" will be determined by the effective integral of the t/A ratio over all the area under stress, on the assumption that the material resistivity ρ does not change. If the material is not homogeneous and materials of different resistivities appear in parallel, the system can be treated as parallel resistors: $R = R_a R_b/(R_a + R_b)$. In this case, the lower-resistivity material usually controls the overall behavior. But if materials of different resistivities appear in series in the electric field, the higher-resistivity material will generally control the current and a majority of the voltage will appear across it, as in the case of series resistors.

The resistance of dielectrics and insulation is usually time-dependent and (for the same reason) frequency-dependent. The dc behavior of dielectrics under stress is an extension of the low-frequency behavior. The ac and dc resistance and permittivity can, in principle, be related for comparable times and frequencies.

Current flow in dielectrics can be divided into parts: (*a*) the true dc current, which is constant with time and would flow indefinitely, is associated with a transport of charge from one electrode into the dielectric, through the dielectric, and out into the other electrode; and (*b*) the polarization or absorption current, which involves, not charge flow through the interface between the dielectric and the electrode, but rather the displacement of charge within the dielectric. This is illustrated in Fig. 5-3, where it is shown that the displaced or absorbed charge is responsible for a reverse current when the voltage is removed.

Polarization current results from any of the various forms of limited charge displacement which can occur in the dielectric. The displacement occurring first (within less than nanoseconds) is the electronic and intramolecular charged atom displacement responsible for the very-high-frequency permittivity. The next slower displacement is the rotation of dipolar molecules and groups which are relatively free to move. The displacement most commonly observed in dc measurements, that is, currents changing in times of the order of seconds and minutes, is due to the very slow rotation of dipolar molecules and ions moving up to internal barriers in the material or at the conductor surfaces. When those slower displacement polarizations occur, the dielectric constant declines with increasing frequency and approaches the square of the optical refractive index η^2 at optical frequencies.

In composite dielectrics (material with relatively lower resistance intermingled with a material of relatively higher resistance) a large interfacial or "Maxwell-Wagner" type of polarization can occur.[5] A circuit model of such a situation can be represented by placing two of the circuits of Fig. 5-1 in series and making the parallel resistance of one much lower than the other. To get the effect, it is necessary that the time constant $R_p C$ be different for each material.

A simple model of the polarization current predicts an exponential decline of the current with time: $I_p = Ae^{-\alpha t}$, similar to the charging of a capacitor through a resistor. Composite materials are likely to have many different time constants, $\alpha = 1/RC$, superimposed. It is found empirically that the polarization or absorption current decreases inversely as a simple negative exponent of the time

$$I = At^{-n} \tag{5-8}$$

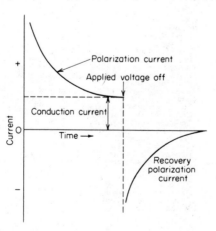

FIG. 5-3 Typical dc dielectric current behavior.

The ratio of the current at 1 min to that at 10 min has been called the polarization

GENERAL PROPERTIES OF INSULATING MATERIALS

index and is used to indicate the quality of composite machine insulation. A low polarization index associated with a low resistance sometimes indicates parallel current leakage paths through or over the surface of insulation (for example, in adsorbed water films).

The level of the conduction current which flows essentially continuously through insulation is an indication of the level of the ionic concentration and mobility in the material. Frequently, as with salt in water, the ions are provided by dissolved, adsorbed, or included impurity electrolytes in the material, rather than by the material itself. Purifying the material will therefore often raise the resistivity. If it is a liquid, purification can be done with adsorbent clays or ion-exchange resins.

The conductivity of ions in an insulation is given by the equation[6]

$$\sigma = \mu e c \quad \Omega^{-1} \cdot cm^{-1} \tag{5-9}$$

where μ is the ion mobility, e is the ionic charge in coulombs, and c is the ionic concentration per cubic centimeter. The mobility, expressed in centimeters per second-volt per centimeter, decreases inversely with the effective internal viscosity and is very low for hard resins, but it increases with temperature and with softness of the resin, or fluidity of liquids. The ionic conductivity also varies widely with material purity. Among the polymers and resins, nonpolar resins such as polyethylene are likely to have high resistivities, of the order of 10^{16} or greater, since they do not readily dissolve or dissociate ionic impurities. Harder or crystalline polar resins have higher resistivity than do similar softer resins of similar dielectric constant and purity. Resins and liquids of higher dielectric constant usually have higher conductivities because they dissolve ionic impurities better, and the impurities dissociate to ions much more readily in a higher dielectric constant medium. Ceramics and glasses have lower resistivity if they contain alkali ions (sodium and potassium), since these ions are highly mobile.

Water is particularly effective in decreasing the resistivity by increasing the ionic concentration and mobility of materials, on the surface as well as internally. Water associates with impurity ions or ionizable constituents within or on the surface or interfaces. It helps to dissociate the ions by virtue of its high dielectric constant and provides a local environment of greater mobility, particularly as surface water films. Electrolyte conduction is discussed in Ref. 6. Table 5-2 indicates the effect of water on the surface resistivity of some insulating materials.

The ionic conductivity σ, exclusive of polarization effects, can be expected to increase exponentially with temperature according to the relation

$$\sigma = \sigma_0 e^{-B/T} \tag{5-10}$$

where T is the Kelvin temperature and σ_0 and B are constants. This relation, log σ versus $1/T$, is shown in Fig. 5-4. It is often observed that at lower temperatures, where the resistivity is higher, the resistivity tends lower than the extrapolated higher temperature line would

TABLE 5-2 Surface Resistivities of Insulation and Effect of Humidity

Material	Ω/square at 100% RH	% RH/decade decrease of Ω/square
Hydrocarbon wax	$>20 \times 10^{12}$	
Silicone rubber	10×10^{12}	
Polytetrafluoroethylene	3.6×10^{12}	
Polystyrene	8.4×10^{11}	
Cellulose acetate	7×10^{9}	6
Polyvinyl chloride acetate	5.7×10^{9}	12
Mica-filled phenolic	5×10^{9}	9
Glazed porcelain	3.7×10^{9}	15
Mica	3×10^{9}	12
Steatite (L-4)	$2-6 \times 10^{8}$	
Cellulose-filled phenolic	2.4×10^{6}	10
Quartz	1.9×10^{6}	

predict. There are at least two possible reasons for this: the effect of adsorbed moisture and the contribution of a very slowly decaying polarization current.

5-5. Variation of Dielectric Properties with Frequency. The permittivity of dielectrics invariably tends downward with increasing frequency, owing to the inability of the polarizing charges to move with sufficient speed to follow the increasing rate of alternations of the electric field. This is indicated in Fig. 5-5. The sharper decline in permittivity is known as a dispersion region. At the lower frequencies the ionic-interface polarization declines first; next the molecular dipolar polarizations decline. With some polar polymers two or more dipolar dispersion regions may occur owing to different parts of the molecular rotation.

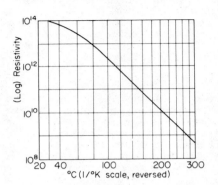

FIG. 5-4 Typical dielectric resistivity-temperature dependence (Corning Glass 7740).

Figure 5-5 is typical of polymers and liquids, but not of glasses and ceramics. Glasses, ceramics, and inorganic crystals usually have much flatter permittivity-frequency curves, similar to that shown for the nonpolar polymer, but at a higher level, owing to their atom-ion displacement polarization, which can follow the electric field usually up to infrared frequencies.

The dissipation factor–frequency curve in Fig. 5-5 indicates the effect of ionic migration conduction at low frequency. It shows a maximum at a frequency corresponding to the permittivity dispersion region. This maximum is usually associated with a molecular dipolar rotation and occurs when the rotational mobility is such that the molecule rotation can just keep up with frequency of the applied field. Here it has its maximum movement in phase with the voltage, thus contributing to conduction current. At lower frequencies the molecule dipole can rotate faster than the field and contributes more to permittivity. At higher frequencies it cannot move fast enough. Such a dispersion region can also occur because of ionic migration and interface polarization if the interfaces are closely spaced, and if the frequency and mobility have the required values.

The frequency region where the dipolar dispersion occurs depends on the rotational mobility. In mobile, low-viscosity liquids it is in the 100- to 10,000-MHz range. In viscous liquids it occurs in the region of 1 to 100 MHz. In soft polymers it may occur in the audio-frequency range, and with hard polymers it is likely to be at very low frequency (indistin-

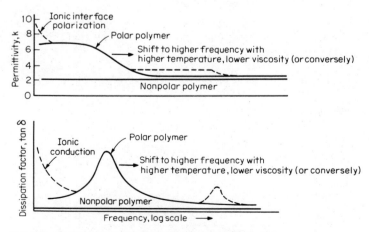

FIG. 5-5 Typical variation in dielectric properties with frequency.

guishable from dc properties). Since the viscosity is affected by the temperature, increased temperature shifts the dispersion to higher frequencies.

5-6. Variation of Dielectric Properties with Temperature. The trend in ac permittivity and conductivity, as measured by the dissipation factor, is controlled by the increasing ionic migrational and dipolar molecular rotational mobility with increasing temperature. This curve, which is indicated in Fig. 5-6, is in most respects a mirror image of the frequency trend shown in Fig. 5-5, since the two effects are interrelated.

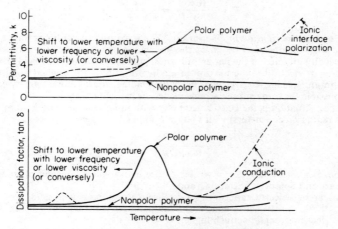

FIG. 5-6 Typical variation in dielectric properties with temperature.

The permittivity-dispersion and dissipation-factor maximum region occurs below room temperature for viscous liquids, and still lower for mobile liquids. In fact mobile liquids may crystallize before they would show dispersion, except at high frequencies. With polymers the dissipation-factor maximum is likely to occur, at power frequencies, at a temperature close to a softening-point or internal second-order transition-point temperature. Dielectric dispersion and mechanical modulus dispersion can usually be correlated at the same temperature for comparable frequencies.

5-7. Composite Dielectrics. The dielectric properties of composite dielectrics are generally a weighted average of the individual component properties, unless there is interaction, such as dissolving (as opposed to intermixing) of one material in another, or chemical reaction of one with another. Interfaces created by the mixing present a special factor, which can often lead to a higher dissipation factor and lower resistivity as a result of moisture and/or impurity concentration at the interface.

The ac properties of sheets of two dielectrics of dielectric constant k_1 and k_2 and of thickness t_1 and t_2 placed in series are related to the properties of the individual materials by the series of capacitance and impedance relations:

$$C = \frac{k_0 k_1 k_2 A}{k_1 t_2 + k_2 t_1} \tag{5-11}$$

$$\tan \delta = \frac{(t_1/t_2)\varepsilon_2' \tan \delta_1 + \varepsilon_1' \tan \delta_2}{\varepsilon_1' + \varepsilon_2'(t_1/t_2)} \tag{5-12}$$

Similarly the properties of two dielectrics in parallel are

$$C = \varepsilon_0 \left(\frac{\varepsilon_1' A_1}{t_1} + \frac{\varepsilon_2' A_2}{t_2} \right) \tag{5-13}$$

$$\tan \delta = \frac{t_2 \varepsilon_1' A_1 \tan \delta_1 + t_1 \varepsilon_2' A_2 \tan \delta_2}{t_2 \varepsilon_1' A_1 + t_1 \varepsilon_2' A_2} \tag{5-14}$$

With steady dc voltages, the resistivities control the current. With equal-area layer dielectrics in series,

$$R = R_1 + R_2 = \frac{1}{A}(\rho_1 t_1 + \rho_2 t_2) \quad (5\text{-}15)$$

When the dielectrics are in parallel and of equal thickness t,

$$R = \frac{R_1 R_2}{R_1 + R_2} = \frac{\rho_1 \rho_2 t}{\rho_1 A_2 + \rho_2 A_1} \quad (5\text{-}16)$$

5-8. Potential Distribution in Dielectrics. The maximum potential gradient in dielectrics is of critical significance insofar as the breakdown is concerned, since breakdown or corona is usually initiated at the region of highest gradient. In a uniform-field arrangement of conductors or electrodes, the maximum gradient is simply the applied voltage divided by the minimum spacing. In divergent fields the gradient must be obtained by calculation (which is possible for some simple arrangements) or by field mapping.

A common situation is the coaxial geometry with inner and outer radii R_1 and R_2. The gradient at radius r (centimeters) with voltage V applied is given by the equation:

$$E = \frac{V}{r \ln (R_2/R_1)} \quad \text{V/cm} \quad (5\text{-}17)$$

The gradient is a maximum at $r = R_1$. Reference books which consider other geometries are Schwaiger and Sorensen,[7] Stratton,[8] and Ollendorff.[9]

When different dielectrics appear in series, the greater stress with ac fields is on the material having the lower dielectric constant. This material will frequently break down first unless its dielectric strength is much higher:

$$\frac{E_1}{E_2} = \frac{\varepsilon_2'}{\varepsilon_1'} \quad \text{and} \quad E_1 = \frac{V}{t_1 + t_2 \varepsilon_1'/\varepsilon_2'} \quad (5\text{-}18)$$

The effect of the insulation thickness and dielectric constant (as well as the sharpness of the conductor edge) to create sufficient electric stress for local air breakdown (partial discharges) is shown in Fig 5-7.[17] With dc fields the stress distributes according to the resistivities of the materials, the higher stress being on the higher-resistivity material.

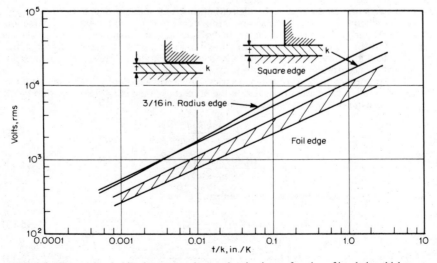

FIG. 5-7 Corona threshold voltage at conductor edges in air as a function of insulation thickness.

GENERAL PROPERTIES OF INSULATING MATERIALS

5-9. Dielectric Strength. This is defined by the ASA as the maximum potential gradient that the material can withstand without rupture. Practically, the strength is often reported as the breakdown voltage divided by the thickness between electrodes, regardless of electrode stress concentration.

Breakdown appears to require not only sufficient electric stress but also a certain minimum amount of energy. It is a property which varies with many factors such as thickness of the specimen, size and shape of electrodes used in applying stress, form or distribution of the field of electric stress in the material, frequency of the applied voltage, rate and duration of voltage application, fatigue with repeated voltage applications, temperature, moisture content, and possible chemical changes under stress.

The practical dielectric strength is decreased by defects in the material, such as cracks and included conducting particles and gas cavities. As will be shown in more detail in later sections on gases and liquids, the dielectric strength is quite adversely affected by conducting particles.

To state the dielectric strength correctly, the size and shape of specimen, method of test, temperature, manner of applying voltage, and other attendant conditions should be particularized as definitely as possible.

ASTM standard methods of dielectric strength testing should be used for making comparison tests of materials, but the levels of dielectric strength measured in such tests should not be expected to apply in service for long times. It is best to test an insulation in the same configuration in which it would be used. Also the possible decline in dielectric strength during long-time exposure to the service environment, thermal aging, and partial discharges (corona), if they exist at the applied service voltage, should be considered. ASTM has thermal life test methods for assessing the long-time endurance of some forms of insulation, such as sheet insulation (D2304), wire enamel (D2307) and others. There are IEEE thermal life tests for some systems such as random wound motor coils (IEEE-117).

The dielectric strength varies as the time and manner of voltage application as indicated in Fig. 5-8. With unidirectional pulses of voltage, having rise times of less than a few microseconds, there is a time lag of breakdown, which results in an apparent higher strength for

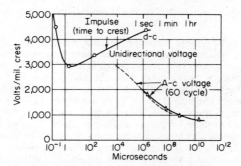

FIG. 5-8 Dielectric strength of 0.032-in pressboard in oil as a function of time of voltage application.

very short pulses. In testing sheet insulation in mineral oil there is usually observed a higher strength for pulses of slow rise time and a still somewhat higher strength for dc voltages.

The trend in breakdown voltage with time, which is illustrated in Fig. 5-8, is typical of many solid insulation systems.

With ac voltages, the apparent strength declines steadily with time as a result of partial discharges[17] (in the ambient medium at the conductor or electrode edge). These penetrate the solid insulation. The discharges result from breakdown of the gas or liquid prior to the breakdown of the solid. See Fig. 5-8. The long-time strength with ac voltage declines, as shown in Fig. 5-9, and levels off at the partial discharge threshold (usually offset) voltage. Mica in particular, as well as other inorganic materials, is more resistant to such discharges. Organic resins should be used with caution where the ac voltage gradient is high and partial

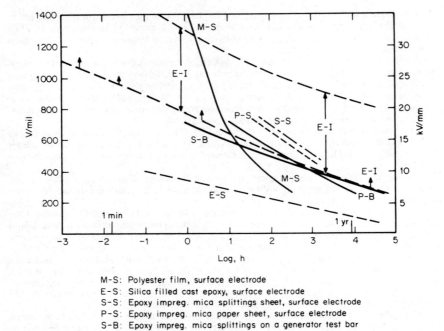

M-S: Polyester film, surface electrode
E-S: Silica filled cast epoxy, surface electrode
S-S: Epoxy impreg. mica splittings sheet, surface electrode
P-S: Epoxy impreg. mica paper sheet, surface electrode
S-B: Epoxy impreg. mica splittings on a generator test bar
P-B: Epoxy impreg. mica paper on a generator test bar
E-I: Silica filled cast epoxy, vacuum cast in electrodes

FIG. 5-9 Alternating-current dielectric strength of various insulations vs. time under stress.

discharges (corona) may be present. Since the presence of partial discharges on insulation is so important to the long-time voltage endurance, their detection and measurement has become a very important quality control and design tool. See Ref. 27 for an extensive review of this subject. If discharges continuously strike the insulation within internal cavities, or on the surface, the time to failure usually varies inversely as the applied frequency, since the number of discharges per unit time increases almost in direct proportion to the frequency. But in some cases, ambient conditions prevent continuous discharges.

With ac voltages, when there are partial discharges either at the surface or internally in cavities or at local points of high stress concentration, there is a steady decline in dielectric strength and eventual breakdown at electric stresses extending down to the partial discharge threshold stress. An empirical inverse power law dependence of failure time t on voltage gradient E, $t = AE^{-n}$, has been used to graph such data. A semilog relation between stress and log time has also been used. In the latter case, the data show an asymptotic approach to the partial discharge threshold stress as indicated in Fig. 5-9. The failure time appears to approach infinity at the partial discharge threshold. Below this level, discharges should have no effect on the failure time, but the material could of course fail by thermal or mechanical deterioration.

Figure 5-9 graphs voltage endurance data of several different types of insulation, including impregnated mica paper and mica splitting insulation, such as is used on high-voltage generator coils. Tests by Wichmann on this mica-filled insulation compare the insulation for two conditions: as sheets (S-S and P-S) with cylinder-to-plate electrodes on the surface (ASTM D2275), and wrapped on a high-voltage conductor bar, as the material would be used in service (S-B and P-B).

Figure 5-9 also compares tests on polyester organic film (M-S), which has a very high short time strength, but a steep decline in strength with time[17] when tested with surface electrodes (ASTM D2275).

GENERAL PROPERTIES OF INSULATING MATERIALS

When organic resin insulation is fabricated to avoid partial discharges using conductors or electrodes intimately bonded to the insulation, as in extruded polyethylene cables with a plastic semiconducting interface between the resin and the coaxial inner and outer metal conductors, respectively, the voltage endurance is greatly extended. Imperfections, however, in this "semicon"-resin interface, or at conducting particle inclusions in the resin, can lead to local discharges and the development of "electrical tree" growth.[28] Vacuum impregnating and casting electrodes or conductors into resin also tends to avoid cavities and surface discharges and greatly improves the voltage endurance at high stresses. This is illustrated by curves in Fig. 5-9 for silica-filled epoxy tested with vacuum cast-in electrodes (E-I) and with electrodes on the surface of a sheet of the epoxy (E-S).[29] The wide band of test values for the cast-in electrode system probably indicates varied degrees of success in avoiding partial discharge sites and imperfections. In this latter system, differences in the expansion coefficient of the resin and the metal conductor (or electrode) are important. In this case the percentage of silica filler in the resin was adjusted so that it matched the metal. See Ref. 28 for a more extensive review of "electrical tree" development.

In practical tests of the dielectric strength, the measured strength usually declines with increasing thickness of material. The breakdown gradient decreases approximately as the inverse half power of the thickness, or conversely, the breakdown voltage increases as the half power of the thickness. This is illustrated in Fig. 5-10. The value of the exponent may

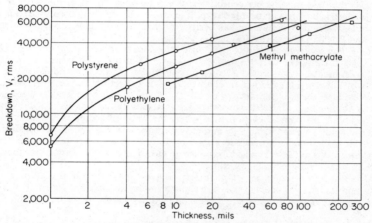

FIG. 5-10 Variation in breakdown strength with thickness, 60-Hz ac voltage, 2 kV/s rate of rise.

vary somewhat with conditions. An exception to this behavior is noted with sheet or wrapper insulation having defects, which are covered by increased thickness. Figure 5-10 illustrates this effect for 1-mil film, which has small defects, giving it a somewhat reduced strength.

The dc strength of solid insulation is usually higher and declines much less with time than the ac strength, since corona discharges are infrequent.

The dielectric strength is much higher where surface discharges are avoided and when the electric field is uniform. This can be achieved with solid materials by recessing spherical cavities into the material and using conducting paint electrodes.

The "intrinsic" electric strength of solid materials measured in uniform fields, avoiding surface discharges, ranges from levels of the order of 0.5 to 1 MV/cm for alkali halide crystals, which are about the lowest, upward to somewhat more than 10 MV/cm. Polymers and some inorganic materials such as mica, aluminum oxide, etc., have strengths of 2 to 20 MV/cm for thin films. The strength decreases with increasing thickness and with temperature above a critical temperature (which is usually from 1 to 100°C), below which the strength has a level value or a moderate increase with increasing temperature. Below the critical

ELECTRICAL ENGINEERING MATERIALS REFERENCE GUIDE

temperature, the breakdown is believed to be strictly electronic in nature and is constant or increases slightly with temperature. Above this temperature, it declines owing to dielectric thermal heating.

The breakdown voltage of thin insulating materials containing defects, which give the minimum breakdown voltage, declines as the area under stress increases. The effect of area on the strength can be estimated from the standard deviation S of tests on smaller areas by applying minimum value statistics:[10] $V_1 - V_2 = 1.497S \log (A_1/A_2)$, where V_1 and V_2 are the breakdown voltages of areas A_1 and A_2.

If the ac or dc conductivity of a dielectric is high, or the frequency is high, breakdown can occur as a result of dielectric heating, which raises the temperature of the material sufficiently to cause melting or decomposition, formation of gas, etc. This effect can be detected by measuring the conductivity as a function of applied electric stress. If the conductivity rises with time, with constant voltage, and at constant ambient temperature, this is evidence of an internal dielectric heating. If the heat transfer to the electrodes and ambient surroundings is adequate, the internal temperature may eventually stabilize, but if this heat transfer is inadequate, the temperature will rise until breakdown occurs. The criterion of this sort of breakdown is the heat balance between dielectric heat input and loss to the surroundings.

The dielectric heat input is given by the equation

$$\sigma E^2 = (5/9\varepsilon' f \tan \delta \times 10^{-12})E^2 \qquad W/cm^3 \qquad (5-19)$$

where E is the field in volts per centimeter. When this quantity is of the order of 0.1 or greater, dielectric heating can be a problem. It is much more likely to occur with thick insulation and at elevated temperatures. Dielectric breakdown is covered by Whitehead,[11] Peek,[12] and Roth[13] and in a review chapter by Mason.[14]

5-10. Water penetration into electrical insulation also degrades the dielectric strength by several mechanisms. The effect of water to increase the insulation conductivity, which has already been mentioned in Sec. 5-4, contributes thereby to a decreased dielectric strength, probably by a thermal breakdown mechanism. Another effect noticed recently, particularly in polyethylene cables, is the development of "water" or "electrochemical trees." Water (and/or a similar high dielectric constant chemical) can diffuse through polyethylene and collect at tiny hygroscopic inclusion sites, where the water or chemical is adsorbed. Then, the electric field causes an expansion and growth of the adsorbed water or chemical in the electric field direction. This may completely bridge the insulation or possibly increase the local electric stress at the site so as to produce an electric tree and eventual breakdown. See Ref. 28 for a review of this phenomenon.

5-11. Ionizing radiation, as from nuclear sources, may degrade insulation dielectric strength and integrity by causing polymer chain scission, and cracking of some plastics, as well as gas bubbles in liquids. Also the conductivity levels in solids and liquids are increased.

5-12. Arc Tracking of Insulation. High current arc discharges between conductors across the surface of organic resin insulation may carbonize the material and produce a conducting track. In the presence of surface water films, formed from rain or condensation, etc., small arc discharges form between interrupted parts of the water film, which is fairly conducting, and conducting tracks grow progressively across the surface, eventually bridging between conductors and causing complete breakdown. Materials vary widely in their resistance to tracking, and there are a variety of dry and wet tests for this property. Table 5-3, from a survey paper by Mandelcorn and Sommerman,[15] indicates the difference between materials and the correlation or lack of correlation between the tests. With proper fillers, some organic resins can be made essentially nontracking. Some resins such as polymethyl methacrylate and polymethylene oxide burst into flame under arcing conditions.

Review references are by Mandelcorn and Sommerman[15] and Olyphant.[16]

5-13. Thermal Aging. Organic resinous insulating materials in particular are subject in varying degrees to deterioration due to thermal aging, which is a chemical process involving decomposition or modification of the material to such an extent that it may no longer function adequately as the intended insulation.[25] The aging effects are usually accelerated by increased temperature, and this characteristic is utilized to make accelerated tests to failure or to an extent of deterioration considered dangerous. Such tests are made at appreciably higher than normal operating temperatures, if the expected life is to be several years or more, since useful accelerated tests should reasonably be completed in less than a year.

TABLE 5-3 Comparison of Tracking Resistance of Various Materials Measured with Seven Test Procedures

	A(15) D495-61 Equiv s/10 Liii(9)	B(25, 26) IEC 113; VDE Drops, 0.9 kV Nekal L .. (27)	C(9, 28) D2132-62T Std Dust-Fog h, 1.5 kV L...(9)	D(9, 27) Lin.-Accel. Dust-Fog H, 1.5 kV L...(27)	E(2) Differential Wet Track W·min W....(2, 32)	F(11) Inclined plane I V, kV V .(11)	G(30) Inclined plane II H, 2.5 kV L .(32, 33)
Column heading (reference)	A(15)	B(25, 26)	C(9, 28)	D(9, 27)	E(2)	F(11)	G(30)
Test method designation	D495-61	IEC 113; VDE	D2132-62T	Lin.-Accel.	Differential	Inclined	Inclined
Units	Equiv s/10	Drops, 0.9 kV Nekal	Std Dust-Fog h, 1.5 kV	Dust-Fog H, 1.5 kV	Wet Track W·min	plane I V, kV	plane II H, 2.5 kV
Symbol (data reference)	Liii(9)	L .. (27)	L...(9)	L...(27)	W....(2, 32)	V .(11)	L .(32, 33)
Polyvinyl chloride	0.5 Tr		0.5 Tr	0.5 Tr			
Phenolic laminate, paper base	0.5 Tr		0.5 Tr		0.2* Tr	1.5 Tr	
Epoxy resin, unfilled	1.7 Tr	60 + No Tr	0.5 Tr		1.6 Tr		
Polyamide resin	58 + Er	5 Tr	0.5 Tr		1.3 Int		
Silicone resin, glass cloth	54 Tr	10 Tr	1.0 Tr	1.0 Tr	1.8 Tr	1.5 Tr	
Melamine resin, glass cloth	47 Tr	6 Tr	3.5 Tr	2.5 Tr	2.3 Tr	2.3 Tr	0.2 Tr
Polyethylene	13 Tr	60 + No Tr	27 Tr	10 Er + Tr			
Polyester, glass mat, h-m-f, 1	25 Tr	No Tr	50 Tr	12 Tr	3.7† Tr	2 Tr	1.1 Tr
Polymethylmethacrylate	100 + Er	No Tr	90 Er	33 Tr	8.1 + Er	6 F	
Polypropylene	310 Er	No Tr	180 Er	40 Tr	8.1 + Er	3.8 Tr	
Epoxy resin, h-m-f	100 + Er	No Tr	200 Er				
Polyester, glass mat, h-m-f, 2	51 Tr	No Tr	350 Er	90 Tr	6.4 Tr	3 Tr	11 Tr
Butyl rubber, h-m-f	100 + Er	No Tr	450 Er	100 Er + Tr	8.1 + Er	6 F	
Silicone rubber, n-m-f	5 Tr		750 Er	120 Er + Tr		3.7 Tr	
Polytetrafluoroethylene	310 + Er	No Tr	2,700 Er	330 Tr	8.1 + Er	7 F	

h-m-f = hydrated-mineral filled; n-m-f = nonhydrated-mineral-filled; Tr = tracked; No Tr = no tracking; Er = eroded; Int = internal; C = carbonized; F = flame.

*Failed 1.3 W, 1 s.

†Failed 5.5 W, 18 s.

5-14 ELECTRICAL ENGINEERING MATERIALS REFERENCE GUIDE

Frequently other environmental factors influence the life, in addition to the temperature. These include presence or absence of oxygen, moisture, and electrolysis. Mechanical and electrical stress may reduce the life by setting a required level of performance at which the insulation must perform. If this level is high, less deterioration of the insulation is required to reach this level.

Sometimes a complete apparatus is life-tested, as well as smaller specimens involving only one insulation material or a simple combination of these in a simple model. New tests are being devised continually,[18] but there has been some standardization of tests by the IEEE and ASTM and internationally by the IEC.

It is important to note that frequently materials are assigned temperature ratings based on tests of the material alone. Often that material, combined with others in an apparatus or system, will perform satisfactorily at appreciably higher temperatures. Conversely, because of incompatibility with other materials, it may not perform at as high a temperature as it would alone. For this reason it is considered desirable to make functional operating tests on complete systems. These can also be accelerated at elevated temperatures and environmental exposure conditions such as humidification, vibration, cold-temperature cycling, etc., introduced intermittently.

The basis for temperature rating of apparatus and materials is discussed thoroughly in *IEEE Standards Publ.* 1. Tests for determining ratings are described in *IEEE Publs.* 98, 99, and 101.

5-14. Application of Electrical Insulation. In applying an insulating material it is necessary to consider not only the electrical requirements but also the mechanical and environmental conditions of the application. Mechanical failure often leads to electrical failure, and mechanical failure is frequently the primary cause for failure of an aged insulation.

The initial properties of an insulation are frequently more than adequate for the application, but the effects of aging and environment may degrade the insulation rapidly to the point of failure. Thus, the thermal and environmental stability should be considered of equal importance. The effects of moisture and surface dirt contamination should be particularly considered, if these are likely to occur.

The application of insulation to shipboard insulation and rotating machinery generally is reviewed by Moses.[19] Application to a variety of apparatus is covered by Jackson.[20] A reference book[21] by Clark surveys the properties of a wide variety of materials and their application characteristics.

References which should be consulted for further details include the following: The ASTM Standards on Insulating Materials[22] are continually revised, and a single-volume collection of the electrical tests and specifications is published every two years. "Progress in Dielectrics" reviews the literature of the field.[23] The annual "Digest of Literature on Dielectrics"[24] is a reference source for each year's published papers in classified form. The "Reports of the Annual National Research Council Conference on Electrical Insulation" and the reports of the approximately biennial NEMA-IEEE "Electrical Insulation Conference" offer collections of papers covering the recent developments.

5-15. General References on Insulating Materials

1. Smyth, C. P.: "Dielectric Behavior and Structure"; New York, McGraw-Hill Book Company, 1955.

2. von Hippel, A.: "Dielectric Materials and Applications"; Cambridge, Mass., and New York, The Technology Press of the Massachusetts Institute of Technology and John Wiley & Sons, Inc., 1954.

3. von Hippel, A.: "Dielectrics and Waves"; New York, John Wiley & Sons, Inc., 1954.

4. Tables of Dielectric Materials, Vols. I–VI, *MIT Lab. Insulation Research Tech. Repts.,* 1944–1958.

5. Miner, D. F.: "Insulation of Electrical Apparatus"; New York, McGraw-Hill Book Company, 1941.

6. MacInnes, D. A.: "The Principles of Electrochemistry"; New York, Reinhold Publishing Corporation, 1939.

7. Schwiger, A., and Sorensen, R. W.: "The Theory of Dielectrics"; New York, John Wiley & Sons, Inc., 1932.

8. Stratton, J. A.: "Electromagnetic Theory"; New York, McGraw-Hill Book Company, 1941.

9. Ollendorff, F.: "Potential Felder der Elektrotechnik"; Berlin, Springer-Verlag, OHG, 1932.

10. Weber, K. H., and Endicott, H. S.: Area Effect and Its Extremal Basis for the Electric Breakdown of Transformer Oil; *Trans. AIEE,* 1956, Vol. 5–III, p. 371.

GENERAL PROPERTIES OF INSULATING MATERIALS 5-15

11. Whitehead, S.: "Dielectric Breakdown of Solids"; New York, Oxford University Press, 1951.

12. Peek, F. W.: "Dielectric Phenomena in High Voltage Engineering"; New York, McGraw-Hill Book Company, 1929.

13. Roth, A.: "Hochspannungstechnik"; Vienna, Springer-Verlag, OHG, 1959.

14. Mason, J. H.: "Progress in Dielectrics"; London, Heywood & Co., 1959, Vol. 1, Chap. 1, Breakdown of Solid Dielectrics.

15. Sommerman, G. M. L.: Electrical Tracking Resistance of Polymers; *Trans. AIEE,* 1960, Vol. 70–III, pp. 69–74. Mandelcorn, L., and Sommerman, G. M. L. "Collected Papers of the 1963 NEMA-IEEE Electrical Insulation Conference"; Chicago, Ill.

16. Olyphant, M.: "Arc Resistance I & II"; *ASTM Bull.,* 1952, Vol. 181, p. 60, Vol. 185, p. 31.

17. Dakin, T. W., Philofsky, H. M., and Divens, W. C.: Effect of Electrical Discharges on the Breakdown of Solid Insulation; *Trans. AIEE,* 1954, Vol. 73–I, pp. 155–162.

18. A Bibliography on Testing of Insulating Materials and Systems for Thermal Degradation, *AIEE Spec. Publ.* S–87.

19. Moses, G. L.: "Electrical Insulation, Its Application to Shipboard Electrical Equipment"; New York, McGraw-Hill Book Company, 1951.

20. Jackson, W.: "The Insulation of Electrical Equipment"; London, Chapman & Hall, Ltd., 1954.

21. Clark, Frank M.: "Insulating Materials for Design and Engineering Practice"; New York, John Wiley & Sons, Inc., 1962.

22. ASTM Standards, Pt. 29, Electrical Insulating Materials, 1964 and succeeding even years.

23. "Progress in Dielectrics"; 1959–1962, London and New York, Heywood & Co. and Academic Press, Inc., Vols. 1 to 4 issued.

24. "Digest of Literature on Dielectrics"; National Academy of Sciences—National Research Council, Conference on Electrical Insulation, Washington, D.C.

25. Dakin, T. W.: Electrical Insulation Deterioration Treated as a Chemical Rate Phenomenon; *Trans. AIEE,* 1948, Vol. 67, p. 113.

26. Field, R. F.: The Formation of Ionized Water Films on Dielectrics under Conditions of High Humidity; *J. Appl. Phys.,* 1946, Vol. 17, p. 318.

27. "Engineering Dielectrics," Vol. 1, "Corona Measurement and Interpretation," R. Bartnikas and E. J. McMahon, (eds); ASTM, Philadelphia, Pa., 1979.

28. "Engineering Dielectrics," Vol. 2, "Electrical Properties of Solid Insulating Materials, Molecular Structure and Electrical Behavior," R. Bartnikas and R. M. Eichhorn, (eds); ASTM, Philadelphia, Pa., 1983.

29. Studniarz, S. A., and Dakin, T. W.: "The Voltage Endurance of Cast Epoxy Resin—II"; Conference Record of 1982 IEEE International Symposium on Electrical Insulation, 82CH1780-6-EI, June 7–9, Philadelphia, pp. 19–25.

30. Wichmann, A., and Gruenwald, P.: "Proc. IEEE International Symposium on Electrical Insulation"; IEEE Conference Record, 76CH1088-4-EI, Montreal, 1976, pp. 88–92.

Chapter 6

THERMAL CONDUCTIVITY OF
ELECTRICAL INSULATING MATERIALS

By CHARLES A. HARPER

6-1 General 6-1
6-2 Basic Thermal-Conductivity Data 6-2
6-3 Use of Fillers 6-3
6-4 Thermal-Conductivity Measurements 6-8
6-5 References 6-8

6-1. General. One of the general characteristics of electrical insulating materials is that they are also good thermal insulating materials. This is true, in varying degrees, for the entire spectrum of insulating materials, including air, fluids, plastics, glasses, and ceramics. While the thermal insulating properties of electrical insulating materials are not especially important for electrical and electronic designs which are not heat-sensitive, modern designs are increasingly heat-sensitive. This is often because higher power levels are being dissipated from smaller part volumes, thus tending to raise the temperature of critical elements of the product design. This results in several adverse effects, including degradation of electrical

6-2   ELECTRICAL ENGINEERING MATERIALS REFERENCE GUIDE

performance and degradation of many insulating materials, especially insulating papers and plastics. The net result is reduced life and/or reduced reliability of the electrical or electronic part. To maximize life and reliability, much effort has been devoted to data and guidelines for gaining the highest possible thermal conductivity, consistent with optimization of product design limitations, such as fabrication, cost, and environmental stresses.[1,2,3,4,*] This section will present data and guidelines which will be useful to electrical and electronic designers in selection of electrical insulating materials for best meeting thermal design requirements. Also, methods of determining thermal conductivity K will be described.

6-2. Basic Thermal-Conductivity Data.   The thermal-conductivity values for a range of materials commonly used in electrical design are shown in Table 6-1. These data show the ranking of the range of materials, both conductors and insulating materials, from high to low. The magnitude of the differences in conductor and plastic thermal-conductivity values can be seen. Note that one ceramic, 95% beryllia, has a higher thermal-conductivity value than some metals—thus making beryllia highly considered for high-heat-dissipating designs which allow its use. Thermal-conductivity values for a range of plastic materials are shown in Table 6-2. Thermal conductivity is variously reported in many different units, and convenient conversions are shown in Table 6-3.

Values of thermal conductivity do not change drastically up to 100°C or higher, and hence only a single value is usually given for plastics. For higher-temperature applications,

---

*Superscripts refer to bibliography, Par. **6-5**.

TABLE 6-1   Thermal Conductivity of Materials Commonly Used for Electrical Design

Material	Thermal conductivity	
	W/(in)(°C)	Btu/(h)(ft)(°F)
Silver	10.6	241
Copper	9.6	220
Eutectic bond	7.50	171.23
Gold	7.5	171
Aluminum	5.5	125
Beryllia 95%	3.9	90.0
Molybdenum	3.7	84
Cadmium	2.3	53
Nickel	2.29	52.02
Silicon	2.13	48.55
Palladium	1.79	40.46
Platinum	1.75	39.88
Chromium	1.75	39.88
Tin	1.63	36.99
Steel	1.22	27.85
Solder (60-40)	0.91	20.78
Lead	0.83	18.9
Alumina 95%	0.66	15.0
Kovar	0.49	11.1
Epoxy resin, BeO-filled	0.088	2.00
Silicone RTV, BeO-filled	0.066	1.5
Quartz	0.05	1.41
Silicon dioxide	0.035	0.799
Borosilicate glass	0.026	0.59
Glass frit	0.024	0.569
Conductive epoxy	0.020	0.457
Sylgard resin	0.009	0.21
Epoxy glass laminate	0.007	0.17
Doryl cement	0.007	0.17
Epoxy resin, unfilled	0.004	0.10
Silicone RTV, BeO-filled	0.004	0.10
Air		0.016

THERMAL CONDUCTIVITY OF INSULATING MATERIALS

TABLE 6-2 Thermal Conductivity for Various Types of Plastics

Material	Specific gravity	Btu/(h)(ft²)(°F)(ft)*
Thermoplastics:		
Polyethylene	0.92–0.96	0.1–0.4
Polytetrafluoroethylene	2.15–2.25	0.1–0.2
Molded thermosets:		
Phenolic, wood-flour-filled	1.32–1.45	0.10–0.19
Phenolic, mineral-filled	1.65–1.92	0.24–0.34
Diallyl phthalate, acrylic fiber	1.31–1.45	0.18–0.19
Laminates, perpendicular-to-face:		
XXXP, paper-phenolic	1.3–1.4	0.04–0.12
G-7, silicone-glass	1.6–1.8	0.07–0.17
G-10, epoxy-glass	1.7–1.8	0.10–0.17
PTFE glass-cloth	2.1–2.2	0.02–0.05
Casting resins and foams:		
Epoxy, unfilled	1.16	0.13–0.20
Epoxy, 73% alumina by weight		0.82
Epoxy, 50–55% silica by weight	1.6–1.7	0.29–0.53
Epoxy, hollow phenolic spheres	0.86	0.16
Epoxy, hollow glass spheres	0.95	0.38
Polyester, unfilled	1.23	0.10–0.15
Polyester, 50% silica by weight	1.6	0.19
Polyurethane foam (10 lb/ft³)	0.16	0.02–0.03

*To obtain: Btu/(h)(ft²)(°F)(in), multiply by 12.

TABLE 6-3 Thermal-Conductivity Conversion Factors

	To			
From	$\dfrac{(cal)(cm)}{(s)(cm^2)(°C)}$	$\dfrac{(W)(cm)}{(cm^2)(°C)}$	$\dfrac{(W)(in)}{(in^2)(°C)}$	$\dfrac{(Btu)(ft)}{(h)(ft^2)(°F)}$
$\dfrac{(cal)(cm)}{(s)(cm^2)(°C)}$	1	4.18	10.62	241.9
$\dfrac{(W)(cm)}{(cm^2)(°C)}$	2.39×10^{-1}	1	2.54	57.8
$\dfrac{(W)(in)}{(in^2)(°C)}$	9.43×10^{-2}	3.93×10^{-1}	1	22.83
$\dfrac{(Btu)(ft)}{(h)(ft^2)(°F)}$	4.13×10^{-3}	1.73×10^{-2}	4.38×10^{-2}	1

such as with ceramics, the temperature effect should be considered. This is shown for ceramics in Fig. 6-1. In the case of ceramics, the composition of the ceramic also strongly influences thermal conductivity, as shown in Table 6-4 and Fig. 6-2.

In addition to bulk insulating materials, insulating coatings are frequently used. Thermal-conductivity values for coatings are given in Table 6-5.

6-3. Use of Fillers to Increase Thermal Conductivity. Many of the plastics used for electrical insulation are liquid casting resins, such as epoxies, polyesters, silicones, and urethanes. Being liquid, and easy to formulate, much work has been done with fillers to increase the thermal conductivity of these materials. As the many available unfilled casting resins are not broadly different in their thermal conductivities, it is evident that a study of fillers and filled compounds will be the key to obtaining the highest possible thermal conductivity.

A survey of fillers with high thermal conductivities shows the following thermal-conductivity data, expressed in calories per centimeter per second per square centimeter per degree Celsius: mica, 0.0012; sand, 0.0028; aluminum, 0.497; copper, 0.918.

It might be deduced from these data that the use of powdered metallic fillers such as copper or aluminum would, when compounded into a casting-resin system, yield a compound that would give the highest possible thermal conductivity. Such is not entirely the case, as can be seen from Table 6-6, which shows the thermal conductivity of a group of filled compounds. For instance, the compounds prepared with coarse-grain sand and tabular alumina have a slightly higher thermal conductivity than does the compound that uses fine-mesh aluminum as filler, despite the much better thermal conductivity of aluminum. On the other hand, the compound filled with 30-mesh aluminum has an appreciably improved thermal conductivity over the sand- and fine-mesh aluminum-filled compounds. This might raise the question of how one aluminum filler can be better than another, when they are both basically the same material. Also note that the copper-filled compound has a lower thermal conductivity than the 30-mesh aluminum compound, even though copper has a much higher thermal conductivity than does aluminum.

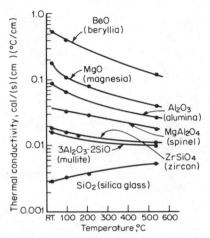

FIG. 6-1 Effect of temperature on thermal conductivity of ceramics (Ref. 5).

The reason for these variations, from what might be expected from a comparative study of filler thermal conductivities alone, lies in the particle type and size of the filler. This is because the particle type and size determine the amount of filler that can be mixed with a given resin and still give a compound of sufficiently low viscosity to be pourable and fluid for embedded-packaging applications. Also involved in the considerations is the specific gravity of the filler; it will be explained below that volume concentration of filler is more meaningful than weight concentration in determining how much the thermal conductivity can be increased. The end result is that maximum thermal conductivity is achieved through the use of the highest-thermal-conductivity filler that will allow the highest volume concentration of filler in the filled compound. This leads to what might be considered the bulk effect.

The Bulk Effect. Perhaps the primary reason why the sand-filled compound has a better thermal conductivity than the aluminum-filled compound in Table 6-6 is the bulk effect. This is best described by the statement that, in general, increase in thermal conductivity of a filled epoxy compound depends more on the quantity of filler added to the compound than on the type of filler used, so long as the filler is of the same particle type. It is observed from Table 6-6 that the mica-filled compound contains 40% of mica by weight, as does the aluminum-filled compound. As can be seen, the weight concentration being the same, the aluminum-filled material has a much higher thermal conductivity than does the mica-filled material. Note, however, that the sand-filled compound contains 70% of sand by weight; thus, there is a much higher concentration of filler in the sand-filled compound on the basis of both weight and volume.

TABLE 6-4 Effect of Alumina Content on the Thermal Conductivity of Alumina Substrates

Alumina, %	Thermal conductivity, cal/(s)(cm²)(°C/cm)	Change in thermal conductivity, %
99	0.070	
98	0.061	−13
96	0.043	−39
85	0.035	−50

THERMAL CONDUCTIVITY OF INSULATING MATERIALS   6-5

Practical considerations enter at this point. Fine-mesh fillers, such as fine mica powder and fine-mesh aluminum powder, give a high-viscosity compound with poor flow properties of 40% weight concentration. (Viscosity considerations are at 70°C for all cases given here, as the base epoxy viscosity is too high for use at room temperature.) However, a sand-filled resin is still very workable at a 70% weight concentration. Actually, a 40% concentration of fine-mesh aluminum filler makes the compound almost unworkable for embedment use, as the compound is approaching a paste at this point. On the other hand, sand can be used in concentrations up to 80% or so. The reason for this is that sand has a much larger particle size than mica or fine-mesh aluminum. Thus, results similar to those obtained with a high concentration of sand filler can also be obtained with a high concentration of such fillers as tabular alumina and other high-conductivity, large-grain-size fillers. An electrical grade of sand is usually much more economical, however. Although Table 6-6 shows that the copper-filled compound has a relatively high thermal conductivity, copper powder is not so conveniently used because of the higher density of the copper filler. The reason again is a practical one. It is often quite difficult to keep powdered copper in suspension in normal embedding compounds long enough for the compound to cure. The powdered copper will rapidly settle to the bottom. The lower the compound viscosity and the longer the pot life, the greater this settling problem becomes.

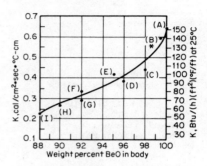

A. Theoretical density 100% BeO
B. Berlox® 99.5% BeO, 95% density
C. Commercial 98% BeO body
D. Commercial 96% BeO body
E. BeO body N*
F. BeO body A4
G. BeO body N4
H. BeO body O
I. BeO body N8

*Reference: J. Amer. Ceramic Soc., vol.33, No. 4, 1950

® Berlox is the registered trademark of National Beryllia Corp. (beryllium oxide)

FIG. 6-2 Thermal conductivity of beryllia as a function of BeO content at 25°C (Ref. 6).

Probably the most likely explanation for the above-described bulk effect is that the casting resins, being good insulators, form an insulating barrier between each particle and the thermal conductivity through the filler particles contributes less than the lower conductivity through the resin layers. Thus, overall thermal conductivity is best increased by adding more filler particles or, more specifically, by increasing the volume concentration of the filler in the compound. The result of the increased concentration is a shorter conductivity path through the resin, a longer conductivity path through the filler, and an overall reduction in the resistance to conductive heat flow through the material. Therefore, in general, improved thermal conductivity is obtained by adding an additional amount of filler material. The upper limitation on filler concentration is controlled by the practical working viscosity of the material. With any filler, at some concentration, the compound will be too thick for pouring and flowing around components in electronic packages.

Typical curves showing the effect of filler content on the thermal conductivity of liquid resin systems are shown in Fig. 6-3.

There is one case, however, where addition of more filler decreases the thermal conductivity. This is where the filler is a hollow, spherical type rather than a solid-particle type. Note from Table 6-6 the extremely low thermal conductivity of the spheroid-filled epoxy. The hollow, spherical fillers are usually filled with air or gas. Here, added spheroid filler essentially has the effect of reducing the conductivity through the insulator, when we consider air and gas to be optimum insulators. Thus, in this case, increasing the quantity of the spherical filler actually increases the resistance to conductive heat flow. Spherical fillers are generally used for reducing the specific gravity of an embedment compound, rather than increasing the thermal conductivity. Also shown in Table 6-6 for comparison are data for the thermal conductivity of an unfilled epoxy resin and for a low-density urethane foam. As would be expected, a very low-density foam material has an extremely low thermal conductivity.

TABLE 6-5 Thermal Conductivity of Various Plastic Coatings

Material	k value,* cal/(s)(cm²)(°C/cm) $\times 10^4$	Source of information
Unfilled plastics:		
Acrylic	4–5	[a]
Alkyd	8.3	[a]
Depolymerized rubber	3.2	H. V. Hardman, DPR Subsidiary
Epoxy	3–6	[b]
Epoxy (electrostatic spray coating)	6.6	Hysol Corp., DK-4
Epoxy (electrostatic spray coating)	2.9	Minnesota Mining & Mfg., No. 5133
Epoxy (Epon† 828, 71.4% DEA, 10.7%)	5.2	
Epoxy (cured with diethylenetriamine)	4.8	
Fluorocarbon (Teflon TFE)	7.0	Du Pont
Fluorocarbon (Teflon FEP)	5.8	Du Pont
Nylon	10	[d]
Polyester	4–5	[a]
Polyethylenes	8	[a]
Polyimide (Pyre-M.L. enamel)	3.5	[e]
Polyimide (Pyre-M.L. varnish)	7.2	[f]
Polystyrene	1.73–2.76	[g]
Polystyrene	2.5–3.3	[a]
Polyurethane	4–5	[n]
Polyvinyl chloride	3–4	[a]
Polyvinyl formal	3.7	[a]
Polyvinylidene chloride	2.0	[a]
Polyvinylidene fluoride	·3.6	[h]
Polyxylylene (Parylene N)	3	Union Carbide
Silicones (RTV types)	5–7.5	Dow Corning Corp.
Silicones (Sylgard types)	3.5–7.5	Dow Corning Corp.
Silicones (Sylgard varnishes and coatings)	3.5–3.6	Dow Corning Corp.
Silicone (gel coating)	3.7	Dow Corning Corp.
Silicone (gel coating)	7 (150°C)	Dow Corning Corp.
Filled plastics:		
Epon 828/diethylenetriamine = A	4	[b]
A + 50% silica	10	[b]
A + 50% alumina	11	[b]
A + 50% beryllium oxide	12.5	[b]
A + 70% silica	12	[b]
A + 70% alumina	13	[b]
A + 70% beryllium oxide	17.8	[b]
Epoxy, flexibilized = B	5.4	[i]
B + 66% by weight tabular alumina	18.0	[i]
B + 64% by volume tabular alumina	50.0	[i]
Epoxy, filled	20.2	Emerson & Cunning, 2651 ft
Epoxy (highly filled)	15–20	Wakefield Engineering Co.
Polyurethane (highly filled)	8–11	International Electronic Research Co.

*All values are at room temperature unless otherwise specified.

†Trademark of Shell Chemical Co., New York, N.Y.

[a]*Mater. Eng.,* Materials Selector Issue, vol. 66, no. 5, Chapman-Reinhold Publication, mid-October 1967.

[b]Wolf, D. C.: *Proc. Nat. Electron. and Packag. Symp.,* New York, June 1964.

[c]Lee, H., and Neville, K.: *Handbook of Epoxy Resins;* New York, McGraw-Hill Book Company, 1966.

[d]Davis, R.: *Reinf. Plast.,* October 1962.

[e]*Du Pont Tech. Bull.* 19, Pyre-M.L. Wire Enamel, August 1967.

[f]*Du Pont Tech. Bull.* 1, Pyre-M.L. Varnish RK-692, April 1966.

[g]Teach, W. C., and Kiessling, G. C.: *Polystyrene;* New York, Reinhold Publishing Corporation, 1960.

[h]Barnhart, W. S., Ferren, R. A., and Iserson, H.: *17th ANTEC of SPE,* January 1961.

[i]Gershman, A. J., and Andreotti, J. R.: *Insulation,* September 1967.

TABLE 6-6 Thermal Conductivity of Embedding Compounds Using Various Fillers

Filler	Filler in compound by weight, %	Thermal conductivity of filled compound, $W/(in^2)(°C)(in)$
Copper powder (Venus A, U.S. Bronze)	90	0.040
30-mesh aluminum	80	0.064
Fine-mesh aluminum	40	0.022
Coarse-grain sand	70	0.025
Tabular alumina	80	0.026
325-mesh mica	45	0.013
325-mesh silica	55	0.019
None (epoxy resin)*	0	0.005
Hollow phenolic spheres	15	0.003
Unfilled urethane foam, 5 lb/ft² density*		0.001

*Presented for comparison.
NOTE: $1 \text{ lb/ft}^2 = 4.882 \text{ kg/m}^2$.

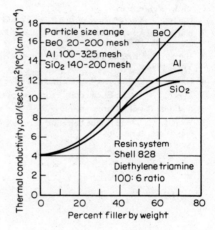

FIG. 6-3 Effect of filler content on thermal conductivity for three filled liquid-resin systems (Ref. 7).

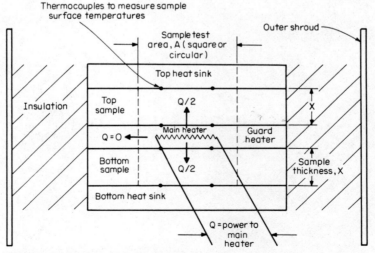

FIG. 6-4 Schematic assembly of guarded hot plate.

6-4. Thermal-Conductivity Measurements. The recognized primary technique for measuring thermal conductivity of insulating materials is the guarded-hot-plate method (ASTM C177). A schematic of the apparatus is shown in Fig. 6-4. The purpose of the guard heater is to prevent heat flow in all but the axial (up and down in the schematic) direction by establishing isothermal surfaces on the specimen's hot side. With this condition established and by measuring the temperature difference across the sample, the electrical power to the main heater area, and the sample thickness, the K factor can be calculated as

$$K = \frac{QX}{2A\,\Delta T} \tag{6-1}$$

Instruments are available for this test which use automatic means to control the guard temperature and record the sample ΔT. Unfortunately this test is fairly expensive.

Another technique uses a heat-flow sensor, which is a calibrated thermopile, in series with the heater, specimen, and cold sink. This method avoids the guard heater and requires only one specimen. This secondary technique is described in ASTM C518.

6-5. References

1. Harper, C. A.: *Handbook of Electronic Packaging,* McGraw-Hill Book Company, New York, 1969.

2. Harper, C. A.: *Handbook of Materials and Processes for Electronics,* McGraw-Hill Book Company, New York, 1970.

3. Harper, C. A.: *Handbook of Thick Film Hybrid Microelectronics,* McGraw-Hill Book Company, New York, 1974.

4. Harper, C. A.: *Handbook of Plastics and Elastomers,* McGraw-Hill Book Company, New York, 1975.

5. Lynch, J. F., et al.: Engineering Properties of Ceramics, *U.S. Dept. Comm. Bull.* AD803-765.

6. National Beryllia Corporation: National Beryllia Technical Bulletin on Berlox BeO.

7. Wolf, D. C.: Trends in the Selection of Liquid Resins for Electonic Packaging, *Proceedings, NEPCON,* New York, June 1964.

Chapter 7

INSULATED CONDUCTORS

By E. J. CROOP

7-1 Magnet Wire Insulation   7-1
7-2 Magnet Strip and Foil Conductors   7-2
7-3 Wire and Cable Insulation   7-2
7-4 References   7-7

7-1. Magnet-Wire Insulation. The term magnet wire includes an extremely broad range of sizes of both round and rectangular conductors used in electrical apparatus. Common round sizes for copper are AWG No. 42 (0.0025 in) to AWG No. 8 (0.1285 in). Ultrafine sizes of round wire, used in very small devices, range as low as AWG No. 60 for copper and AWG No. 52 for aluminum.

Approximately 30 different "enamels" are used commercially at present in insulating magnet wire. Enamel insulations generally are lowest in cost and best in space factor. The most widely used materials are based on polyvinyl acetals, polyesters, polyester amide-imides, and epoxy resins. The polyvinyl acetal and polyester types of materials possess good mechanical properties and good flexibility and perform well in automatic winding machines. Where low cost is important and winding conditions are not too severe, oleoresinous types and modified oleoresinous types are still used. Polyurethanes are employed where ease of solderability, without solvent or mechanical stripping, is required. These do not have high cut-through resistance, however. Epoxy enamels are used where resistance to chemicals and to moisture is important. Polyimide and other aromatic polymer types are employed for operation in the 200 to 220°C range. However, the aromatic types are among the most expensive of the enamels.

7-2   ELECTRICAL ENGINEERING MATERIALS REFERENCE GUIDE

Table 7-1 lists some of the commonly used enameled wires by temperature class. It should be understood that this temperature rating is based on a thermal test[1]* and does not include other environmental factors such as exposure to high humidity or use with a varnish which may impair its thermal stability. Cycling tests which include humidification greatly reduce the lives at temperature of many insulation systems capable of undergoing hydrolytic as well as thermal and oxidative degradation.

Table 7-2 lists some fibrous insulations commonly used for insulating magnet wire conductors.[2] Fibrous insulations are employed where positive separation and high reliability are required. These are generally higher in cost and poorer in space factor than enamel insulations.

7-2. Magnet-Strip and Foil Conductors. Magnet strip is a term generally employed to describe conductors, both copper and aluminum, with a width-to-thickness ratio greater than 50:1, while smaller ratios place the conductors in the category of rectangular magnet wire. If the thickness of the strip is less than 0.008 in, it is often referred to as "foil." Strip conductors are used in many electromagnetic devices including transformers, choke coils, welders, motor and generator fields, lift magnet coils, and electric clutches and brakes. Some of the advantages[3] of strip conductors are more uniform voltage distribution under surge or impulse conditions, better heat transfer, improved space factor, and stronger coil structure. For insulation, paper and polyester film (0.0005 in or less in thickness) have been used as interleaving materials. The width of the interleave is usually about 0.125 in wider than the strip. Other available interleaving materials are asbestos, polytetrafluoroethylene film, mica, and glass cloth. The most widely used insulation is enamel, which provides the best space factor and lowest cost. Many of the enamels used for insulating wire can be used also for strip, but the most widely used are epoxy and a modified polyester type. The enamel thickness generally ranges from 0.00025 to 0.0005 in on each side, or a build of 0.0005 to 0.0010 in.

7-3. Wire and Cable Insulation. Many materials are used in wire and cable insulations. Some general references (6,7,8,11) which review the recent state of the art are included at the end of this section.

Polyvinyl chloride (PVC) is widely used for primary insulation or jacketing on communication wires, control cable, bell wire, building wire, hookup wire, fixture wire, appliance cords, power cables, motor leads, etc. Many formulations are available, including those with flame resistance. Dielectric strength is excellent, and flexibility is very good. PVC is one of the most versatile of the lower-cost conventional insulations. A conductive PVC can be used for both shielding and jacketing.

Butyl rubber, when properly compounded, is characterized by excellent resistance to oxidation and aging, exceptional ozone resistance, and very good electrical properties. Resistance to moisture and chemicals is also very good. Applications include low- and high-power cables, apparatus leads, and control cables. Ethylene-propyelene terpolymer rubbers (EPT) are replacing butyl in some applications.

Neoprene has been used as a cable-jacketing material for more than 30 years. Its application over lead-sheathed and rubber-insulated cables has grown rapidly during this time. Although the electrical properties of neoprene are inferior to many other insulations, they are adequate for low-voltage work.

Nitrile-butadiene rubber (NBR) offers excellent resistance to oils and solvents but has low electrical resistivity.

Polyethylene (PE) is used in wires and cables in very large amounts. Polyethylene has excellent electrical properties plus good abrasion resistance and solvent resistance (at temperatures below 50°C). It is employed for hookup wire, coaxial cable, communication cable, line wire, lead wire, high-voltage cable, etc. The high-molecular-weight polyethylene has been the predominant cable insulation for many years. Also available for primary insulation are propylene/ethylene copolymers which are highly compressive and abrasion-resistant. Their electrical characteristics are very similar to polyethylene. These materials are covered in REA Specifications PE-22 and PE-23. Chemically cross-linked filled polyethylene is growing in usage for hookup and lead wire. Properties are similar to those of conventional PE except for a marked improvement in heat resistance, mechanical properties, aging charac-

---

*Superscripts refer to bibliography, Par. **7-4**.

TABLE 7-1 Some Typical Enamel-Insulated Wires

Temperature class, °C	Type	NEMA Std.	Federal Spec.* J-W-1177/sub:	Advantages
105	Acrylic	MW-37	/7, types SA	Resists refrigerants, low cost
105	Nylon	MW-6	/3, types N	Excellent windability, solderable
105	Oleoresinous (plain enamel)	MW-1	/1, types E	Low cost
105	Polyvinyl formal (Formvar)	MW-15, MW-18	/4, /16, types T	Excellent windability
105	Polyvinyl formal, isocyanate modified, for hermetic use (Formetic)	None	None	Excellent resistance to R-22
105	Polyvinyl formal with nylon overcoat	MW-17	/5, types TN	Improved windability over Formvar and resistance to hot solvents
105	Polyvinyl formal with polyvinyl butyral overcoat	MW-19	None	Can be self-bonded by heat or solvent
105	Cellulose lacquer	None	None	Can be applied in thin coatings to very fine wires. Bonds by solvent activation
105	Polyurethane	MW-2	/2, types U	Solderable. Can be coated at high speeds
130	Epoxy	MW-14, MW-9	None	Resistance to solvents, chemicals, hydrolysis
130	Epoxy with self-bonding overcoat	None	None	Used in making coils self-supporting. Eliminates varnish dip
155–180	Polyester	MW-5	/10, types L	Good windability, heat shock, thermal stability
200	Modified polyester	MW-74	None	Good windability, heat shock, thermal stability

TABLE 7-1 Some Typical Enamel-Insulated Wires (*Continued*)

Temperature class, °C	Type	NEMA Std.	Federal Spec.* J-W-1177/sub:	Advantages
180	Polyester amide-imide	MW-72	/12, types H	Good windability, heat shock, thermal stability, hermetic refrigerant resistance
>180	Polytetrafluoroethylene		None	Good thermal stability to 250°C. Solvent resistance
220	Polyimide, aromatic	MW-16, MW-20,	/15, /18, types M	Excellent thermal stability, solvent resistance, flexibility, scrape resistance, cut-through
220	Polamide-imide, aromatic (overcoat)	Used in MW-35	None, used in /13, /14, types K	Somewhat lower cost than polyimide at some sacrifice in properties. Currently used chiefly as an overcoating to improve windability and hermetic refrigerant resistance
220	Ceramic with polytetrafluoroethylene overcoat	None	None	High cut-through resistance
180	Ceramic with silicone overcoat	None	None	High thermal stability. Radiation resistant
>220	Ceramic with polyimide overcoat	None	None	High thermal stability
650	Ceramic	None	None	High thermal stability. Radiation resistant

*See complete Federal Spec. J-W-1177/GEN. It supersedes the former MIL-W-583C, March 6, 1963, and Int. Fed. Spec. J-W-001177 (Navy-Ships) September 21, 1973.

TABLE 7-2 Some Typical Fibrous-Covered Wires

Temperature class, °C	Type	NEMA Std.	Federal Spec. J-W-1177/sub:	Advantages
90–105	Paper	MW-31, MW-33	None	High electric strength when impregnated with oil (oil-filled transformers)
90–105	Cotton yarn	MW-11, MW-12	None	Positive separation of conductors, good varnish absorption and bonding, good abrasion resistance
—	Cellulose-acetate fiber	None	None	Can be self-bonded by solvent activation
130	Asbestos with organic bond	None	None	High compressive strength. Being replaced*
155	Glass fibers, organic bond	MW-41, MW-42	/19 types GV	Positive separation of conductors
155	Glass and polyester fibers, organic bond	MW-45, MW-46	/20, types DgV	Positive separation of conductors, greatly improved abrasion resistance over glass alone
200	Glass and polyester fibers, silicone bond	MW-47	/24, /25, types DgH	Positive separation. High-temperature capability
200	Polyamide fiber paper (Du Pont NOMEX)	MW-60	None	Tough, high-temperature and moisture-resistant paper
180	Asbestos with silicone bond	None	None	High compressive strength. Being replaced*
650	Glass fibers, organic bond with dispersed ceramic filler	None	None	Windability and high thermal stability

*Check with latest OSHA and EPA regulations.

TABLE 7-3 Some Properties of Common Wire-Insulation Materials

Physical properties	Unplasticized PVC	Plasticized PVC	Silicone rubber	Nylon	TFE fluorocarbon	FEP fluorocarbon	Polyethylene	Irradiated polyolefin
Specific gravity*	1.40	1.2–1.5	1.9	1.13	2.15	2.15	0.930	1.2
Tensile strength, lb/in^2	6,000–9,000	1,000–3,000	4,200	4,000–7,000	2,500	2,500	1,900–2,600	2,500
Elongation, %	2–40	200–400	...	300–600	200–300	250–330	200	250
Abrasion resistance*	Good	Good	Poor	Excellent	Good	Fair	Good	Good
Maximum continuous operating temperature, °C*	105	105	200	150	260	260	80	135
Melting point, °C	200	200	>375	300	327	275	120	Not thermoplastic
Flexibility at −180°C*	Cracks	Cracks	Cracks	...	Good	Good	Cracks	Fair
Cut-through resistance*	Good	Fair	Fair	Excellent	Fair	Fair	Good	Good
Flammability, in/min	Self-extinguishing	Self-extinguishing	10–78	Self-extinguishing	Nonflammable	Nonflammable	1.0	Self-extinguishing
Dielectric strength, V/mil (short time)	425–1,300	1,000	375	385	480	550	480	1,000
Dielectric constant, 1,000 c/s	5–10	2–4	4.2	4–10	2.0	2.1	2.3	2.6
Volume resistivity, Ω·cm	2×10^{12}	2×10^{14}	$>3 \times 10^{13}$	4.56×10^{16}	Approx. 10^{19}	$>2 \times 10^{13}$	10^{16}	$>10^{16}$

*These properties are of particular importance in aerospace applications.
NOTE: Data compiled by Hughes Aircraft Company.
1 lb/in^2 = 6.895 kPa.
SOURCE: From Adams, H. S.: *Electrotechnol.,* 1963, vol. 72, no. 3, p. 133.

INSULATED CONDUCTORS

teristics, and freedom from environmental stress cracking. Flame resistance can be provided by proper compounding. Uses include building wire, control cable, automotive wiring, and lead wire for motors and appliances. Polyethylene can also be cross-linked by irradiating the insulation on the wire. Advantages are similar to those of the chemically cross-linked material, but the process is generally limited to thin wall insulations, such as hookup wire (5 to 12 mils wall thickness). Foamed or cellular polyethylene represents a small but important part of the wire and cable insulation field. Dielectric constants of the order of 1.5 can be attained in this manner. In coaxial cables for community antenna television and closed-circuit television, the trend has been away from solid polyethylene to foamed polyethylene cable. Coaxial cables for military applications have either a solid low-density polyethylene insulation or polytetrafluoroethylene (TFE) in solid, semisolid, or tape-wrap form. PE and TFE have dielectric constants and dissipation factors which vary little over wide frequency and temperature ranges.

Polypropylene is the lightest of all plastics. It is similar to polyethylene in electrical properties but offers better heat resistance, tensile strength, and abrasion resistance. The material may be extruded, foamed, and made into cast and biaxially oriented films. Polypropylene film is being used as a cable wrap.

Fluorinated ethylene propylene (FEP) and TFE are used in critical applications where heat resistance, solvent resistance, and reliability are important, for example, wiring in jet aircraft, military electronic equipment, and supervisory wiring for steam-turbine generators.

Polyimide film laminated to FEP film (HF film) is a heat-sealable material which offers possibilities of savings in weight and space for wire insulation. It is rated for continuous use at 200°C.

Some properties of typical wire insulation materials are shown in Table 7-3 (from Ref. 5).

For a review of insulation for integrated microelectronic circuits see Ref. 6.

7-4. References on Insulated Conductors.

1. Tentative Method of Test for Relative Thermal Endurance of Film Insulated Round Magnet Wire, ASTM D2307 (based on former IEEE No. 57).

2. Saums, H. L.: Magnet Wire, Strip, Hollow Conductors and Superconductors; *Insulation Directory/ Encyclopedia Issue,* May 1965, pp. 332–352, Lake Publishing Corp., Libertyville, Ill.

3. Edge Conditioned Aluminum Strip Conductor, *Publ.* 731-1-8(5–665), 1965, Reynolds Metals Co., Richmond, Va.

4. Noble, M. G., and Savage, R. M.: A Status Report on Silicone Rubber for Wire and Cable Insulation; *Insulation,* November 1965, vol. 11, no. 12, p. 51.

5. Adams, H. S.: Problems in Insulated Wire and Cable in Space-Vehicle Systems; *Electrotechnol.,* 1963, vol. 72, no. 3, pp. 133–135.

6. Staff Report, "Where Does Insulation Technology Stand Today for Integrated Microelectronic Circuits?"; *Insulation,* September 1965, vol. 11, no. 10, p. 108.

7. Harper, C. A.: *Handbook of Wiring, Cabling, and Interconnecting for Electronics;* McGraw-Hill, 1972, Chaps. 5 and 7.

Chapter 8

INSULATING GASES

By T. W. DAKIN

8-1 General Properties of Gases 8-1
8-2 Dielectric Properties at Low Electric Fields 8-2
8-3 Dielectric Breakdown 8-3
8-4 Relative Dielectric Strengths of Gases 8-4
8-5 Corona and Breakdown in Nonuniform Fields between Conductors 8-6
8-6 Corona Discharges on Insulator Surfaces 8-7
8-7 Flashover on Solid Surfaces in Gases 8-7
8-8 High Pressure Gas Breakdown 8-8
8-9 Breakdown at High Frequency 8-9
8-10 Vacuum Breakdown 8-10
8-11 References 8-10

8-1. General Properties of Gases. A gas is a highly compressible dielectric medium, usually of low conductivity and with a dielectric constant only a little greater than unity, except at high pressures. In high electric fields the gas may become conducting as a result of impact ionization of the gas molecules by electrons accelerated by the field, and by secondary processes, which produce partial breakdown (corona) or complete breakdown. Conditions which ionize the gas molecules, such as very high temperatures and ionizing radiation (ultraviolet rays, x-rays, gamma rays, high-velocity electrons, and ions, such as alpha particles), will also produce some conduction in a gas.

The gas density d (grams per liter) increases with pressure p (torrs or millimeters of mercury) and gram-molecular weight M and decreases inversely with the absolute temperature

T (degrees Celsius + 273), according to the relation

$$d = \frac{M}{22.4} \frac{p}{760} \frac{273}{T} \qquad \text{g/L} \qquad (8\text{-}1)$$

The above relation is exact for ideal gases but is only approximately correct for most common gases. For exact values, tables should be consulted as well as more exact equations such as the Van der Waals equation.[1]*

If the gas is a vapor in equilibrium with a liquid or solid, the pressure will be the vapor pressure of the liquid or solid. The logarithm of the pressure varies as $-\Delta H/RT$, where ΔH is the heat of vaporization in calories per mole and R is the molar gas constant, 1.98 cal/(mol)(°C).[1] This relation also applies to all common atmospheric gases at low temperatures, below the points where they liquify.

8-2. Dielectric Properties at Low Electric Fields. *Dielectric Constant.* The dielectric constant k of gases is a function of the molecular electrical polarizability and the gas density. It is independent of magnetic and electric fields except when a significant number of ions is present. Values of the dielectric constant of some common gases are given in Table 8-1 at

TABLE 8-1   Dielectric Constant of Gases, 20°C, 1 atm gas

Air*	1.000536	He	1.000065
N_2	1.000547	A	1.000517
O_2	1.000495	SF_6	1.002084
CO_2	1.000921	H_2	1.000254

*Dry, CO_2 free.

atmospheric pressure and 20°C. The increment above unity $(k - 1)$ may be estimated[2] approximately by assuming that it varies proportionally to the pressure and inversely to the Kelvin temperature.

Conduction. The conductivity of a pure molecular gas at moderate electric stress and moderate temperature can be assumed, in the absence of any ionizing effect such as ionizing radiation, to be practically zero. Ionizing radiation induces conduction in the gas to a significant extent, depending on the amount absorbed and the volume of gas under stress.[3] The energy of the radiation must exceed, directly or indirectly, the ionization energy of the gas molecules and thus to produce an ion pair (usually an electron and positive ion). The threshold ionization energy is of the order of 10 to 25 electronvolts (eV)/molecule for common gases (10.86 eV for methyl alcohol, 12.2 for oxygen, 15.5 for nitrogen, and 24.5 for helium). Only very short wavelength ultraviolet light is effective directly in photoionization, since 10 eV corresponds to a photon of ultraviolet with a wavelength of 1240 Å. Since the photoelectric work function of metal surfaces is much lower (2 to 6 eV, for example, copper about 4 eV) the longer-wavelength ultraviolet commonly present is effective in ejecting electrons from a negative conductor surface. Such cathode-ejected electrons give the gas apparent conductivity.

High-energy radiation from nuclear disintegration is a common source of ionization in gases. Nuclear sources usually produce gamma rays of the order of 10^6 eV energy. Only a small amount is absorbed in passing through a low-density gas. A flux of 1 R/h produces ion pairs corresponding to a saturation current (segment ab of Fig. 8-1) of 0.925×10^{-13} A/cm^3 of air at 1 atm pressure, if all the ions formed are collected at the electrodes. The effect is proportional to the flux and the gas density.

At a voltage stress below about 100 V/cm, some of the ions formed will recombine before being collected and the current will be correspondingly less (segment oa of Fig. 4-53). Higher stresses do not increase the current if all the ions formed are collected. A very small current, of the order of 10^{-21} A/cm^3 of air, is attributable to cosmic rays and residual natural radioactivity.

---

*Superscripts refer to bibliography, Par. **8-11**.

Electrons (beta rays) produce much more ionization per path length than gamma rays, because they are slowed down by collisions and lose their energy more quickly. Correspondingly, the slower alpha particles (positive helium nuclei) produce a very dense ionization in air over a short range. For example, a 3-million-eV (MeV) alpha particle has a range in air of 1.7 cm and creates a total of 6.8×10^5 ion pairs. A beta particle (an electron) of the same energy creates only 40 ion pairs per centimeter and has a range of 13 m in air.

It should be noted that ionizing radiation of significant levels has only a small effect on gas dielectric strength. For example, the ionization current produced by a corona discharge from a needle point is typically much higher than that produced by a radiation flux of significant level, 10^{11} gamma photons per square centimeter.

Thermally induced conductivity occurs in gases, at very high temperatures, as a result of impact ionization by the very high velocity molecules in the gas. This ionization can be calculated from the Saha equation[4] if the ionization energy is known. Such conductivity, in air, becomes significant only above 2000°C. Introduction of quantities of "seed" atoms such as sodium and potassium, which have low ionization energies, has been used in MHD generators to increase the gas conductivity substantially at high temperatures. The chemical reactions in flames also produce significant quantities of ions, and these can carry currents.

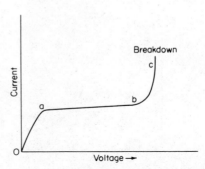

FIG. 8-1 Current-voltage behavior of a slightly ionized gas.

At temperatures increasing above 600°C, it has been shown that thermionic electron emission from negative conductor surfaces produces significant currents compared with levels typical of electrical insulation. The order of magnitude of this effect can be estimated from the Richardson thermionic-emission equation.[4]

Since the rate of production of ions by the various sources mentioned above is limited, the current in the gas does not follow Ohm's law, unless the rate of collection of the ions at the electrodes is small compared with the rate of production of these ions, as in the initial part of segment *oa* in Fig. 8-1.

8-3. Dielectric Breakdown. *Uniform Fields.* The dielectric breakdown of gases is a result of an exponential multiplication of free electrons induced by the field. It is indicated by segment *bc* of Fig. 8-1. It is generally assumed that the initiation of breakdown requires only one electron. However, if only a few electrons are present prior to breakdown, it is not easily possible to measure the trend of current shown in Fig. 8-1. If the breakdown is completed between metal electrodes, the spark develops extremely rapidly into an arc, involving copious emission of electrons from the cathode metal and, if the necessary current flow is permitted, vaporization of metal from the electrodes. Table 8-2 gives the dielectric strength of typical gases. References 17–22 are general references on gas breakdown.

In uniform electric fields breakdown occurs at a critical voltage which is a function of the product of the pressure p and spacing d (Paschen's law), as illustrated in Fig. 8-2 for

TABLE 8-2 Relative Dielectric Strengths of Gases
(0.1 in gap)

Air	0.95	CF_4	1.1
N_2	1.0	C_2F_6	1.9
CO_2	0.90	C_3F_8	2.3
H_2	0.57	C_4F_8 cyclic	2.8
A	0.28	CF_2Cl_2	2.4
Ne	0.13	C_2F_5Cl	2.6
He	0.14	$C_2F_4Cl_2$	3.3
SF_6	2.3–2.5		

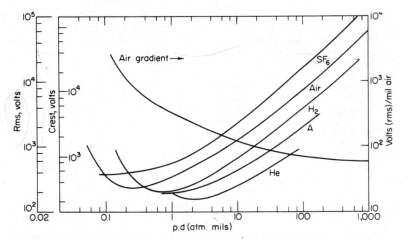

FIG. 8-2 Pressure-spacing dependence of the dielectric strength of gases (Paschen's curves).

several gases at 20°C. New international summary curves have been published[23] by CIGRE for air, N_2, and SF_6, and curves for H_2, CO_2, and He are given in Ref. 24.

It would be more accurate to consider the gas density-spacing product, since the dielectric strength varies with the temperature only as the latter affects the gas density. It will be noted that the electric field at breakdown decreases as the spacing increases. This is typical of all gases and is due to the fact that a minimum amount of multiplication of electrons must occur before breakdown occurs. A single electron accelerated by the field creates an avalanche which grows exponentially as $e^{\alpha x}$, where x is the distance and α is the Townsend ionization coefficient (electrons formed by collision per centimeter), which increases rapidly with electric field. At small spacings α and the field must be higher for sufficient multiplication. In divergent electric fields or large spacings, it has been found that when the integral $\int \alpha_E \, dx$ increases to about 18.4 (10^8 electrons), sufficient space charge develops to produce a streamer type of breakdown. It seems to be apparent that the final step in gas breakdown before arc development is the development of a branched filamentary streamer which proceeds more easily from the positive electrode toward the negative electrode.

The dielectric strength of air for larger sphere gaps is given in Table 8-3, selected from IEEE Standard 4 (revision of AIEE Standard 4). These values are used as voltage standards, but they indicate also the trend of the breakdown stress downward with increasing spacing. They also indicate that the impulse strength is almost indentical with the crest 60-Hz strength for smaller spacings but is a little greater for larger spacings. The positive (high-terminal) dc strength is the same as the positive-impulse strength. The higher values than the crest 60-Hz on the larger spacings are because of a slight asymmetry of the field due to the ground plane.

Pressure and (moderate) temperature affect the dielectric strength only as they affect the gas density according to Paschen's law (Fig. 8-2). IEEE Standard 4 gives correction factors for the relative pressure effect on sphere gap breakdown (see Table 8-4).

8-4. Relative Dielectric Strengths of Gases. The relative dielectric strength, with few exceptions, tends upward with increasing molecular weight. There are a number of factors other than molecular or atomic size which influence the retarding effect on electrons. These include ability to absorb electron energy on collision and trap electrons to form negative ions. The noble atomic gases (helium, argon, neon, etc.) are poorest in these respects and have the lowest dielectric strengths. Table 8-2 gives the relative dielectric strengths of a variety of gases at 1 atm pressure at a $p \cdot d$ value of 1 atm $\times$ 0.25 cm. The relative strengths vary with the $p \cdot d$ value, as well as gap geometry, and particularly in divergent fields where corona begins before breakdown. It is best to consult specific references with regard to divergent field breakdown values.

TABLE 8-3 Sphere Gap with One Sphere Grounded*

Peak values of disruptive-discharge voltages in kilovolts (50 % values for impulse tests)
Valid for:
  alternating voltages
  full negative standard impulses and impulses with longer tails
  direct voltages of either polarity
Atmospheric reference conditions: 25°C and 101.3 kN/m² (760 mm Hg)

Sphere-gap spacing, cm	Sphere diameter, cm							
	6.25	12.5	25	50	75	100	150	200
1	31.4	31.2						
2	57.5	58.0						
3	78.0	84.0	84.5					
4	(93.5)	106	110					
5	(105)	127	135	136	136			
6	(114)	144	158					
8		(171)	203					
10		(192)	240	259	260	262	262	262
12		(208)	271					
15			(309)	367	380	383		
20			(362)	452	483	500	500	500
25			(393)	520	575	605		
30				(575)	655	700	730	735
40				(660)	(785)	862	940	960
50				(720)	(880)	1000	1110	1160
75					(1025)	(1210)	1420	1510
100						(1340)	(1630)	1810
130							(1840)	(2070)
150							(1930)	(2210)
180								(2370)
200								(2450)

*Condensed from Table 1 in IEEE No. 4, 1969.
NOTE: The figures in parentheses, which are for spacings of more than $0.5D$, will be within ±5% if maximum clearances are met. On errors for direct current see paragraph 2.5.2.2 in IEEE No. 4.
  For full positive standard impulses and impulses with longer tails, the values are zero to 7% higher, depending on the gap. For those values and for intermediate gaps, consult IEEE No. 4, 1969.

TABLE 8-4 Air-Density Correction Factors for Sphere Gaps

Relative air density	Sphere diameter, cm		
	6.25	12.5	25
0.50	0.547	0.535	0.527
0.55	0.595	0.583	0.575
0.60	0.640	0.630	0.623
0.65	0.686	0.677	0.670
0.70	0.732	0.724	0.718
0.75	0.777	0.771	0.766
0.80	0.821	0.816	0.812
0.85	0.866	0.862	0.859
0.90	0.910	0.908	0.906
0.95	0.956	0.955	0.954
1.00	1.000	1.000	1.000
1.05	1.044	1.045	1.046
1.10	1.090	1.092	1.094

8-5. Corona and Breakdown in Nonuniform Fields between Conductors. In nonuniform fields, when the ratio of spacing to conductor radius of curvature is about 3 or less, breakdown occurs without prior corona. The breakdown voltage is controlled by the integral of the Townsend ionization coefficient α across the gap.[6] At larger ratios of spacing to radius of curvature, corona discharge occurs at voltage levels below complete gap breakdown, as shown in Fig. 8-3, Ref. 28.

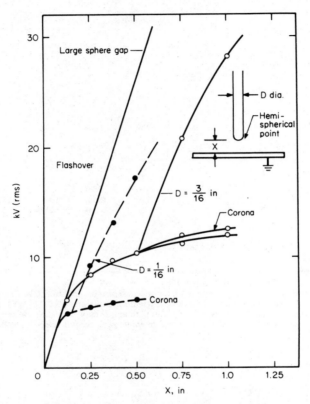

FIG. 8-3 Alternating current breakdown and corona starting voltages for a hemispherical point to plane in air as a function of gap length (in air). See Ref. 28.

According to Peek,[7] corona in air at atmospheric pressure occurs before breakdown when the ratio of outer to inner radius of coaxial electrodes exceeds 2.72, or where the ratio of gap to sphere radius beween spheres exceeds 2.04. These discharges project some distance from the small-radii conductor but do not continue out into the weaker electric field region, until a higher voltage level is reached.

Such partial breakdowns are often characterized by rapid pulses of current and radio noise. With some conductors at intermediate voltages between onset and complete breakdown, they blend into a pulseless glow discharge around the conductor. When corona occurs before breakdown, it creates an ion space charge around the conductor, which modifies the electric field, reducing the stress at sharp conductor points in the intermediate voltage range. At higher voltages, streamers break out of the space-charge region and cross the gap.

The surface voltage stress at which corona begins increases above that for uniform field breakdown stress, since the field to initiate breakdown must extend over a finite distance. An empirical relation developed by Peek[7] is useful for expressing the maximum surface stress for corona onset in air for several geometries of radius r cm:

For concentric cylinders: $\quad E = 31\delta(1 + 0.308/\sqrt{\delta r}) \quad$ kV/cm $\qquad$ (8-1)

For parallel wires: $\quad E = 29.8\delta(1 + 0.301/\sqrt{\delta r}) \quad$ kV/cm $\qquad$ (8-2)

For spheres: $\quad E = 27.2\delta(1 + 0.54\sqrt{\delta r}) \quad$ kV/cm $\qquad$ (8-3)

where δ is the density of air relative to that at 25°C and 1 atm pressure.

8-6. Corona Discharges on Insulator Surfaces. It has been shown by a number of investigators that the discharge-threshold voltage stress on or between insulator surfaces is the same as between metal electrodes.[8] Thus, the threshold voltage for such discharges can be calculated from the series dielectric-capacitance relation for internal gaps of simple shapes, such as plane and coaxial gaps, insulated conductor surfaces, hollow spherical cavities, etc.

The corona-initiating voltage at a conductor edge on a solid barrier depends on the electric stress concentration and generally on the ratio of the barrier thickness to its dielectric constant[9] (see Fig. 5-7), except with low surface resistance. Any absorbed water or conducting film raises the corona threshold voltage by reducing stress concentration at the conductor edge on the surface.

It is sometimes possible to overvolt such gaps considerably prior to the first discharge, and the offset voltage may be below the proper voltage due to surface-charge concentration. With ac voltages, pulse discharges occur regularly back and forth each half cycle, but with dc voltage the first discharge deposits a surface charge on the insulator surface which must leak away, before another discharge can occur. Thus, corona on or between insulator surfaces is very intermittent with steady dc voltages, but discharges occur when the voltage is raised or lowered.

8-7. Flashover on Solid Surfaces in Gases. As has been mentioned in the previous section on partial discharges, the breakdown in gases is influenced by the presence of solid insulation between conductors. This insulation increases the electric stress in the gas. A particular case of this is the complete breakdown between conductors across or around solid insulator surfaces. This can occur when the conductors are on the same side of the insulation or on opposite sides. A significant reduction in flashover voltage can occur whenever a significant part of the electric field passes through the insulation. The reduction is influenced by the percentage of electric flux which passes through the solid insulation and the dielectric constant of the insulation.

In the application of insulation in the outdoor environment, such as on transmission and distribution lines, or inside in severely polluted or wet locations, it is also important to recognize that conducting layers on the surface (such as from rain and fog with dissolved salts) can greatly reduce the surface flashover voltage in air (or any gas). This is illustrated by Fig. 8-4 for an epoxy strip.[29] The flashover voltage reduction increases with increasing

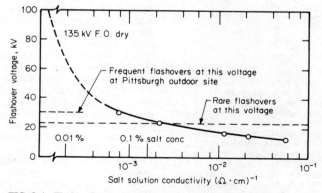

FIG. 8-4 Flashover voltage of a 15-in long × 2-in wide insulator strip in a salt-solution fog.

surface conductivity. Suspension or post insulators and outdoor bushings are similarly affected as is the flat strip of Fig. 8-4. Reference 25 outlines standard methods of polluting insulator surfaces with controlled conducting layers to assess the flashover voltage in service. Several methods are used: salt fog exposure, clean fog with preapplied conducting coatings, and others. Nonwettable coatings such as silicone grease (which must be periodically replaced) help to maintain a higher flashover voltage in seacoast and polluted areas. More permanent silicone coatings are being tested. Cast epoxy and silicone insulators and bushings made with resins which are nontracking and durable (which are essential properties in outdoor and polluted applications) have performed as well as, or better than, ceramic or glass insulators in seacoast applications.

8-8. High-Pressure Gas Breakdown. The dielectric strength of gases can be increased very considerably by increasing the pressure (and hence the density). At moderate pressures the increase in strength is slightly less than proportional to the pressure. At higher pressures the increase becomes appreciably less than proportional to the pressure, as indicated[5] in Fig. 8-5. In several cases such as CO_2, N_2, SF_6, and hexane vapor, the compressed gas strength[6] has been shown to approach that of the pure liquid. At very high pressures and the corresponding high stresses for breakdown, the breakdown voltage declines below that predicted by Paschen's law, Fig. 8-2. The departure from Paschen's law seems to begin for gases at stresses on the order of 2×10^5 volts/per centimeter, and the breakdown level becomes much more sensitive to surface roughness and conducting particles.

With nonuniform fields when corona occurs before breakdown, maxima are observed in the breakdown voltage-pressure curve for electronegative gases like SF_6.[26] With increasing pressure and an electric stress concentration point, the lower pressure part of the curve increases somewhat like that for the fixed wire particle in Fig. 8-6, but as the pressure is increased still further, the breakdown curve goes over a maximum and decreases toward a level nearly the same as the partial discharge (corona) threshold.

It should be noted that the high maximum in breakdown voltage shown in Fig. 8-6 for the fixed particle projection from the central conductor is typical for points in SF_6 and other electron-attaching gases. In this pressure region, corona discharges from a point create a cloud of ions about the point which electrostatically shield the point, raising the breakdown voltage with ac and 60 Hz. Because the ion space charge takes some time to develop, the impulse voltage breakdown is lower than the 60-Hz voltage breakdown in the pressure range. This is an important consideration in practice. This phenomenon also influences the breakdown with moving particles, which have a lower breakdown in this region, as illustrated in the Fig. 8-6.

The increased use of high-pressure gas (particularly SF_6) insulation for enclosed coaxial transmission lines, substations, and power circuit breakers has led to increased problems with conducting particles in the system.[27] These may drastically reduce the dielectric strength, as is indicated by the data in Fig. 8-6. Three situations are shown: (1) a fixed particle projecting from the central conductor, multiple (16) free particles with (2) the test voltage raised slowly and (3) the test voltage held for 3 min. Delayed breakdown may occur in gas insulation with particles. See Ref. 27 for more details on the effect of particles. Free particles will be moved about by the electric field, producing greatly increased electric stress concentration, partial discharges, and reduced breakdown voltage. The effect increases greatly with the high applied average electric stresses used in high gas pressure systems.

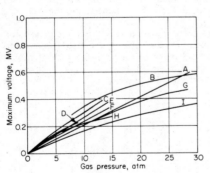

FIG. 8-5 Breakdown of N_2, CO_2, and air at high-pressure, 12.7-mm gap. Philip (*Trans. AIEE*, 1963, vol. 82, p. 356): 64-mm sphere facing negative high-voltage terminal in N_2 + CO_2 (50%) (*A*); Kusko: Uniform field gap in N_2 + CO_2 (*B*); Trump, Stafford, and Cloud: Uniform field gap in air (*C*); Ganger: 50-mm-diameter sphere-to-plane gap in N_2 (*D*); Finkelmann: Uniform field gap in N_2 (*E*); Finkelmann: Uniform field gap in CO_2 (*F*); Trump, Cloud, Mann, and Hanson (*Elec. Eng.*, 1950, vol. 69, p. 961): Uniform field gap in N_2 + CO_2 (*G*); Palm: Uniform field gap in N_2 (*H*); Howell: Uniform field gap in air (*I*).

INSULATING GASES

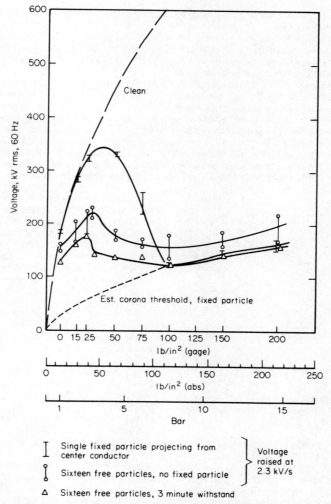

FIG. 8-6 Effect of wire particles (0.64 × 0.45 cm diameter) on the breakdown voltage of compressed SF_6 in a coaxial system (7.6 cm inside diameter × 25 cm outside diameter) over a range of pressures.

One corrective measure, which was not used in Ref. 27 tests, is the introduction of "particle traps," which are regions of low electric stress arranged so that moving particles will eventually fall into them and remain "trapped." Conductor particles may also cling to solid insulator surfaces which support high-voltage conductors and thereby produce stress concentration and reduced breakdown voltage.

8-9. Breakdown at High Frequency. The ac dielectric strength of gases declines only slightly (6 to 10% at 600 kHz) as the frequency is increased, until the time of a half cycle is about the same as the transit time, first of positive ions, and then of electrons across the gap.[10] At these critical frequencies (10^5 to 10^7 Hz) small maxima have been observed. Above the critical frequency, cumulative ionization occurs in the gap, and there is a sharp drop in breakdown voltage. At these high frequencies, the breakdown voltage is set by the equilibrium between production of electrons by electron impact ionization and loss by diffusion to the walls or electrodes.

8-10. Vacuum Breakdown. When the pressure and gas density in a system are so low that the electron mean free path is much larger than the spacing of conductors, electron multiplication by impact ionization of the gas molecules cannot take place. This occurs at pressures well below the Paschen's minima.

In the absence of direct gaseous ionization, breakdown can occur, at high stresses, from electrode effects. While the exact mechanism of vacuum breakdown has not been determined, there are several phenomena which can lead to breakdown. One of these is cathode field emission, which may be enhanced by imperfections, in or on the cathode surface, which increase the local stress, or even heating by the high current density. Steady cathode emission currents, which can lead to breakdown at elevated stresses, have been observed.

Another process which also seems likely to occur is a cathode-anode regeneration process of elementary particles. Electrons strike the anode with enough energy to create photons and positive ions which return to the cathode to generate more electrons and ions, etc.

At larger spacings, breakdown seems to be controlled by the total voltage rather than the gradient. The breakdown voltage increases approximately as the square root of the spacing. One mechanism which can account for this, together with other aspects of vacuum breakdown, is the Cranberg clump hypothesis, which presumes that a microscopic particle of many atoms is accelerated by the field from one electrode to the other, gaining enough kinetic energy to vaporize itself and some atoms from the electrode when it strikes the electrode. The vapor formed by this impact then leads to breakdown by a gas discharge process. Figure 8-7 indicates the range of breakdown voltages of vacuum, from a review paper by Hawley.[11] Breakdown in vacuum is very sensitive to residual particulate matter on the electrodes or in the system. Frequently, initial breakdown values on fresh systems are quite low, and electrodes can be "conditioned" to higher breakdown levels by repeatedly breaking down the system with limited current discharges.[13-16]

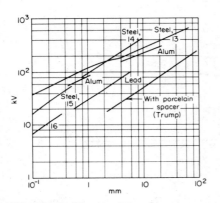

FIG. 8-7 Breakdown voltage of vacuum gaps (numbers correspond with those in bibliography).

Supporting insulators between electrodes in vacuum may reduce the breakdown voltage drastically below the level of breakdown in clear gaps.[12] It has been shown that flashover of such insulators in vacuum is initiated at the contact between the insulator and the cathode. If this region is shielded from the field and the insulator properly shaped, much higher breakdown voltage values can be obtained.

8-11. References on Insulating Gases

1. MacDougall, F. H.: "Physical Chemistry"; New York, The Macmillan Company, 1952.
2. Smyth, C.: "Dielectric Behavior and Structure"; New York, McGraw-Hill Book Company, 1955.
3. Curran, S. C., and Craggs, J. D.: "Counting Tubes"; London, Butterworth Scientific Publications, 1949.
4. Cobine, J. D.: "Gaseous Conductors"; New York, Dover Publications, Inc., 1958.
5. Trump, J. G., Cloud, R. W., Mann, J. G., and Hanson, E. P.: Influence of Electrodes on D-C Breakdown in Gases at High Pressures; *Electr. Eng.,* 1950, vol. 69, p. 961. Philp, S. F. Compressed Gas Insulation in the Million Volt Range, SF_6, N_2, CO_2; *Trans. IEEE,* 1963, vol. 82, p. 356.
6. Loeb, L. B.: Electrical Breakdown of Gases, "Encyclopedia of Physics"; Berlin, Springer-Verlag, OHG, 1956, vol. XXII.
7. Peek, F. W.: "Dielectric Phenomena in High Voltage Engineering"; New York, McGraw-Hill Book Company, 1929.
8. Hall, H. C., and Russek, R. M.: Discharge Inception and Extinction in Dielectric Voids; 1954, *Proc. IEE,* vol. 101-2, p. 47.

INSULATING GASES 8-11

9. Dakin, T. W., Philofsky, H., and Divens, W.: Effect of Electrical Discharges on the Breakdown of Solid Insulation; *Trans. AIEE*, 1954, vol. 73–I, pp. 155–162.

10. Brown, S. C.: Breakdown in Gases, Alternating and High Frequency Fields, "Encyclopedia of Physics"; Berlin, Springer-Verlag, OHG, 1956, vol. XXII.

11. Hawley, R.: Vaccum as an Insulator; *Vacuum*, 1960, vol. 10, p. 310.

12. Kofoid, M. J.: Effect of Metal-Dielectric Junction Phenomena on High Voltage Breakdown over Insulators in Vacuum; 1960, *Trans. AIEE*, vol. 79–III, p. 999.

13. Trump, J. G., and Van de Graaf, R.: The Insulation of High Voltages in Vacuum; *J. Appl. Phys.*, 1947, vol. 18, p. 327.

14. Slivkov, I. N.: Mechanism for Electrical Discharge in Vacuum; *Sov. Phys. Tech. Phys.*, 1957, vol. 2, p. 1919.

15. Denholm, A. S.: The Electrical Breakdown of Small Gaps in Vacuum; *Can. J. Phys.*, 1958, vol. 36, p. 476.

16. Leader, D.: Electrical Breakdown in Vacuum; *Proc. IEE*, 1953, vol. 100–2A, p. 138.

17. Meek, J. M., and Craggs, J. D.: "Electrical Breakdown of Gases"; New York, Oxford University Press, 1953.

18. Llewellyn Jones, F.: "Ionization and Breakdown in Gases"; London, Methuen & Co., Ltd.

19. Ganger, B.: "Der Elektrische Durchschlag von Gasen"; Berlin, Springer-Verlag, OHG, 1953.

20. Gas Discharges I, "Encyclopedia of Physics"; Berlin, Springer-Verlag, OHG, 1956, vols. XXI and XXII.

21. Dakin, T. W., and Berg, D. Theory of Gas Breakdown, chapter in "Progress in Dielectrics"; London and New York, Heywood & Co., Ltd., and Academic Press, Inc., 1962, vol. 4.

22. Roth, A.: "Hochspannungstechnik," 4th ed.; Berlin, Springer-Verlag, OHG, 1959.

23. Dakin, T. W., with German and French members of CIGRE Group 15-03: Breakdown of Gases in Uniform Fields—Paschen Curves for Nitrogen, Air, and SF_6; *Electra* (published by CIGRE, Paris), No. 32, p. 61, January 1974.

24. Winkelnkemper, H., Krasucki, Z., Gerhold, J., and Dakin, T. W.: Breakdown of Gases in Uniform Fields, Paschen Curves for Hydrogen, Carbon Dioxide and Helium; *Electra*, No. 92, p. 67, May 1977 (published together with Ref. 23 as a booklet by CIGRE, Paris, France.)

25. "Artificial Pollution Tests on High Voltage Insulators to Be Used on A-C Systems"; IEC Standard 507, 1975.

26. Works, C. N. and Dakin, T. W.: Dielectric Breakdown of SF_6 in Non-uniform Fields; 1953, *Trans. AIEE*, vol. 72, pt. 1, pp. 682–687.

27. Wootton, R. E., Cookson, A. H., Emery, F. T., and Farish, O.: "Investigation of High Voltage Particle Initiated Breakdown in Gas Insulated Systems"; Electric Power Research Institute Report EL-1007.

28. Narbut, P., Berg, D., Works, C. N., and Dakin, T. W.: Factors Controlling Electric Strength of Gaseous Insulation; *Trans. AIEE*, 1959, vol. 78–III, pp. 59–74.

29. Dakin, T. W., and Mullen, G. A.: Continuous Recording of Outdoor Insulator Surface Conductance; *IEEE Trans.* on Electrical Insulation, EI-7, December 1972, pp. 169–175.

Chapter 9

INSULATING OILS AND LIQUIDS

By T. W. DAKIN

9-1 General Considerations 9-1
9-2 Mineral Insulating Oils 9-2
9-3 Dielectric Properties of Mineral Oils 9-3
9-4 Dielectric Strength of Mineral Oils 9-3
9-5 Deterioration of Oil 9-4
9-6 Servicing, Filtering, and Treating 9-5
9-7 Synthetic Liquid Insulation 9-6
9-8 Fluorcarbon Liquids 9-6
9-9 Silicone Fluids 9-6
9-10 Ester Fluids 9-6
9-11 References 9-7

9-1. General Considerations. Typical insulating liquids are natural or synthetic organic compounds and frequently consist of mixtures of essentially isomeric compounds with some range of molecular weight. The mixture of very similar but not exactly the same molecules, with a range of molecular size and with chain and branched hydrocarbons, prevents crystallization and results in a low freezing point, together with a relatively high boiling point. Typical insulating liquids have permittivities (dielectric constants) of 2 to 7 and a wide range of conductivities, depending upon their purity. The dc conductivity in these liquids is usually due to dissolved impurities, which are ionized by dissociation. Higher ionized impurity and conductivity levels occur in liquids having higher permittivities and lower viscosities.

The function of insulating liquids is to provide electrical insulation and heat transfer. As insulation, the liquid is used to displace air in the system and provide a medium of high electric strength to fill pores, cracks, and gaps in insulation systems. It is usually necessary to fill and impregnate systems with liquid under vacuum, so that all air bubbles are eliminated. If air is completely displaced in all high electric field regions, the corona threshold voltage and breakdown voltage for the system are greatly increased. The viscosity selected for a liquid insulation is often a compromise to provide the best balance between electrical insulation and heat transfer and other limitations such as flammability, solidification at low temperatures, and pressure development at high temperatures in sealed systems.

The most commonly used insulating liquids are natural hydrocarbon mineral oils refined to give low conductivity and selected viscosity and vapor-pressure levels for transformer, circuit-breaker, and cable applications.

A variety of synthetic fluids are also used for particular applications where the higher cost above that of mineral oil is warranted by the requirements of the application or by the improved performance in relation to the apparatus design.

9-2. Mineral Insulating Oils. Mineral insulating oils are hydrocarbons (compounds of hydrogen and carbon) refined from crude petroleum deposits from the ground.[1]* They consist partly of aliphatic compounds with the general formula C_nH_{2n+2} and C_nH_{2n}, comprising a mixture of straight and branched chain and cyclic or partially cyclic compounds. Many oils also contain a sizable fraction of aromatic compounds related to benzene, naphthalene, and derivatives of these with aliphatic side chains. The ratio of aromatic to aliphatic components depends on the source of the oil and its refining treatment. The percent aromatics is of importance to the gas-absorption or evaluation characteristics under electrical discharges[2] and to the oxidation characteristics.[3]

The important physical properties of a mineral oil (as for other insulating liquids as well) are listed in Table 9-1 for three types of mineral oils. In addition to these properties, mineral oils which are exposed to air in their application have distinctive oxidation characteristics which vary with type of oil and additives and associated materials.[4]

TABLE 9-1 Characteristic Properties of Insulating Liquids

Type of liquid	Mineral oil		
	Transformer	Cable and capacitor	Solid cable
Specific gravity.....................	0.88	0.885	0.93
Viscosity, Saybolt sec at 37.8°C......	57–59	0.100	100
Flash point, °C....................	135	165	235
Fire point, °C.....................	148	185	280
Pour point, °C....................	−45	−45	−5
Specific heat......................	0.425	0.412	....
Coefficient of expansion.............	0.00070	....	0.00075
Thermal conductivity, cal/(cm)(s)(°C)	0.39	....	....
Dielectric strength,* kV~...........	30	....	....
Permittivity at 25°C................	2.2	....	....
Resistivity, $\Omega \cdot$cm $\times 10^{12}$...........	1–10	50–100	1–10

*ASTM D877.

Many manufacturers now approve the use of any of several brands of mineral insulating oil in their apparatus provided that they meet their specifications which are similar to ASTM D1040, values from which are tabulated in Table 9-1. Low values of dielectric strength may indicate water or dirt contamination. A high neutralization number will indicate acidity, developed very possibly from oxidation, particularly if the oil has already been used. Presence of sulfur is likely to lead to corrosion of metals in the oil.

The solubility of gases and water in mineral oil is of importance in regard to its function

---

*Superscripts refer to bibliography, Par. **9-11.**

INSULATING OILS AND LIQUIDS

in apparatus. Solubility is proportional to the partial pressure of the gas above the oil:

$$S = S_0(p/p_0) \tag{9-1}$$

where S is the amount dissolved at pressure p if the solubility is expressed as the amount S_0 dissolved at pressure p_0.

The solubility is frequently expressed in volume percent of the oil. Values for solubility of some common gases in transformer oil[5] at atmospheric pressure (760 torr) and 25°C are air 10.8%; nitrogen 9.0%; oxygen 14.5%; carbon dioxide 99.0%; hydrogen 7%; methane 30% by volume. The solubilities of all the gases, except CO_2, increase slightly with increasing temperature. Water is dissolved in new transformer oil to the extent of about 60 to 80 ppm at 100% relative humidity and 25°C. The amount dissolved is proportional to the relative humidity. Solubility of water increases with oxidation of the oil and the addition of polar impurities, with which the water becomes associated. Larger quantities of water can be suspended in the oil as fine droplets.

9-3. Dielectric Properties of Mineral Oils. The permittivity of mineral insulating oils is low, since they are essentially nonpolar, containing only a few molecules with electric dipole moments. Some oils possess a minor fraction of polar constituents, which have not been identified. These contribute a dipolar character to the dielectric properties at low temperature and/or high frequency, similar to the trends shown in Figs. 5-5 and 5-6. A typical permittivity for American transformer oil at 60 Hz is 2.19 at 25°C, declining almost linearly to 2.11 at 100°C. At low temperatures and high frequencies, values of permittivity as high as 2.85 have been noted in oils with a relatively high level of polar constituents.

The dc conductivity levels of mineral oils range from about $10^{-15} \Omega^{-1} \cdot cm^{-1}$ for pure new oils up to $10^{-12} \Omega^{-1} \cdot cm^{-1}$ for contaminated used oils.[6] This conductivity is due to dissociated impurity ions or ions developed by oil oxidation.[7] It increases approximately exponentially with temperature about 1 decade in 80°C.

Alternating-current dissipation-factor values are nearly proportional to the dc conductivity, $10^{-13} \Omega^{-1} \cdot cm^{-1}$, corresponding to a tan δ of 0.008. If no electrode polarization or interfacial polarization at solid barrier surface effects are present, the dc conductivity σ should be related to the ac conductivity (tan δ) by

$$\sigma = \tfrac{5}{9}\varepsilon' f \tan \delta \times 10^{-12}$$

where ε' is the dielectric permittivity (Table 5-1) and f is the frequency.

9-4. Dielectric Strength of Mineral Oils. The dielectric strength of mineral oils, as with all liquids, varies considerably with the state of purity, particularly with respect to particulate matter and moisture. Typical values (ASTM D877, D1816 standard test gaps) are shown in Fig. 9-1 as a function of moisture content.[8]

The dielectric strength of mineral oil has been shown by Weber and Endicott[10] to decrease with increasing area under stress according to the relation

$$V_1 - V_2 = 1.497S \log (A_2/A_1) \tag{9-2}$$

where S is the standard deviation of tests with the smaller area. This relation is derived by application of minimum value statistics, assuming that the largest defect controls the breakdown. A_1 is the smaller area.

Typical ac values of the dielectric strength versus spacing and electrode geometry, which affects the maximum stress, are shown[9] in Fig. 9-2. It must not be assumed that the levels shown can be maintained indefinitely, since particulate matter may move into the field and reduce the strength. The dielectric strength is thus dependent upon the time of voltage application. This is illustrated in Fig. 9-3, where the dielectric strength is plotted as a function of the number of 60 Hz cycles using electrodes like those in ASTM D877. The single-cycle breakdown voltage is usually close to the impulse breakdown voltage; most of the decline in dielectric strength occurs in the time range from one cycle to a few thousand cycles, or about 1 min. Typical impulse breakdown voltages of transformer oil are shown in Fig. 9-4. Usually the impulse strength is about two to three times the crest 60-Hz 1-min strength. The difference decreases as the oil purity from particles increases.

The covering of metal conductor surfaces has been known to increase the ac strength of oil gaps.[11]

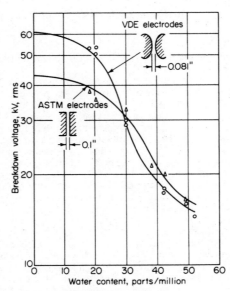

FIG. 9-1 Electric strength of transformer oil vs. water content with ASTM and VDE electrodes (rate of voltage rise, 2 kV rms/s).

Corona or partial breakdown can occur in mineral oil, as with any liquid or gas, when the electric stress is locally very high and complete breakdown is limited by a solid barrier or large oil gap (as with a needle point in a large gap). Such discharges produce hydrogen and methane gas, and sometimes carbon with larger discharges. Dissolved air is also sometimes released by the discharge. If the gas bubbles formed are not ejected away from the high field, they will reduce the subsequent discharge threshold voltage as much as 80%. The resistance of insulating oils to partial discharges is measured by two ASTM gassing tests: D2298 (Merrill test) and D2300 (modified Pirelli test). These tests measure the amount of decomposition gas evolved under specified conditions of exposure to partial discharges. A minimum amount of gas is, of course, preferred, particularly in applications for cables or capacitors. In fact, conventional mineral oils are inadequate in this respect for application in modern 60-Hz power capacitor designs.

9-5. Deterioration of Oil. Deterioration of oil in apparatus partially open or "breathing" is subject to air oxidation. This leads to acidity and sludge. There is no correlation between the amount of acid and the likelihood of sludging or the amount of sludge. Sludge clogs the ducts, reduces the heat transfer, and accelerates the rate of deterioration. ASTM tests for oxidation of oils are D1904, D1934, D1313, and D1314.

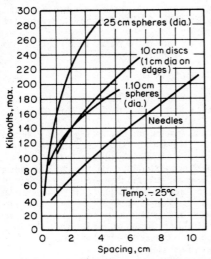

FIG. 9-2 Sparkover of various shaped electrodes in oil at 60 Hz.

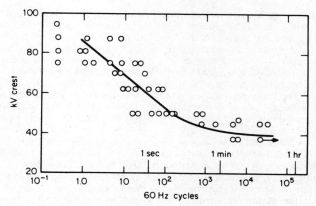

FIG. 9-3 Time to breakdown of a 0.1-in oil gap with 1-in square-edge electrodes (ASTM D877). (See Ref. 15.)

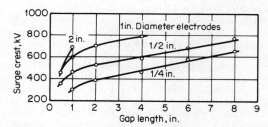

FIG. 9-4 Curve showing relation of gap length to minimum surge crest voltage required for breakdown between cylindrical electrodes with hemispherical ends immersed in oil, 1½- × 40-μs wave.

Copper and lead and certain other metals accelerate the oxidation of mineral oils. Oils are considerably more stable in nitrogen atmospheres.

Inhibitors are now commonly added both to new and to used oils to delay the oxidation. Ditertiary butyl paracresol (DBPC) is the inhibitor most commonly used at present.

9-6. Servicing, Filtering, and Treating. Oil in service is usually maintained by testing for acidity, dielectric strength, inhibitor content, interfacial tension, neutralization number, peroxide number, pour point, power factor, refractive index and specific optical dispersion, resistivity, saponification, sludge, corrosive sulfur, viscosity, and water content, as outlined in ASTM D117. These properties indicate various types of contamination or deterioration which might affect the operation of the insulating oil. It has been suggested that interfacial tension below a certain value indicates that sludging is imminent or has started (see *ASTM Spec. Tech. Publ.* 135, 1952).

Depending on the voltage rating of the apparatus, the oil is maintained above 16 to 22 kV (ASTM Test D877). The usual contaminants are water, sludge, acids, and, in circuit-breaker oils, carbon. The centrifuge is best suited for removing large quantities of water, heavier solid particles, etc. The blotter filter press is used for the removal of minute quantities of water, fine carbon, etc. In another method, after removing the larger particles the oil is heated and sprayed into a vacuum chamber, where the water and volatile acids are removed. Sludge and very fine solids are then taken out by a blotter filter press. All units are assembled together so that the process is completed in a single pass. Some work has been done in reclaiming oil by treating it to reduce acidity. One process is similar to the

9-6 ELECTRICAL ENGINEERING MATERIALS REFERENCE GUIDE

later stages in refining. Another treatment uses activated alumina, fuller's earth, or silica gel. The IEEE guide for Maintenance of Insulating Oil is published as *IEEE Standards Publ.* 64.

It has been found that analysis of the dissolved gas in oil or above the oil in oil-insulated transformers and cables is a good diagnostic tool to detect electrical faults, particularly, or deterioration, generally. For example, continuing or intermittent partial discharges produce hydrogen and low-molecular-weight hydrocarbons such as methane, ethane, ethylene, which accumulate in the oil and can be accurately measured to assess the magnitude of the fault. Higher-current arc faults produce acetylene in addition to H_2 and other low-molecular-weight hydrocarbons. Thermal deterioration of cellulosic or paper insulation is indicated by elevated concentrations of CO and CO_2 in the oil. Reference 16 should be consulted for more details on this subject.

9-7. Synthetic Liquid Insulation. Synthetic chlorinated diphenyl and chlorinated benzene liquids (askarels) have been widely used from the mid-1930s up to the mid-1970s and are still in service in many power capacitors and transformers, where they were adopted for their nonflammability, as well as good electrical characteristics.[6] Since the mid-1970s their use has been banned in most countries due to their alleged toxicity and resistance to biodegradation in the environment. Now, when apparatus containing these fluids, which are commonly referred to as "PCBs," are taken out of service, environmental regulations in the United States require that the fluid not be released into the environment. Waste fluid should be incinerated at high temperature with HCl reactive absorbent scrubbers in the stack, since this acid gas is a product of the combustion. Methods of disposing of solid material (such as paper) saturated with PCBs have not been completely agreed upon.

New synthetic fluids have been developed and are now widely applied in power capacitors, where the electrical stresses are very high. These fluids include aromatic (containing benzene rings) hydrocarbons, some of which have excellent resistance to partial discharges. They are not fire-resistant, however.[17]

Very high boiling, low-vapor-pressure, high-flash-point ($>300°C$) hydrocarbon oils are being tried for power transformers with some fire resistance. Methods for assessing the risk of fire with such liquids, as well as with silicones, are still being debated.

Perchloroethylene (tetrachloroethylene), a nonpolar liquid, is now in use in sealed medium-power transformers, where nonflammability is required. With a boiling point at atmospheric pressure of $121°C$, this fluid is completely nonflammable. It is also widely used in dry cleaning. Other important classes of synthetic insulating fluids are discussed in the following sections.

9-8. Fluorocarbon Liquids. A variety of nonpolar nonflammable perfluorinated aliphatic compounds, in which the hydrogen has been completely replaced by fluorine, are available with different ranges of viscosity and boiling point from below room temperature to more than $200°C$. These compounds have low permittivities (near 2.0) and very low conductivity. They are inert chemically and have low solubilities for most other materials. The chemical formula for these compounds is one of the following: C_nF_{2n}, C_nF_{2n+2}, and $C_nF_{2n}O$. The presence of the oxygen in the latter formula does not seem to reduce the stability. These compounds have been used for filling electronic apparatus[13] and large transformers to give high heat-transfer rates together with high dielectric strength. The vapors of these liquids also have high dielectric strengths.[14]

9-9. Silicone Fluids. These fluids, chemically formed from $Si-O$ chains with organic (usually methyl) side groups, have a high thermal stability, low temperature coefficient of viscosity, low dielectric losses, and high dielectric strength. They can be obtained with various levels of viscosity and correlated vapor pressures. Rated service temperatures extend from -65 to $200°C$, some having short-time capability up to $300°C$. Their permittivity is about 2.6 to 2.7, declining with increasing temperature. These fluids have a tendency to form heavier carbon tracks than other insulating liquids when breakdown occurs. They cannot be considered fireproof but will reduce the risk of fire due to their low vapor pressure.

9-10. Ester Fluids. There are a few applications, mostly for capacitors, where organic ester compounds are used. These liquids have a somewhat higher permittivity, in the range of about 4 to 7, depending on the ratio of ester groups to hydrocarbon chain lengths. Their conductivities are generally somewhat higher than those of the other insulating liquids discussed here. The compounds are easily subject to hydrolysis with water to form acids and alcohols and should be kept dry, particularly if the temperature is raised. Their thermal stability is poor. Specifically dibutyl sebacate has been used in high-frequency capacitors and castor oil in energy-storage capacitors.

INSULATING OILS AND LIQUIDS

9-11. References on Insulating Oils and Liquids

1. Gruse, W. A., and Stevens, D. R.: "Chemical Technology of Petroleum," 3d ed.; New York, McGraw-Hill Book Company, 1960.

2. Berberich, L. J.: Influence of Gaseous Electric Discharge on Hydrocarbon Oils; *Ind. Eng. Chem.,* 1938, vol. 30, p. 280. Blodgett, R. B, and Bartlett, S. C.: *Trans. AIEE,* 1961, vol. 80, p. 528. Olds, W. F., Feich, G., and Eich, E.: *Ann Rept. NRC 1960 Conf. on Electrical Insulation,* p. 93.

3. Berberich, L. J.: Oxidation Inhibitors in Electrical Insulating Oils; *ASTM Bull.* 149, pp. 65–73, 1947. Ford, J. G., and Sloat, T. K.: Inhibitors Lengthen Life of Transformer Oil; *Westinghouse Eng.,* 1950, p. 250.

4. Symposium on Insulating Oils: *ASTM Bulls.* 146 and 149, 1947. (Several authors.)

5. Kaufman, R. B., Shimanski, E. J., and MacFadyen, K. W.: Gas and Moisture Equilibriums in Transformer Oil; *Trans. AIEE,* 1955, vol. 74–III, p. 312.

6. Clark, F. M.: "Insulating Materials for Design and Engineering Practice"; New York, John Wiley & Sons, Inc., 1962.

7. Piper, J. D.: Chapter in "Dielectric Materials and Applications," A. von Hippel (ed.); New York, John Wiley & Sons, Inc., 1954.

8. Rohlfs, A. F., and Turner, F. J.: Correlation between the Breakdown Strength of Large Oil Gaps and Oil Quality Gauges; *Trans. AIEE,* 1956, vol. 75–III.

9. Peek, F. W.: "Dielectric Phenomena in High Voltage Engineering"; New York, McGraw-Hill Book Company, 1929.

10. Weber, K. H., and Endicott, H. S.: Area Effect and Its Extremal Basis for the Electric Breakdown of Transformer Oil; *Trans. AIEE,* 1956, vol. 75–III, p. 371.

11. Roth, A.: "Hochspannungstechnik," 4th ed.; Vienna, Springer-Verlag, OHG, 1959.

12. White, A. H., and Morgan, S. O.: The Dielectric Properties of Chlorinated Diphenyl; *J. Franklin Inst.,* 1933, vol. 216, p. 635.

13. Kilham, L. F., and Ursch, P. R.: *Proc. Natl. Electronics Computer Conf.,* Los Angeles, May 1955.

14. Berberich, L. J., Works, C. N., and Lindsay, E. W.: *Trans. AIEE,* 1955, vol. 74–I, p. 660.

15. Dakin, T. W., Studniarz, S. A., and Hummert, G. T.: *Annual Report, NRC-NAS Conference on Electrical Insulation and Dielectric Phenomena,* 1972.

16. *IEEE Guide for the Determination of Generated Gases in Oil Immersed Transformers and Their Relation to the Serviceability of the Equipment;* ANSI-IEEE C57, 104, 1978.

17. Mandelcorn, L., Dakin, T. W., Miller, R. L., and Mercier, G.: High-Voltage Power Capacitor Dielectrics, Recent Developments; *Proc. 14th IEEE Electrical/Electronics Insulation Conference,* Boston, Mass., October 1979, IEEE Publication No. 79CH1510-7-EI, p. 250

Chapter 10

MICA AND MICA PRODUCTS

T. W. DAKIN

10-1	Mica Sources 10-1		10-4	Dielectric Strength of Mica 10-2	
10-2	Composition and Physical Properties 10-1		10-5	Synthetic Mica 10-3	
			10-6	Mica Paper 10-3	
10-3	Dielectric Properties of Mica 10-2		10-7	References 10-3	

10-1. Mica Sources. Mica insulation is derived from a class of minerals of finely laminated structure and very easy cleavage, the flakes of which are very flexible and tough and extremely resistant to heat. Reference 1 discusses the structure, properties, and sources in detail.[1]

The two classes most commonly used for electrical purposes are (1) the ferromagnesia mica phlogopite, also known as "amber" or "silver mica," and (2) the potassium mica muscovite, known as "India," "white," "ruby," or potash mica. The phlogopite mica is produced principally in Madagascar and Canada. The muscovite is produced principally in India, with quantities also obtained in the United States, Africa, and South America. Mica for electrical purposes must be carefully selected owing to mineral and vegetable inclusions occurring in the slabs. ASTM Designation D351 describes the standard methods for grading and classification of natural muscovite mica in blocks and films. Blocks range from 0.007 inches in thickness; films range from 0.0008 to 0.0040 in. NEMA specifies the thickness of 10 splittings at 0.006 to 0.0011 in. NEMA phlogopite splittings range from 0.007 to 0.012 in for 10 splittings.

10-2. Composition and Physical Properties. The micas are complex and variable in composition. Muscovite mica is translucent; is white, ruby, green, or brown; and is harder and less flexible than the phlogopite, which is opaque and ranges from pale yellow (amber) and silver to brown and green.

The idealized formula for muscovite mica is $K_2Al_4Al_2Si_6O_{20}(OH)_4$ and for phlogopite mica $K_2Mg_6Al_2Si_6O_{20}(OH)_4$. The structure consists of firmly bonded double sheets of aluminum silicate (muscovite) or magnesium aluminum silicate (phlogopite). The individual layers of the double sheets are internally held together with hydroxyl groups and other atoms of aluminum and magnesium. The cleavage plane occurs between these double sheets, which are more loosely bonded by potassium atoms, which lie in the cleavage plane. If clear, thin sheets of muscovite or phlogopite mica are heated to a temperature of 400 to 600°C, no perceptible alteration occurs, and their clearness and elasticity are retained, but muscovite starts to lose H_2O at about 500°C, phlogopite at 1000°C. At higher temperatures, between 900 and 1000°C, muscovite becomes silver-white, with a pronounced metallic appearance, loses considerably in clearness, and becomes rather brittle, so that it can be pulverized into a thin white dust. Phlogopite withstands this temperature much better; it loses but little in transparency and does not become so brittle. The melting points are in the range of 1200 to 1300°C. The safe maximum temperature is about 500°C for muscovite and 1000°C for phlogopite. (Also see results of heating tests.[2]) The specific heat varies from 0.206 to 0.209. Mica will withstand great mechanical pressures perpendicular to the plane of lamination, but the laminations have very weak cleavage and are easily split into very thin flakes or leaves. It resists to a high degree the attack of gases such as combustion products but is attacked by warm hydrofluoric acid, molten potassium hydrate, warm alkaline carbonates, and water containing carbon dioxide. Magnesia mica is attacked by concentrated sulfuric acid, but the potassium micas are not attacked by either hydrochloric acid or sulfuric acid. Contact with oil is injurious to mica, as it penetrates the laminae and seriously impairs their cohesion.

10-3. Dielectric Properties of Mica. The resistivity of mica at 25°C ranges from about 10^{12} to 10^{16} Ω·cm depending on inclusions, etc., muscovite usually having a higher resistivity. The dissipation factors (tan δ) of muscovite splittings range from 0.0001 to 0.0008 and phlogopite from 0.003 to 0.09, over the frequency range from 60 Hz to 1 MHz. The high-temperature resistivity of muscovite and phlogopite splittings is illustrated in Fig. 10-1, from data of Dakin and coworkers. Changes in permittivity and tan δ with temperature are illustrated in Fig. 10-2.[3]

Reference 2 gives data for frequencies of 100 to 1000 kHz, at a test temperature of 25°C. All samples were muscovite, except two phlogopite samples from Madagascar.

The presence of stains or inclusions so seriously affects the tan δ as to render such stained micas unsuited for radio purposes. The tan δ of phlogopite mica was found to be so high as to render it also unsuitable for radio purposes. See ASTM D748 for measurement methods and acceptable values. Table 12 lists the properties of mica of different origins.

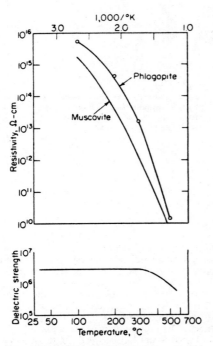

FIG. 10-1. Resistivity (Dakin et al.) and apparent dielectric strength (Hackett and Thomas) of mica splittings at elevated temperature.

10-4. Dielectric Strength of Mica. The dielectric strength of India ruby mica was investigated by Moon and Norcross with dc potentials, and the breakdown was found to have linear relationship to thickness, independent of temperature up to at least 100°C. The

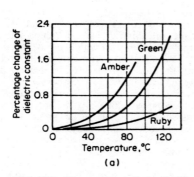
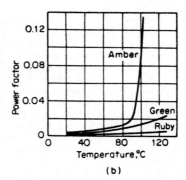

FIG. 10-2 Characteristics of mica. (*a*) Change in dielectric constant with temperature. (*b*) Change in power factor with temperature.

MICA AND MICA PRODUCTS

TABLE 10-1 Properties of Mica

Origin	Kind	No. of samples	Permittivity		tan δ	
			Average	Spread	Average	Spread
India	Clear	4	7.32	7.90–7.07	0.02	0.02–0.01
United States except North Carolina	Clear	4	7.02	7.36–6.62	0.02	0.03–0.01
North Carolina	Clear	8	7.22	8.69–6.57	0.02	0.04–0.01
Haywood County, N.C.	Clear	4	7.31	8.69–6.57	0.02	0.04–0.01
United States and India	Stained	17	——	9.64–5.83	——	8.36–0.06
Madagascar	——	2	——	6.07–5.41	——	7.12–0.38

breakdown gradients ranged from 3520 kV/cm in air to 6960 kV/cm in transformer oil and 10,600 kV/cm in a mixture of xylene and aniline, at 20°C. Higher temperature-strength values are shown in Fig. 10-1.

In these tests the edge effect of the test electrodes was not eliminated except in the bath of xylene and aniline. Dielectric strength tests[2] made in transformer oil, at 60 Hz, gave results on 19 lots of domestic clear ruby samples of 5.7 kV/cm for 1-mil thickness, 9.9 kV for 5-mil thickness, and 13.2 kV for 9-mil thickness. Reference 2 also gives similar results for mica from other sources. Mica is particularly noted for its resistance to corona discharges, thus maintaining its ac dielectric strength much better than any other thin dielectric material, when exposed to discharges either on the surface or within internal gas spaces in built-up insulation.

10-5. Synthetic Mica. A modified mica is now made synthetically. Its structure is similar to phlogopite mica with the hydroxyl groups (OH) replaced by fluorine. Its permittivity is 6.5 and its tan δ and resistivity similar to those of muscovite. Its principal advantage in properties is its thermal stability, particularly in the absence of air, where it will withstand heating to 1100 to 1200°C. In the presence of atmospheric moisture at elevated temperatures it will very slowly change, losing fluorine. It is not available in sizable splittings, but only as a compacted aggregate.

10-6. Mica Paper. Several methods have been developed for disintegrating rough mica into very thin, small-area splittings and re-forming it into a continuous paperlike sheet. Such sheets are of themselves weak mechanically and must usually be supported by resin impregnation or backing with plastic films or fibrous material. The properties of these composite sheets, which are also available as tapes, vary widely depending on the impregnating and supporting material. The sheets are now used widely in many applications where mica splittings were formerly employed. Some forms, when properly applied, have a high dielectric strength, in excess of 100 V/mil, which is maintained under corona discharges much better than that of organic resins. Mica paper is also available with inorganic (commonly phosphate) impregnation which can be cured up hard by heating.

10-7. References on Mica

1. Skow, M. L.: Mica, a Material Survey; *U.S. Dept. Interior Information Circ.* 8125, 1962.

2. Lewis, A. B., Hall, E. L., and Caldwell, F. R.: NBS *J. Res.,* vol. 7, p. 403, 1931.

3. Dannatt, C., and Goodall, S. E.: The Permittivity and Power Factor of Micas; *J. IEEE,* vol. 69, pp. 490, 1034, 1931.

4. Moon, P. H., and Norcross, A. S.: *Trans. AIEE,* vol. 49, p. 755, 1930.

5. Hackett, W., and Thomas, A. M.: The Electric Strength of Mica and Its Variation with Temperature; *Proc. IEE,* vol. 88, p. 295, 1940.

Chapter 11

PLASTICS

R. R. DIXON

11-1	Definitions 11-1		11-6	Cold-Molded Compounds 11-9
11-2	Molded Materials 11-1		11-7	Laminated Sheet 11-12
11-3	Thermoplastic Molding Compounds 11-1		11-8	Filament-Wound Materials 11-25
			11-9	Films 11-25
11-4	Reinforced Thermoplastics 11-9		11-10	References 11-25
11-5	Thermosetting Compounds 11-9			

11-1. Definitions. Plastics are materials which consist of organic substances of high molecular weight, which at some stage in their processing can be shaped by flowing under pressure, and which are solid in the finished state. The organic substances are called resins or polymers and are derived from natural gas, petroleum, coal, or in a few cases, natural materials such as cellulose. The polymers are frequently mixed with other materials to modify their properties or their processing characteristics. For example, inorganic fillers may be added to improve physical properties, fluids may be added to change flow parameters, reinforcements may be included to increase strength, and antioxidants, stabilizers, pigments, and lubricants may be added. Plastics are available in a number of forms; those of concern in this section are molded materials, laminates, and films.

There has been considerable interest recently in electromagnetic-induction (EMI) and radio frequency-interference (RFI) shielding for plastics housings, and many different techniques have been developed to achieve this goal.

11-2. Molded Materials. Molded materials are formed by causing the plastics to flow in a mold, usually with the application of heat and pressure. The plastics materials may be pure polymer or heterogeneous mixtures of polymers, fillers, reinforcements, and other additives. Some orientation of the ingredients occurs during molding. The most common molding processes are compression molding, transfer molding, injection molding, extrusion, blow molding, casting, rotational molding, and laminating. Molding materials can be broadly divided into three categories: thermoplastic materials, thermosetting materials, and cold-molding compounds.

Standards. Various bodies, both governmental and private, maintain standards and specifications for plastics materials. These include U.S. government (Fed, MIL, CS/PS), American Society for Testing and Materials (ASTM), Society of Automotive Engineers (Aerospace Materials Specifications), National Electrical Manufacturers Association (NEMA), Underwriters' Laboratories (UL), and International Organization for Standardization (IDO). A complete compilation of standards can be found in the World Index of Plastics Standards, National Bureau of Standards Special Publication 352.

11-3. Thermoplastic Molding Compounds. Thermoplastic molding compounds are materials capable of being repeatedly softened or melted at elevated temperatures.

The polymers consist of long linear molecules which may have some branching but are not interconnected (cross-linked) to other molecules. The change of state in molding is physical rather than chemical. A typical thermoplastic structure is paraffin. Thermoplastic materials can be molded quickly but must be used below their softening temperatures.

ASTM and government standards for thermoplastics are listed in Table 11-1. Typical properties of most of the following thermoplastics are shown in Table 11-2.

11-1

TABLE 11-1 Thermoplastics Standards

Plastics	ASTM	Government	Plastics	ASTM	Government
ABS	D1788	Fed L-P-1183	Polypropylene	D2146	Fed L-P-394
	D3011			D2853	MIL-C-19978
Acetal	D2133	Fed L-P-392			MIL-C-39022
	D2948	MIL-P-46137			MIL-P-46109
Acrylic	D788	Fed L-P-380			MIL-C-52472
	D1547	Fed L-P-507	Polystyrene	D1892	Fed L-P-396
Butylene	D2581			D3011	Fed L-P-398
Cellulose acetate	D706	Fed L-P-397			Fed L-P-506
Cellulose acetate	D707	Fed L-P-349			MIL-P-21347
butyrate					MIL-P-60312
Cellulose acetate	D1562		Polysulfone		MIL-P-46120
propionate					MIL-P-46133
CTFE	D1430	Fed L-P-385	PTFE	D1457	Fed L-P-403
		MIL-P-55028		D2902	
Ethyl cellulose	D787		Polyurethane		MIL-I-4997
FEP	D2116	Fed L-P-389			MIL-C-5898
	D2902	Fed L-P-523	PVC	D1047	Fed L-P-375
	D3368			D1755	Fed L-P-535
Nylon	D789	Fed L-P-395		D1784	Fed L-P-1035
	D2897	Fed L-P-410		D2287	Fed HH-I-595
	D4066	MIL-M-19887		D2474	MIL-C-17
		MIL-M-20693		D2740	MIL-W-76
		MIL-P-22096		D3150	MIL-I-631
PFA	D2902			D3915	MIL-C-915
	D3307				MIL-C-3432
Polyallomer		Fed L-P-394			MIL-C-3884
Polycarbonate	D2473	Fed L-P-393			MIL-I-3930
	D2848	MIL-P-81390			MIL-I-5086
	D3935				MIL-P-6264
Polyester	D3220	MIL-P-46161			MIL-W-7072
	D3221				MIL-C-7078
Polyethylene	D1248	Fed L-P-378			MIL-W-7444
	D2103	Fed L-P-390			MIL-C-15452
	D2853	Fed L-P-512			MIL-C-15479
		Fed L-P-530			MIL-W-16878
		MIL-C-17			MIL-C-23437
		MIL-C-9660			MIL-C-27072
		MIL-C-19547			MIL-C-55036
		MIL-C-27072	PVDC	D729	Fed L-P-1041
Polyethylene	D4020		PVDF	D3222	
(UHMW)			PVF		Fed L-P-1040
PE-CTFE	D3275				MIL-W-5086
PE-TFE	D3159				MIL-C-7078
Polyolefin	D2853				MIL-W-16878
Polyphenylene	D2874				MIL-W-22759
oxide–based					MIL-I-23053
					MIL-P-46122
					MIL-W-81044
					MIL-W-81822
			SAN	D1432	Fed L-P-399
Polyphenylene	D3646			D3011	
sulfide	D4067				

TABLE 11-2 Thermoplastics Properties

Properties	ASTM test	ABS	Acetal	Acrylic	Butylene	Cellulosic	CTFE	FEP	PE-CTFE
Electrical									
Arc resistance, s	D495	85	180	——	——	200	>360	>300	18
Dielectric constant	D150								
60 Hz		5.0	3.7	3.9	——	7.5	2.8	2.1	2.3
10^6 Hz		4.5	3.7	3.6	2.5	7.0	2.7	2.1	2.3
10^9 Hz		3.8	3.7	3.2	——	7.0	2.5	2.1	2.3
Dissipation factor	D150								
60 Hz		0.008	0.005	0.06	——	0.06	0.0012	0.0002	0.005
10^6 Hz		0.007	0.005	0.05	0.0004	0.07	0.027	0.0002	0.0015
10^9 Hz		0.015	0.005	0.03	——	0.10	0.017	0.002	0.015
Dielectric strength, V/mil, step by step	D149	460	320	400	500	520	550	500	450
Volume resistivity, $\Omega \cdot$ cm	D257	5×10^{16}	10^{15}	10^{15}	10^{16}	10^{14}	10^{18}	10^{18}	10^{16}
Mechanical									
Tensile strength, psi	D638	7,000	10,000	11,000	3,000	9,000	6,000	3,100	7,000
Tensile modulus, 10^5 psi	D638	3.5	5	4	0.3	4	3	0.5	0.2
Elongation, %	D638	110	75	10	300	70	250	330	200
Compressive strength, psi	D695	10,000	18,000	18,000	——	5,000	7,400	2,200	——
Flexural strength, psi	D790	14,000	14,000	19,000	2,000	16,000	9,300	——	7,000
Impact strength, ft · lb/in	D256	12	2.3	0.5	5	5.2	2.7	——	——
Hardness	D785	R100	M94	M105	D60 (shore)	R125	R75	D60 (shore)	R95
Thermal									
Deflection temperature, °F	D648	218	255	210	135	195	258	158	170
Continuous-use temperature, °F		210	195	200	180	160	390	400	355
Coefficient of expansion, 10^{-5} in (in · °C)	D696	1.3	8	9	13	18	7	10	8
Thermal conductivity, 10^{-4} (cal · cm)/(cm^2 · s · °C)	C177	8	5.5	6.0	20	8	5	6	4
Flammability									
Oxygen index	D2863	18	18	18	——	——	——	——	——

TABLE 11-2 Thermoplastics Properties (*Continued*)

Properties	ASTM	ABS	Acetal	Acrylic	Butylene	Cellulosic	CTFE	FEP	PE-CTFE
Chemical									
Water absorption, % in 24 h	D570	0.45	0.25	0.4	0.01	2.0	0	0.01	0.01
Radiation resistance		10^8	10^6	—	10^8	10^6	10^4	10^4	10^4
Not resistant to:	D543	Oxidizing acids, ketones, esters, chlorinated solvents	Alkalies, acids	Ketones, esters, aromatic and chlorinated solvents		Ketones, esters, aromatic and chlorinated solvents	Chlorinated solvents	—	—

Properties	ASTM test	PE-TFE	PFA	PTFE	PVDF	Nylon	Ether sulfone	Aryl ester	Aryl ether
Electrical									
Arc resistance, s	D495	72	—	>300	60	120	110	120	180
Dielectric constant	D150								
60 Hz		2.6	2.1	2.1	8.4	4.3	3.5	3.0	3.1
10^6 Hz		2.6	2.1	2.1	6.4	5.3	3.5	3.0	2.9
10^9 Hz		2.6	2.1	2.1	—	—			3.1
Dissipation factor	D150								
60 Hz		0.0006	0.0002	0.0002	0.049	0.020	0.001	0.001	0.006
10^6 Hz		0.0008	0.0002	0.0002	0.17	0.029	0.005	0.022	0.004
10^9 Hz		0.005	0.0002	0.0002	—	0.04			0.007
Dielectric strength, V/mil, step by step	D149	350	400	430	260	280	400	425	390
Volume resistivity, $\Omega \cdot$cm	D257	10^{16}	10^{18}	10^{18}	10^{14}	10^{14}	10^{17}	10^{16}	10^{16}
Mechanical									
Tensile strength, psi	D638	6,500	4,300	5,000	5,200	11,000	12,000	10,000	7,500
Tensile modulus, 10^5 psi	D638	1	1	0.6	2	4	3.5	—	3
Elongation, %	D638	400	300	400	100	300	60	50	90
Compressive strength, psi	D695	7,100	—	1,700	8,000	5,000	—	12,000	—
Flexural strength, psi	D790	—	—	—	8,600	6,100	18,600	12,000	11,000
Impact strength, ft·lb/in	D256	—	—	3	3	2	1.6	5	8
Hardness	D785 D2240	D75 (shore)	D64 (shore)	D60 (shore)	D80 (shore)	R120	M88	M90	R117
Thermal									
Deflection temperature, °F	D648	160	—	250	195	167	390	340	300
Continuous-use temperature, °F		360	500	500	300	250	360	265	250

	D696	9	12	10	8.5	8	5.5	6	4
Coefficient of expansion, 10^{-5} in/(in·°C)	D696	9	12	10	8.5	8	5.5	6	4
Thermal conductivity, 10^{-4} (cal·cm)/(cm²·s·°C)	C177	—	—	6	6	6	7	4	7

Flammability

Oxygen index	D2863	—	—	>90	44	24	—	34	—

Chemical

Water absorption, % in 24 h	D570	0.03	0.03	0	0.04	1.5	0.43	0.25	0.25
Radiation resistance		10^4	10^4	10^4	—	10^5	5×10^7		
Not resistant to:	D543	—	—	—	—	Acids, phenols	Ketones, esters aromatic and chlorinated solvents	Alkalies, aromatic and chlorinated solvents	Acids, esters, ketones, chlorinated solvents

Properties	ASTM test	Polycarbonate	Ether-ether-ketone	Ether-imide	Ethylene (med. density)	Imide	Phenylene oxide	Phenylene sulfide	Methyl pentene
Electrical									
Arc resistance, s	D495	100	—	120	235	80	75	180	—
Dielectric constant	D150								
60 Hz		3.2	3.0	3.0	2.35	3.4	2.7	—	2.1
10^6 Hz		3.0	3.2	3.0	2.35	3.4	2.7	3.1	2.1
10^9 Hz		2.9	3.3	3.0	2.35	—	2.7	3.2	2.1
Dissipation factor	D150								
60 Hz		0.001	0.001	0.001	0.0005	0.0055	0.0004	—	0.0001
10^6 Hz		0.002	0.003	0.001	0.0005	0.0018	0.0007	0.0004	—
10^9 Hz		0.010	0.003	0.0028	0.0005	—	0.0024	0.0007	0.0001
Dielectric strength, V/mil, step by step	D149	380	480	620	700	500	400	400	700
Volume resistivity, Ω·cm	D257	10^{16}	5×10^{16}	10^{17}	10^{16}	10^{16}	10^{16}	10^{16}	10^{16}
Mechanical									
Tensile strength, psi	D638	9,500	13,000	15,000	3,500	17,000	7,800	10,000	3,600
Tensile modulus, 10^5 psi	D638	3.5	5	4	0.5	2	4	5	1.9
Elongation, %	D638	130	100	60	600	10	50	3	13
Compressive strength, psi	D695	12,000	—	21,000	7,000	30,000	12,000	—	5,500
Flexural strength, psi	D790	12,500	—			29,000	12,800	20,000	4,500
Impact strength, ft·lb/in	D256	18	1.6	1.0	16	0.7	5	0.3	1.0
Hardness	D785 D2240	M70	—	M109	D50	E99	R115	R124	R93

TABLE 11-2. Thermoplastics Properties (*Continued*)

Properties	ASTM test	Polycarbonate	Ether-ether-ketone	Ether-imide	Ethylene (med. density)	Imide	Phenylene oxide	Phenylene sulfide	Methyl pentene
Thermal									
Deflection temperature, °F	D648	285	320	390	120	520	212	275	105
Continuous-use temperature, °F		250	390	340	250	550	203	340	250
Coefficient of expansion, 10^{-5} in/(in·°C)	D696	7	——	5.6	16	5	6	5.5	12
Thermal conductivity, 10^{-4} (cal·cm)/(cm^2·s·°C)	C177	5	——	2.6	10	2.6	4	6.8	4
Flammability									
Oxygen index	D2863	25	35	47	18	44	30	44	——
Chemical									
Water absorption, % in 24 h	D570	0.18	0.14	0.25	0.01	0.3	0.06	0.02	0.01
Radiation resistance		10^8	10^9	10^8	10^8	10^9	10^6	10^6	10^8
Not resistant to:	D543	Acids, alkalies, aromatic and chlorinated solvents	Ketones	Strong bases, chlorinated solvents	Acids, chlorinated solvents	Alkalies	Aromatic and chlorinated solvents	——	

Properties	ASTM test	Propylene	Styrene	SAN	Styrene/ maleic	Sulfone	PVC, rigid	PVC, flexible
Electrical								
Arc resistance, s	D495	140	130	120	60	90	10	15
Dielectric constant	D150							
60 Hz		2.3	2.6	3.0	2.6	3.1	4.0	9.0
10^6 Hz		2.3	2.7	2.8	2.6	3.1	3.8	8.0
10^9 Hz		2.3	2.7	——	——	3.1	3.1	4.5

Dissipation factor	D150							
60 Hz		0.0005	0.0006	0.006	0.003	0.0008	0.020	0.15
10^6 Hz		0.0007	0.0003	0.008	——	0.0010	0.017	0.16
10^9 Hz		0.0018	0.0004	——	——	0.0034	0.019	0.14
Dielectric strength, V/mil, step by step	D149	600	600	——	——	400	——	290
Volume resistivity, $\Omega\cdot$cm	D257	10^{16}	10^{16}	10^{16}	——	10^{16}	10^{16}	10^{15}
Mechanical								
Tensile strength, psi	D638	5,000	10,000	11,000	5,000	10,000	7,000	2,500
Tensile modulus, 10^5 psi	D638	2	5	4.5	3	4	5	——
Elongation, %	D638	500	2	3	30	100	60	——
Compressive strength, psi	D695	——	14,000			14,000	11,000	300
Flexural strength, psi	D790	8,000	10,000	11,000	8,400	15,000	13,000	1,200
Impact strength, ft·lb/in	D256	1.0	0.4	0.4	5	1.3	8	——
Hardness	D785	R80	M70	M82	R109	M69	D75 (shore)	A75 (shore)
	D2240							
Thermal								
Deflection temperature, °F	D648	130	210	215	235	345	150	150
Continuous-use temperature, °F		250	160	175	——	320	160	150
Coefficient of expansion, 10^{-5} in/(in·°C)	D696	9	8	6.5	9	5.6	8	12
Thermal conductivity, 10^{-4} (cal·cm)/(cm²·s·°C)	C177	5	3	——		6	4	4
Flammability								
Oxygen index	D2863	18	18	18	18	31	40	28
Chemical								
Water absorption, % in 24 h	D570	0.03	0.10	0.3	0.1	0.22	0.2	0.5
Radiation resistance		10^6	10^6	10^8		10^8	10^7	10^7
Not resistant to:	D543	——	Aromatic and chlorinated solvents	Aromatic and chlorinated solvents	Alkalies, aromatic and chlorinated solvents	Aromatic and chlorinated solvents	Ketones, esters aromatic solvents	Ketones, esters aromatic solvents

ELECTRICAL ENGINEERING MATERIALS REFERENCE GUIDE

ABS. These polymers are made from acrylonitrile, butadiene, and styrene. Since the ratio of each ingredient and the chain configuration can be varied, a large number of polymers are available. They are characterized by toughness.

Acetals. Derived from formaldehyde, two polymers exist, a homopolymer and a copolymer. Acetals are stiff materials with a low coefficient of friction.

Acrylics. Acrylics are a large family of materials, of which methyl methacrylate is the most common. Their most significant properties are transparency and resistance to weathering.

Cellulosics. Cellulose esters and ethers (cellulose acetate, cellulose acetate butyrate, ethyl cellulose, etc.) have good impact strength and are transparent.

Fluorocarbons. All of this family of polymers contain fluorine, which makes them quite resistant to chemicals and to thermal degradation. They are also resistant to weathering and have low-friction surfaces. The family consists of:

CTFE: chlorotrifluoroethylene

FEP: fluorinated ethylene/propylene

PE-CTFE: poly(ethylene/chlorotrifluoroethylene)

PE-TFE: poly(ethylene/tetrafluoroethylene)

PFA: perfluoroalkoxy

PTFE: polytetrafluoroethylene

PVDF: polyvinylidene fluoride

Nylons. Nylons are polyamides, in which the associated number (e.g. nylon 6/6) defines the polymer chain construction. Nylons are strong, tough materials with a low coefficient of friction.

Polyarylester. This material is a high-temperature polyester, transparent, with thermal stability and good electrical properties.

Polyarylether. This material is an alloy of polysulfone with an ABS. It is thermally stable and is resistant to many chemicals.

Polybutylene. Polybutylene is a flexible material with good stress-crack resistance and good electrical properties.

Polycarbonates. Polycarbonates are transparent polymers with excellent toughness and good temperature stability.

Polyesters. Polyesters are available as polybutylene terephthalate and polyethylene terephthalate. Although the unfilled compound is useful in an oriented form (e.g. blow-molded parts or film), molded parts are generally made from fiber-filled compounds.

Poly(ether-ether-ketone). This polymer is an aromatic polyether with flame resistance. It can be used as a wire coating.

Polyetherimide. A recently marketed material, it has high strength and good electrical properties, and it is flame retardant.

Polyethylenes. This family consists of a large number of materials, with rigidity increasing as specific gravity increases from 0.91 to 0.96. Electrical properties are excellent. For additional polyolefins, see polybutylene, polypropylene, and polymethylpentene.

Polyimides. These polymers are noted for their outstanding thermal stability. They are flame-retardant and have good electrical properties.

Polymethylpentene. The lowest-density plastics material, this polymer has good transparency and excellent electrical properties. It is more heat-resistant than polyethylene or polybutylene.

Polyphenylene Oxides. Resins based on polyphenylene oxide are available in various grades. Generally they have good impact strength and thermal stability.

Polyphenylene Sulfide. This is a rigid resin with excellent chemical, flame, and thermal resistance.

Polypropylenes. Polymers in this hydrocarbon family have low density, good electrical properties, and good chemical resistance. They are sometimes reacted with ethylene to form copolymers with increased toughness.

Polystyrenes. This is a family of materials with good electrical properties. The base

PLASTICS **11-9**

polymer is transparent, but it is frequently modified to improve impact resistance, and the material is then no longer transparent.

Polysulfones. Polysulfone is a transparent, rigid material. Modifications of this polymer are available as polyethersulfone and polyarylsulfone. They all have good thermal stability.

Polyvinylchlorides. A widely known plastics material, these polymers can be made either rigid or flexible and are chemically resistant with good impact resistance.

Styrene Acrylonitriles. These materials are modifications of polystyrene to improve chemical resistance.

Styrene/Maleic Anhydride Copolymers. This is a recent family of polystyrene materials with improved heat distortion. Impact grades are available.

11-4. Reinforced Thermoplastics. Certain thermoplastics benefit from the addition of glass fibers. The addition of glass fibers in the range of 10 to 40% by weight results in an increase in flexural strength, modulus, creep resistance, and dimensional stability. Impact strength is not always enhanced. Table 11-3 shows typical physical properties of reinforced thermoplastics.

11-5. Thermosetting Compounds. Thermosetting molding materials are partly polymerized materials which are cured in a mold, usually with heat and pressure, thereby completing the chemical reaction. The cured polymer has thus been converted into an infusible, insoluble product. Since the change in state is chemical rather than physical, thermosetting plastics cannot be softened to the flow point by heat after curing. In general, thermosetting resins are combined with fillers and/or reinforcements to produce molding compounds. Fillers and reinforcements include wood flour, glass fibers, natural or synthetic fibers, metal fibers, and a wide variety of mineral materials such as calcium carbonate and mica.

Applicable government and ASTM standards for thermosetting molding compounds are shown in Table 11-4. Typical properties of these materials are given in Table 11-5 on pages 11-13 to 11-15.

Alkyd (Polyester). This large family of polyester materials is cured by the use of cross-linking monomers incorporated with unsaturated polyester polymers. The alkyd portion of the family is available as molding pellets, and the polyester compounds are available in forms such as bulk molding compounds (BMC) and sheet molding compounds (SMC). These materials are particularly good in electrical applications and can be made track-resistant and flame-retardant.

DAP. Diallylphthalate (DAP) and diallyl isophthalate are the most common of the allyl polymers. They are dimensionally stable, with low water absorption and excellent electrical properties. They are often selected for electronic applications.

Epoxies. Although most often used in the cast form, epoxies can also be used as molding compounds, with glass fiber or mineral reinforcements.

Melamines. These polymers are the reaction product of melamine and formaldehyde. They are hard and have good surface appearance. They have good arc resistance.

Phenolics. These polymers are the reaction product of phenol and formaldehyde. They have good chemical resistance, and properties can be varied by the wide range of filler possibilities. Electrical properties are good, but arc resistance is poor.

Silicones. Silicones are high-temperature polymers in which the polymer chains are $-Si-O-$ linkages instead of $-C-C-$ linkages. This type of linkage results in good thermal stability and a continuous-use temperature around 215°C.

Ureas. These polymers are the reaction products of urea and formaldehyde. They are usually reinforced with cellulose fibers. They have good chemical resistance and perform well electrically, but they have limited dimensional and thermal stability.

11-6. Cold-Molded Compounds. Cold-molded materials are formed in molds under high pressure at room temperature and are subsequently hardened in ovens. Cold-molding materials are either refractory (inorganic) or nonrefractory (organic), depending on the nature of the binder. Refractory materials consist of cement, lime, water glass, or silica mixed with water and a filler such as asbestos. These materials are more heat-resistant than the nonrefractory compounds and have higher arc resistance. Nonrefractory materials consist of asphalts, coal-tar pitches, oils, gums, resins, or waxes combined with fillers such as asbestos, silica, or magnesia compounds. The cold-molded materials are frequently used in electrical insulation where thermal stability and arc resistance are important. These materials are inexpensive, low in luster, and available in a limited color range.

TABLE 11-3 Properties of Reinforced Thermoplastics

Properties	ASTM test	ABS 20% G.F.	Acetal 30% G.F.	PE-TFE 25% G.F.	Nylon 6/6 20% G.F.	Ether sulfone 20% G.F.	Aryl ester 20% G.F.	Aryl ether 20% G.F.	Polycarbonate 20% G.F.	Ether-ether-ketone 30% G.F.
Electrical										
Dielectric constant	D150	—				—	—	—		—
60 Hz		—	4.0	3.4	4.0	—	—	—	3.2	—
10^6 Hz		3.4	4.0	3.4	3.5	3.9	3.6	—	3.1	—
Dissipation factor	D150	—				—	—	—		—
60 Hz		—	0.004	0.004	0.018	—	—	—	0.0009	—
10^6 Hz		0.009	0.004	0.002	0.017	0.007	0.013	—	0.0073	—
Mechanical										
Tensile strength, psi	D638	12,000	19,000	12,000	18,000	18,000	16,000	14,000	16,000	23,500
Tensile modulus, 10^5 psi	D638	7	15	12	12	10	8	9	10	3
Elongation, %	D638	3	3	8	3	3.5	3	5	5	3
Compressive strength, psi	D695	14,000	11,000	10,000	21,000	20,000	21,000	17,000	18,000	22,400
Flexural strength, psi	D790	16,000	29,000	10,700	29,000	25,000	24,000	18,000	21,000	34,400
Impact strength, ft·lb/in	D256	1.2	1.8	9	2.1	1.4	2	1.8	3	2.6
Thermal										
Distortion temperature, °F	D648	210	330	410	480	410	355	290	300	590
Continuous-use temperature, °F		165	220	390	250	350	—	240	265	—
Thermal coefficient, 10^{-5} in/(in·°C)	D696	36	40	30	41	36	38	36	27	—
Thermal conductivity, 10^{-4} (cal·cm)/(cm^2·s·°C)	C177	—	7.6	5.7	10	8	8	—	6	—

Properties	ASTM test	Ether-imide 30% G.F.	PPO 30% G.F.	PPS 30% G.F.	Propylene 20% G.F.	Styrene 20% G.F.	SAN 20% G.F.	Styrene/maleic 20% R.H.	Sulfone 30% G.F.	Urethane 20% R.H.	Polyester 30% G.F.
Electrical											
Dielectric constant	D250										
60 Hz		—	2.9	3.7	2.4	3.1	3.5	—	3.6	——	3.8
10^6 Hz		—	2.9	3.8	2.4	3.0	3.4	—	3.5	4.8	3.7
Dissipation factor	D150										
60 Hz		—	0.0009	0.004	0.002	0.005	0.005	—	0.0029	——	0.002
10^6 Hz		—	0.0015	0.001	0.003	0.002	0.009	—	0.0049	0.012	0.02
Mechanical											
Tensile strength, psi	D638	24,500	14,500	20,000	9,100	11,500	17,400	13,000	16,000	7,000	17,000
Tensile modulus, 10^5 psi	D638	13	9	16	6	10	12	10	13	1	10
Elongation, %	D638	3	5	3	2.5	2.5	2.5	4	2	30	3
Compressive strength, psi	D695	23,500	17,500	21,000	7,000	16,700	20,000	16,000	20,000	——	19
Flexural strength, psi	D790	33,000	20,000	29,000	10,000	15,000	20,000	17,000	22,000	8,000	27
Impact strength, ft·lb/in	D256	2	2.3	1.4	1.4	0.9	1	2.5	1.5	8	1.8
Thermal											
Distortion temperature, °F	D648	410	280	500	285	200	215	235	360	165	410
Continuous-use temperature, °F		330	265	—	220	180	180	—	350	——	285
Thermal coefficient, 10^{-5} in/(in·°C)	D696	2.0	3.6	2.5	4.3	4.0	3.6	3.0	2.5	7.5	4.5
Thermal conductivity,	C177	——									——
10^{-4} (cal·cm)/(cm²·s·°C)		—	4	10	7	6	7	4	8	9	—

TABLE 11-4 Thermosetting Plastics Standards

Material	ASTM	Government
Alkyd (ester)	D1201	MIL-M-14
DAP	D1636	MIL-M-14
Epoxy	D3013	MIL-M-24325
Melamine	D704	MIL-M-14
Phenolic	D700	FEDL-P-1125
	D3881	MIL-M-14
Silicone	——	MIL-M-14
Urea	D705	

Properties. Typical properties for a refractory compound and a nonrefractory material are given in Table 11-6 on page 11-17.

11-7. Laminated Sheet. Laminated products are made by bonding together two or more layers of material. The layers generally comprise reinforcing fibers such as cellulose, glass fibers, synthetic fibers, or (to a limited extent) asbestos in the form of fabrics, papers, or mats impregnated with a thermosetting resin. The laminates may be parallel-laminated, in which case the reinforcing fibers are oriented generally parallel to the strongest direction in tension, or cross-laminated, wherein the reinforcements are at 90° directions to one another. Laminates with other orientations are also manufactured. Laminates are frequently used for structural applications in electrical devices because of their superior physical properties.

Standards. Standards and specifications for thermosetting laminates have been prepared by ASTM (D709-78 and D1532-77) and NEMA (LI 1-1977). Types include paper, fabric, asbestos, glass, nylon, flame-resistant, copper-clad, thin copper-clad, PTFE, polyester, and postforming composites. The grades are as follows, and contain phenolic resin unless otherwise stated. See Table 11-7 for properties.

Paper-Based Phenolics:

Grade X. Paper-based sheet and tubes intended for mechanical applications where electrical properties are of minor importance. Not equal to fabric-based grade in impact strength.

Grade XP. Paper-based sheet intended for hot punching. More flexible and not so strong as grade X.

Grade XPC. Paper-based sheet for cold punching. More flexible, lower flexural strength and higher cold flow than grade XP.

Grade XX. Paper-based sheet, tube, and rods suitable for electrical applications. Good machinability.

Grade XXP. Paper-based sheet with better electrical and moisture properties than grade XX. More suitable for hot punching.

Grade XXX. Paper-based sheet, tube, and rod for rf work. High humidity and cold-flow resistance.

Grade XXXP. Paper-based sheet with high insulation resistance and low losses at high humidities. More suited for hot punching than grade XXX.

Grade XXXPC. Paper-based sheet similar to grade XXXP but with lower punching temperatures.

Fabric-Based Grades:

Grade C. Cotton fabric (>4 oz/yd^2) sheet, tube, and rod. Mechanical grade with high impact strength.

TABLE 11-5 Thermosetting Molded Materials Properties

Properties	ASTM test	Alkyd (ester) Glass fiber	Alkyd (ester) Mineral	DAP Glass fiber	DAP Mineral	Epoxy Glass fiber	Epoxy Mineral	Melamine Glass fiber	Melamine Cellulose
Electrical									
Arc resistance, s	D495	180	150	130	130	150	150	180	180
Dielectric constant	D150								
60 Hz		5.0	6.0	4.7	4.7	5.0	5.0	11.0	9.5
10^3 Hz		4.7	6.0	4.5	5.0	——	——	——	9.0
10^6 Hz		4.4	5.0	4.4	4.0	4.4	4.4	7.5	8.4
Dissipation factor	D150								
60 Hz		0.03	0.01	0.02	0.02	0.02	0.02	0.10	0.20
10^3 Hz		0.02	0.01	0.02	0.02	——	——	——	0.10
10^6 Hz		0.02	0.01	0.01	0.02	0.02	0.02	0.02	0.03
Dielectric strength, V/mil	D149	375	375	400	400	350	350	300	350
Volume resistivity, $\Omega\cdot$cm	D257	10^{15}	10^{15}	10^{15}	10^{14}	10^{14}	10^{14}	10^{11}	10^{12}
Mechanical									
Tensile strength, psi	D638	8,000	7,000	10,000	7,000	15,000	8,000	8,000	10,000
Tensile modulus, 10^5 psi	D638	20	20	20	20	30	10	20	10
Elongation, %	D638	1	1	4	4	4	1	1	1
Compressive strength, psi	D695	25,000	20,000	30,000	25,000	30,000	30,000	30,000	30,000
Flexural strength, psi	D790	20,000	10,000	15,000	10,000	20,000	15,000	18,000	15,000
Impact strength, ft·lb/in	D256	15	0.4	10	0.5	7	0.4	6	0.4
Hardness	D785	E95	E98	E85	E85	M105	M105	M115	M120
Thermal									
Deflection temperature, °F	D648	>400	400	400	400	>250	>250	390	360
Continuous-use temperature, °F		260	310	260	260	260	260	300	300
Coefficient of expansion, 10^{-5} in/(in·°C)	D696	3	4	3	3	3.5	5	2	4
Thermal conductivity, 10^{-4} (cal·cm)/(cm²·s·°C)	C177	2	2	10	10	7	10	11	15
Oxygen index	D2863	20–40	20–40	>30	>30	30	30	>18	>18

TABLE 11-5. Thermosetting Molded Materials Properties *(Continued)*

Properties	ASTM test	Alkyd (ester)		DAP		Epoxy		Melamine	
		Glass fiber	Mineral	Glass fiber	Mineral	Glass fiber	Mineral	Glass fiber	Cellulose
Chemical									
Water absorption, % in 24 h	D570	0.3	0.3	0.3	0.3	0.1	0.1	0.3	0.6
Radiation resistance		10^8	10^8	10^8	10^8	10^8	10^8	10^6	10^6
Not resistant to:	D543	Alkalies	Alkalies	Acids, alkalies	Acids, alkalies	Acids, oxidizers	Acids, oxidizers	Acids, alkalies	Acids, alkalies

Properties	ASTM test	Phenolic		Silicone		Urea
		Glass fiber	Cellulose	Glass fiber	Mineral	Cellulose
Electrical						
Arc resistance, s	D495	30	30	250	420	120
Dielectric constant	D150					
60 Hz		7.2	7.0	5.2	3.6	9.0
10^3 Hz		6.5	6.5	5.0	——	7.5
10^6 Hz		5.6	5.5	4.7	5.0	6.8
Dissipation factor	D150					
60 Hz		0.10	0.20	0.004	0.004	0.04
10^3 Hz		0.04	0.10	0.004	——	0.03
10^6 Hz		0.02	0.05	0.002	0.002	0.03
Dielectric strength, V/mil	D149	300	350	350	280	350
Volume resistivity, $\Omega \cdot cm$	D257	10^{13}	10^{11}	10^{15}	10^{15}	10^{11}

Mechanical						
Tensile strength, psi	D638	10,000	7,000	5,000	4,500	7,000
Tensile modulus, 10^5 psi	D638	20	10	———	———	10
Elongation, %	D638	0.2	0.6			1
Compressive strength, psi	D695	30,000	30,000	11,000	10,000	30,000
Flexural strength, psi	D790	35,000	10,000	12,000	8,000	12,000
Impact strength, ft·lb/in	D256	12	0.4	12	0.3	0.3
Hardness	D785	M110	E80	M84	M84	M115
Thermal						
Deflection temperature, °F	D648	400	350	>500	>500	260
Continuous-use temperature, °F		350	300	500	500	170
Coefficient of expansion, 10^{-5} in/(in·°C)	D696	1.5	4	4	4	3
Thermal conductivity, 10^{-4} (cal·cm)/(cm²·s·°C)	C177	12	6	7	8	9
Oxygen index	D2863	>30	25	31	31	30
Chemical						
Water absorption, % in 24 h	D570	0.6	0.6	0.2	0.1	0.6
Radiation resistance		10^8	10^6	10^6	10^6	10^7
Not resistant to:	D543	Strong acids and alkalies	Strong acids and alkalies	Strong acids and alkalies	Strong acids and alkalies	Alkalies, acids

11-16 ELECTRICAL ENGINEERING MATERIALS REFERENCE GUIDE

TABLE 11-6 Cold-Molded Materials Properties

Properties	ASTM test	Nonrefractory (organic)	Refractory (inorganic)
Electrical			
Arc resistance, s	D495	200	300
Dielectric constant	D150		
60 Hz		28	——
10^3 Hz		18	——
10^6 Hz		6.0	——
Dissipation factor	D150		
60 Hz		0.20	——
10^6 Hz		0.07	——
Dielectric strength			
Short time	D149	115	45
Step by step		75	80
Volume resistivity, $\Omega \cdot$cm	D257	10^{12}	——
Mechanical			
Tensile strength, psi	D638	3,000	2,200
Compressive strength, psi	D695	15,000	18,000
Flexural strength, psi	D790	10,000	7,500
Impact strength, ft·lb/in	D256	0.4	0.4
Hardness	D785	M85	M85
Thermal			
Resistance to heat, °F		500	1,000
Burn rate		Nil	Nil
Chemical			
Water absorption, % in 24 h	D570	0.5	0.5
Not resistant to:	D543	Strong acids, alkalies, some solvents	Strong acids

Grade CE. Sheet, tube, and rod made from the same fabric as grade C. For electrical applications requiring greater toughness than grade XX. More moisture-resistant than grade C. Not recommended for primary insulation at voltages over 600 V.

Grade L. Cotton fabric (<4 ox/yd^2) sheet, tube, and rod. Mechanical grade not so tough as grade C, and suitable for gears.

Grade LE. Sheet, tube, and rod similar to grade L. Not recommended for primary insulation at voltages over 600 V.

Grade N-1. Nylon cloth-based sheet, tube, and rod with excellent electrical properties under high humidity. Good impact strength, but creeps at elevated temperatures.

Asbestos-Based Grades:

Grade A. Asbestos-paper–based sheet and tube. More resistant to flame and heat than cellulose-based material. Limited to voltages below 250 V.

Grade AA. Asbestos-fabric sheet and tube. More heat-resistant and stronger than grade A. Not recommended for primary insulation.

Glass-Based Grades:

Grade GPO-1. Sheet of random-laid glass-fiber reinforcement (mat) with polyester resin binder. General-purpose material for mechanical and electrical applications.

TABLE 11-7 Properties of Laminated Sheet

Properties	X	XP	XPC	XX	XXP	XXX	XXXP	XXXPC
	ASTM/NEMA grades							
	Phenolics/paper							
Electrical								
Strength perpendicular to laminations, V/mil								
Short time: $\frac{1}{32}$ in	950	900	850	950	950	900	900	900
$\frac{1}{16}$ in	700	650	600	700	700	650	650	650
$\frac{1}{8}$ in	500	470	425	500	500	470	470	470
Step by step: $\frac{1}{32}$ in	700	650	625	700	700	650	650	650
$\frac{1}{16}$ in	500	450	425	500	500	450	450	450
$\frac{1}{8}$ in	360	320	290	360	360	320	320	320
Insulation resistance, MΩ (ASTM C-96/35/90)	——	——	——	60	500	1,000	20,000	50,000
Dielectric constant, 1 MHz	——	——	——	5.5	5.0	5.3	4.6	4.6
Dissipation factor, 1 MHz	——	——	——	0.045	0.040	0.038	0.038	0.038
Mechanical								
Tensile strength, psi								
Lengthwise	20,000	12,000	10,500	16,000	11,000	15,000	12,000	12,000
Crosswise	16,000	9,000	8,500	13,000	8,500	12,000	9,500	9,500
Modulus, 10^5 psi								
Lengthwise	19	12	10	15	9	13	10	10
Crosswise	14	9	8	12	7	10	8	8
Compressive strength, psi								
Flatwise	36,000	25,000	22,000	34,000	25,000	32,000	25,000	25,000
Edgewise	19,000	——	——	23,000	——	25,000	——	——
Thermal								
Coefficient, in/(in·°C)	$\longleftarrow$			2×10^{-5}				$\longrightarrow$
Temperature index, °C								
$\frac{1}{32}$–$\frac{1}{16}$ in	130	130	130	130	130	130	125	125
$\frac{1}{16}$ in and in over	130	130	130	140	140	140	125	125

TABLE 11-7 Properties of Laminated Sheet (*Continued*)

| | ASTM/NEMA grades | | | | | | | |
| | Phenolics/cloth | | | | | Polyester | | |
Properties	C	CE	L	LE	N-1	GPO-1	GPO-2	GPO-3
Electrical								
Strength perpendicular to laminations, V/mil								
Short time: 1/32 in	——	——	——	700	850	——	——	——
1/16 in	——	500	——	700	600	400	400	400
1/8 in	——	360	——	360	450	350	350	350
Step by step: 1/32 in	——	——	——	450	650	——	——	——
1/16 in	——	300	——	300	450	——	——	——
1/8 in	——	220	——	220	300	——	——	——
Insulation resistance, MΩ (ASTM C-96/35/90)	——	——	——	30	50,000	——	——	——
Dielectric constant, 1 MHz	——	——	——	5.8	3.9	——	——	——
Dissipation factor, 1 Mhz	——	——	——	0.055	0.038	——	——	——
Mechanical								
Tensile strength, psi								
Lengthwise	10,000	12,000	14,000	13,500	8,500	9,000	9,000	9,000
Crosswise	8,000	9,000	10,000	9,000	8,000	9,000	9,000	9,000
Modulus, 10^5 psi								
Lengthwise	10	9	12	10	5	11	11	11
Crosswise	9	8	9	9	5	10	10	10
Compressive strength, psi								
Flatwise	37,000	39,000	35,000	37,000	(20,000)	30,000	30,000	30,000
Edgewise	23,500	24,500	23,500	25,000	——	20,000	20,000	20,000
Thermal								
Coefficient in/(in·°C)	←		2×10^{-5}		→	——	——	——
Coefficient in/(in·°C)								
Temperature index, °C								
1/32–1/16 in	85	85	85	85	——	——	——	——
1/16 in and over	115/125*	115/125*	115/125*	115/125*	——	——	130/160*	130/160*

Properties	ASTM/NEMA grades					
	Glass cloth					
	G-3	G-5	G-7	G-9	G-10	G-11
Electrical						
Strength perpendicular to laminations, V/mil						
Short time: $\frac{1}{32}$ in	750	——	450	450	750	750
$\frac{1}{16}$ in	700	350	400	400	700	700
$\frac{1}{8}$ in	600	260	350	350	550	550
Step by step: $\frac{1}{32}$ in	550	——	400	400	500	500
$\frac{1}{16}$ in	500	220	350	350	450	450
$\frac{1}{8}$ in	450	160	250	275	350	350
Insulation resistance, MΩ (ASTM C-96/35/90)	——	100	2,500	10,000	200,000	200,000
Dielectric constant, 1 MHz	——	7.8	4.2	7.2	5.2	5.2
Dissipation factor, 1 MHz	——	0.020	0.003	0.018	0.025	0.025
Mechanical						
Tensile strength, psi						
Lengthwise	23,000	37,000	23,000	37,000	40,000	40,000
Crosswise	20,000	30,000	18,500	30,000	35,000	35,000
Modulus, 10^5 psi						
Lengthwise	18	20	16	24	26	26
Crosswise	15	18	15	20	21	21
Compressive strength, psi						
Flatwise	50,000	70,000	45,000	70,000	60,000	60,000
Edgewise	17,500	25,000	14,000	25,000	35,000	35,000
Thermal						
Coefficient in/(in·°C)	1.8×10^{-5} ⟵		1×10^{-5}		⟶ 0.9×10^{-5}	
Temperature index, °C						
$\frac{1}{32}$–$\frac{1}{16}$ in	140/170*	-/140*	170/220*	-/140*	130/140*	140/160*
$\frac{1}{16}$ in and over	140/170*	-/140*	170/220*	-/140*	130/140*	170/180*

*Elec./Mech.

TABLE 11-7 Properties of Laminated Sheet (*Continued*)

Properties	ASTM/NEMA grades						
	CEM-1	FR-1	FR-2	FR-3	FR-4	FR-5	FR-6
Electrical							
Strength perpendicular to laminations, V/mil							
Short time: $\frac{1}{32}$ in	——	900	800	650	750	750	——
$\frac{1}{16}$ in	——	650	650	600	700	700	400
$\frac{1}{8}$ in	——	470	470	475	550	550	——
Step by step: $\frac{1}{32}$ in	——	650	650	500	500	500	——
$\frac{1}{16}$ in	——	450	450	450	450	450	——
$\frac{1}{8}$ in	——	320	320	325	350	350	——
Insulation resistance, MΩ (ASTM C-96/35/90)	——	——	50,000	100,000	200,000	200,000	20,000
Dielectric constant, 1 MHz	——	——	4.6	4.6	5.2	5.2	4.3
Dissipation factor, 1 MHz	——	——	0.038	0.035	0.025	0.025	0.030
Mechanical							
Tensile strength, psi							
Lengthwise	11,000	12,000	12,000	14,000	40,000	40,000	7,000
Crosswise	9,500	9,000	9,500	12,000	35,000	35,000	7,000
Modulus, 10^5 psi							
Lengthwise	10	12	10	12	26	27	——
Crosswise	9	9	8	10	21	22	——
Compressive strength, psi							
Flatwise	37,000	25,000	25,000	29,000	60,000	60,000	30,000
Edgewise	23,500	——	——	——	35,000	35,000	——
Thermal							
Coefficient, in/(in·°C)	←————— 2×10^{-5} —————→				1×10^{-5}	0.9×10^{-5}	2×10^{-5}
Temperature index, °C							
$\frac{1}{32}$–$\frac{1}{16}$ in	85	130	75	90	130/140*	140/160*	——
$\frac{1}{16}$ in and over	115/125*	130	105	110	130/140*	170/180*	——

*Elec./Mech.

PLASTICS   **11-21**

Grade GPO-2. Sheet of random-laid glass-fiber reinforcement (mat) with polyester resin binder. Similar to GPO-1 except that it is flame-retardant.

Grade GPO-3. Construction and performance similar to grade GPO-2, except that it is also resistant to tracking under electrical arcs.

Grade G-3. Sheet, tube, and rod made of continuous-filament-type glass cloth. General-purpose grade, high impact strength, and low bond strength.

Grade G-5. Sheet, tube, and rod made of continuous-filament glass fabric with melamine resin binder. Highest mechanical strength and hardest grade. Good heat and arc resistance. Excellent electrically when dry. Good dimensional stability.

Grade G-7. Sheet made of continuous-filament-type glass cloth with silicone resin binder. Similar to grade G-6, but bond strength and flame resistance slightly lower. Higher electrical strength.

Grade G-9. Sheet, tube, and rod made of continuous-filament-type glass cloth with heat-resistant melamine resin binder. Similar to grade G-5, except more heat-resistant.

Grade G-10. Sheet, tube, and rod made of continuous-filament-type glass cloth with epoxy resin binder. Extremely high mechanical strength. Good electrical loss properties both wet and dry.

Grade G-11. Sheet, tube, and rod made of continuous-filament-type glass cloth with heat-resistant epoxy resin binder. Similar to grade G-10, but retains 50% of its initial flexural strength at 150°C after 1 h at 150°C.

Flame Resistant Grades:

Grade CEM-1. Sheet made of continuous-filament-glass surface and cellulose-paper core, with flame-retardant epoxy resin. It can be cold punched. Modifications are available as grades CEM-2 through CEM-8.

Grade FR-1. Paper-based sheet similar to grade XP, except that it has a reduced rate of burning.

Grade FR-2. Paper-based sheet, tube, and rod modified to be flame-retardant after the ignition source is removed.

Grade FR-3. Paper-based sheet, tube, and rod with epoxy resin binder. Higher flexural strength than grade XXXPC, low dielectric losses, and stable at high humidity.

Grade FR-4. Sheet, tube, and rod of glass cloth and epoxy resin binder. Similar to grade G-10, but flame-retardant.

Grade FR-5. Sheet, tube, and rod of glass cloth and epoxy resin binder. Similar to grade G-11, but flame retardant.

Grade FR-6. Sheet of random-laid glass-fiber reinforcement (mat) with polyester resin binder and fillers. Used as a printed circuit board.

TABLE 11-8   Filament-Wound Products

	Fiber type		
Property*	S-glass	Graphite	Kevlar-49
Flexural strength, 10^3 psi	240	230	77
Flexural modulus, 10^6 psi	8	18	8
Tensile strength, 10^3 psi	180	220	130
Tensile modulus, 10^6 psi	8	20	9
Short beam shear strength, 10^3 psi	10	14	10
Specific gravity	2.0	1.5	1.4

*Properties are based on undirectional prepreg, fiber volume 60%, and Fiberite 948A1 epoxy resin.

TABLE 11-9 Film Properties

Properties	ASTM	ABS	Acrylic	Cellulose acetate	Ethyl cellulose	Cellophane	CTFE	FEP	PE-CTFE	PE-TFE
Electrical										
Dielectric constant	D150									
10^3 Hz		3.0	3.8	3.6	3.1	3.2	2.6	2.0	2.6	2.6
10^6 Hz		3.0	3.3	3.2	3.0	——	2.4	2.0	2.6	2.6
10^9 Hz		——	2.6	3.2	——	——	2.3	2.0	——	——
Dissipation factor	D150									
10^3 Hz		——	0.05	0.013	0.01	0.015	0.023	0.0002	0.002	0.0008
10^6 Hz		——	0.04	——	0.04	——	0.012	0.004	0.013	0.005
10^9 Hz		——	0.01	0.038	0.05	——	0.004	0.002	——	——
Dielectric strength, V/mil	D149	375	2,400	4,000	1,500	2,200	2,200	7,000	5,000	3,500
Volume resistivity, $\Omega \cdot$ cm	D257	10^{16}	10^{16}	10^{12}	10^{15}	10^{11}	10^{18}	10^{19}	10^{16}	10^{16}
Mechanical										
Tensile strength, 10^3 psi	D882	5–10	5	7–16	8–10	7–18	6–10	2–3	8–10	7–8
Elongation, %	D882	30	80	40	25	30	100	300	200	300
Burst strength	D774	——	——	45	70	40	25	11	——	——
Folding endurance	D2176	——	6,000	1,000	Excell.	500	Good	4,000	Excell.	——
Thermal										
Maximum temperature, °F	D759	200	170	160	250	200	300	400	>300	——
Heat seal range, °F		——	250–350	350–450	——	250–300	450–500	540–700	475–500	525
Burning rate	(CS-192)	0.04	0.1	1.5	0.5	1.5	(NB)	(NB)	(SE)	(SE)
Chemical										
Water vapor transmission, $(cm^3 \cdot mil)/(atm \cdot 24 \text{ h} \cdot 100 \text{ in}^2)$	E96(B)	——	8	20	10	1–100	0.04	0.4	0.6	0.04
CO_2 permeability, g-mil/$(24 \text{ h} \cdot 100 \text{ in}^2)$	D1434	175	——	900	5,000	2	30	2,000	110	250
N_2 permeability, g-mil/$(24\text{h} \cdot 100 \text{ in}^2)$	D1434	8	——	35	600	1	3	320	10	30

Properties	ASTM	TFE	Polyvinyl fluoride	Polyvinylidene fluoride	Ionomer	Polyethylene (med. dens.)	Ethylene/ vinylacetate	Polypropylene (oriented)	Polycarbonate	Polyester (terephthalate)
Electrical										
Dielectric constant	D150									
10^3 Hz		2.1	8.5	7.5	2.4	2.2	2.8	2.2	3.0	3.2
10^6 Hz		2.1	7.4	6.1	2.4	2.2	2.8	2.2	3.0	3.0
10^9 Hz		2.1	——	3.0	2.4	2.2	——	2.2	2.9	2.8
Dissipation factor	D150									
10^3 Hz		0.0002	0.016	0.02	0.007	0.003	0.015	0.0002	0.002	0.005
10^6 Hz		0.0002	——	0.06	0.007	0.0003	0.015	0.0002	0.010	0.016
10^9 Hz		0.0002	——	0.11	0.007	0.0003	——	0.0002	0.012	0.006
Dielectric strength, V/mil	D149	1,200	3,500	300–1000	1,000	500	500	8,000	1,500	7,500
Volume resistivity, $\Omega \cdot$ cm	D257	10^{18}	10^{13}	10^{14}	10^{16}	10^{17}	10^9	10^{18}	10^{16}	10^{18}
Mechanical										
Tensile strength, 10^3 psi	D882	2–4	7–18	5–7	5	2–4	1–3	25–30	8–10	20–35
Elongation, %	D882	200	200	400	350	50–650	600	65	95	100
Burst strength	D774	——	50	——			10		>30	70
Folding endurance	D2176	Very High	20,000	26,000	Very High	Very High	Excellent	Excellent	300	Very High
Thermal										
Maximum temperature, °F	D759	500	225	270	160	220	160	290	270	300
Heat seal range, °F		——	400–425	375–425	200–500	250–400	150–300	300–320	375–430	425–450
Burning rate	(CS-192)	NB	——	(SE)	——	——	——	——	(SE)	——
Chemical										
Water vapor transmission, gm-mil/(24 h $\cdot$ 100 in^2)	E96(B)	——	——	2.6	0.4	0.7	2	0.3	11	2
CO$_2$ permeability, (cm$^3 \cdot$ mil)/(atm $\cdot$ 24 h $\cdot$ 100 in^2)	D1434	——	11	6		2,000	6,000	540	1,200	20
N$_2$ permeability, (cm$^3 \cdot$ mil)/(atm $\cdot$ 24h $\cdot$ 100 in^2)	D2434	——	0.3	9	(100)	200	400	20	50	1

TABLE 11-9 Film Properties (*Continued*)

Properties	ASTM	Nylon 6 (oriented)	Polyimide	Polystyrene (oriented)	Polysulfone	Vinylchloride acetate	PVC (nonplasticized)	PVCD/PVC	Urethane	Rubber hydrochloride
Electrical										
Dielectric constant	D150									
10^3 Hz		3.7	3.5	2.5	3.1	3.2	3.2	4.2	6.	3.0
10^6 Hz		3.4	3.4	2.5	3.0	3.0	3.0	3.5	6.	3.0
10^9 Hz		(dry)	——	2.5	3.0	2.8	2.8	2.7	——	——
Dissipation factor	D150									
10^3 Hz		0.016	0.013	0.0005	0.0008	0.016	0.016	0.06	0.05	0.006
10^6 Hz		0.025	0.010	0.0005	0.0034	0.018	0.016	0.07	——	——
10^9 Hz		(dry)	——	0.0005	0.0041	0.019	0.019	0.02	——	——
Dielectric strength, V/mil	D149	1,400	7,000	5,000	7,500	425–1300	425–1300	4,000	600–1300	——
Volume resistivity, cm	D257	——	10^{18}	10^{16}	10^{16}	10^{16}	10^{16}	10^{14}	10^{11}	10^{13}
Mechanical										
Tensile strength, 10^3 psi	D882	32–35	25	8–12	8–11	3–8	7–10	8–16	5–12	3–5
Elongation, %	D882	100	70	20	90	10–100	40	60	200–700	200–800
Burst strength	D774	——	75	25	——	——	40	30	——	Stretches
Folding endurance	D2176	Excellent	10,000	——	——	Very high	——	Excellent	——	Excellent
Thermal										
Maximum temperature, °F	D759	300	600	190	350	200	200	290	190	200
Heat seal range, °F		380–450	——	250–350	500–550	275–360	350–420	250–300	275–375	225–250
Burning rate	(CS-192)	——	(SE)	——	12	——	1	(SE)	——	(SE)
Chemical										
Water vapor transmission, gm-mil/(24 h · 100 in^2)	E96(B)	10	5	8	18	4	1	0.5	50	Low
CO_2 permeability, (cm^3 · mil)/(atm · 24 h · 100 in^2)	D1434	11	450	900	800	900	50	20	1,000	12,000
N_2 permeability, (cm^3 · mil)/(atm · 24h · 100 in^2)	D1434	1	6	50	40	40	5	1	80	500

PLASTICS

Fluorocarbon-Based Grades:

Grade GT. Continuous-filament glass-fabric-based cloth with polytetrafluoroethylene (PTFE) binder. It will meet the requirements of Class 180.

Grade GX. Similar to GT except dielectric properties are controlled within closer limits and they are used in X-band applications.

Copper-Clad Laminates. These grades consist of grades XXXP, XXXPC, CEM-1, FR-2, FR-3, FR-4, FR-5, FR-6, G-10, G-11, GT, and GX laminates. Known as copper-clad, they have copper foil bonded to one or both surfaces and are used for printed wiring. Standards are maintained by the Institute for Interconnecting and Packaging Electronic Circuits (IPC) and ASTM (D1867-78).

11-8. Filament-Wound Materials. "Filament winding" refers to the process of wrapping resin-impregnated continuous filaments of reinforcing fiber around a mandrel, forming a surface of revolution in various patterns. Then the part is cured and the mandrel removed. Fibers include glass, aramid (Kevlar from E. I. duPont), graphite, and others (including hybrid combinations); resins include epoxies, polyimides, and even thermoplastics.

Because of the wide variation in available resins, types of fibers, fiber content, and winding patterns, properties of the final product vary considerably. One set of properties is shown in Table 11-8.

Filament-wound parts have high hoop strength and have been used as fuse tubes, lightning-arrester tubes, bushings, and bands on rotors for generators and motors. Note that the graphite composites can be electrically conducting.

11-9. Films. Films are made from plastic by extrusion, calendering, or solvent casting. The polymers used for films are generally thermoplastics. Films are available in thicknesses of 0.25 to 10 mils (or more) and in a wide variety of widths. One of the most useful properties of films to an electrical designer is the high dielectric strength usually associated with these materials. Physical properties are also high. The strength properties of some films can be dramatically increased by orientation during manufacture. Orientation is the process of selectively stretching the film in one or two directions, thereby reducing its thickness and causing changes in the crystallinity of the polymer. Usually it is accomplished under conditions of elevated temperature during processing, and the benefits are lost if the processing temperature is exceeded during service.

Most films can be bonded to other substrates with a variety of adhesives. The films which do not readily accept adhesives can be surface-treated for bonding by several chemical means. Some films are combined to obtain bondable surfaces. Examples of these composites are olefins laminated to polyester film and fluorocarbons laminated to polyimide film.

Films are used extensively in wire and cable, fractional-horsepower motors, capacitors, coils, transformers, and batteries.

Properties. Typical properties for films are given in Table 11-9.

11-10. References on Plastics

1. *Modern Plastics Encyclopedia;* McGraw-Hill, New York, 1982–83. (Yearly revision.)

2. *International Plastics Selector;* Codura Publications, LaJolla, CA, 1980.

3. Harper, C. A.: *Handbook of Plastics and Elastomers;* McGraw-Hill, New York, 1975.

4. Schwartz, S. S., and Goodman, S. H.: *Plastics Materials and Processes;* Van Nostrand Reinhold, New York, 1982.

5. *NEMA Standards;* NEMA, Washington, DC.

6. *ASTM Standards;* ASTM, Philadelphia, PA, 1982.

Chapter 12

PAPER, FIBER, WOOD, AND INSULATING FABRICS

G. R. SPRENGLING

12-1 Cellulose 12-1	12-7 References on Paper, Fiber, and Wood 12-4
12-2 Unimpregnated Cellulose Paper 12-1	12-8 Insulating Fabrics 12-4
12-3 Pressboards and Fiber 12-3	12-9 Impregnated Fabrics 12-5
12-4 Impregnated Papers 12-3	12-10 Thermal Endurance of Fabrics 12-7
12-5 Altered Cellulose and Synthetic Fiber Papers 12-3	12-11 Uses of Fabrics 12-7
12-6 Wood 12-4	

12-1. Cellulose. Cellulose is one of the oldest of electrical insulating materials, and it remains today the most widely used. It is employed in the form of papers, fabrics, and pressboards. Pure cellulose, commonly termed alpha cellulose, is essentially a polymer of glucose and is thus a carbohydrate of the formula $(C_6H_{10}O_5)_n$, in which n, the number of units in the polymer, varies from 1100 to 8000, depending on the source. This molecule contains three hydroxyl groups, which cause cellulose to have both a high dielectric constant (6.0 for kraft cellulose, at 60 Hz, decreasing with the degree of crystallinity) and a high affinity for water.[1]* The latter of these properties is the salient cause of difficulty in dealing with cellulose as insulation.

Common insulating papers are made chiefly from coniferous woods but also from cotton and linen rags, rope, and other materials. These are first treated by a chemical process, most commonly the sulfite or the kraft (sulfate) process, and then comminuted to a dispersed pulp by "beating." The pulp suspension is formed into a loose sheet by filtering on a moving wire screen (the Fourdrinier) and compacted into paper by calendering with heated rolls. Accordingly, paper has different properties, chiefly mechanical, in the machine direction (MD) and that transverse to it (CMD). Thus the MD strength is 1.5 to 5 times higher but the elongation 2 to 3 times smaller than the CMD properties.

The paper- and pulp-making process does not produce pure cellulose. Even a rather pure grade of electrical paper will contain about 4% of pentosans, 3% of lignin, and 0.35 to 1.0% of ash impurities. The processing, especially the degree of beating, also determines the proportion of the cellulose present in the form of crystallites or in the much more hygroscopic amorphous form. With modern deionized water washing methods the various processes can produce paper of equivalent electrical quality.

12-2. Unimpregnated Cellulose Paper is used for a variety of purposes in insulation, such as telephone-cable insulation, household electric cable, small transformers and capacitors, and the like. So-called fish paper used in motor-slot cell insulation is a type of vulcanized fiber (which see). The electrical properties of such papers depend chiefly on their pore volume, which may vary from 10 to 60%, and on their moisture content, which is normally 7

---

*Superscripts refer to references Par. **12-7.**

12-1

to 12% in contact with air of average humidity. The apparent dielectric constant for pure, dry paper varies with porosity, as shown[2] in Fig. 12-1. The dielectric constant and loss factor vary with frequency, as shown[3] in Fig. 12-2.

The dielectric loss factor (tan δ) of dry paper may be as low as 0.0009 to 0.004 at 40°C. Both the loss factor and the dielectric constant change sharply in papers exposed to humid

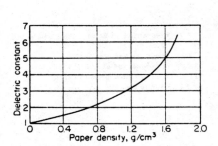

FIG. 12-1 Effect of density on the dielectric constant of cellulose paper. (Sakamoto and Yoshida, *ETJ Japan*, 1956.)

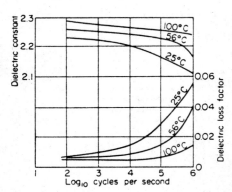

FIG. 12-2 Effect of frequency on dielectric constant and loss in cable paper. (Race, Hemphill, Endicon, *General Electric Review*, 1940, volume 43, p. 492.)

air, as shown[4] in Fig. 12-3. At the same time the volume resistivity of the paper changes in a straight line from 5×10^{15} Ω·cm in equilibrium with dry air to 10^{10} Ω·cm at 84% relative humidity. The loss factor of moist paper also rises with increasing test voltage gradient.

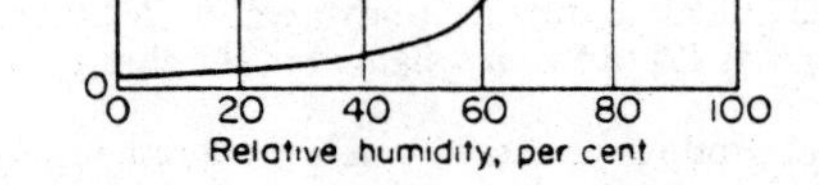

These figures demonstrate the importance of using dry paper for insulation. The drying of cellulose is a difficult operation, since at least a portion of the contained water is so tightly bound that it can be removed only at elevated temperature where the process overlaps with thermal degradation of the cellulose itself.[5] Such degradation is accelerated by the water present, especially if trapped in equipment, and also by the presence of air, some metallic oxides, alkali, and light. Commercial drying of paper at temperatures above 115 to 120°C is usually impractical. Insofar as the electrical properties achievable by drying are of major importance, as, for example, in telephone cable, paper dryness must be carefully maintained by the sealing of the insulation in an impervious sheath (of lead), sometimes even pressurized with inert, dry gas. Otherwise moisture is rapidly regained in contact with air. Because of the above factors the IEEE and NEMA accepted use of unimpregnated paper insulation extends only to apparatus operating at temperatures not above 90°C.

FIG. 12-3 Effect of ambient humidity on dielectric constant and loss in paper cable. (Dieterle, *Bulletin Swiss Electrotech Institute*, 1955, volume 22, p. 3.)

The electric strength of untreated kraft insulating paper may be 160 V/mil of thickness. It is characteristic of paper that this electric strength is a function not only of paper density but also of the distribution of defects in the electrical sense. A study on single-thickness paper has shown these to be invariably either "holes" (low-density areas) or opaque inclusions, which are usually electrically conducting. When two or more layers of paper are tested together, the defects are usually unassignable.[6]

12-3. Pressboards and fiber are other forms in which cellulose is used for insulation. Pressboard is essentially a thicker form of paper, made in thicknesses up to 0.5 in, which is used chiefly impregnated with oil in transformers. Vulcanized fiber is made by chemical treatment of cellulose (usually with zinc chloride) followed by lamination to the desired thickness. Standard grades of fiber (NEMA Standard VU-1-1954) are: (1) Electrical insulation grade, gray in color and made to ⅛-in thickness. Thinner samples of this grade are known as "fish paper" and are used as slot liners in motors. (2) Commercial grade, thicker than the above and red, black, or gray in color. (3) Bone grade, harder than the above, gray, furnished in tubes and rods. (4) Trunk grade, used for suitcases. In general, fiber is of low electric strength (50 to 300 V/mil) and high water absorption. It is used for its toughness and abrasion resistance in low-voltage applications. Fiber is also used in arc-interrupting devices, since it leaves no carbon residue when burned and evolves large quantities of gas when heated, which helps extinguish the arc.

12-4. Impregnated papers are usually impregnated with mineral oils or chlorinated hydrocarbons (Askarels) or with various synthetic resins. So impregnated paper becomes capable of high-voltage applications. Thermal stability is also improved, permitting application in apparatus operating at 105°C. As a general rule of thumb the life of such insulation can be expected to change by a factor of 2 for each 8°C change in the operating temperature. One system of paper additives has been found which will enable a rise in the operating temperature limit for oil-impregnated paper by up to 30°C or which will yield a corresponding eightfold increase in life at equal temperature.[7]

In general, the same factors affecting the electrical properties and thermal life of cellulose as noted above also dictate the properties of impregnated papers, notably the ingress of moisture. These effects are retarded but not prevented by the impregnants. Where optimum properties are needed, as in capacitors and transformers, such insulation is therefore often used in sealed enclosures. In dry, impregnated paper the effective dielectric constant of the composite can be approximated by calculation from the known constants of the impregnant and cellulose and the paper density. Impregnated paper is used at higher voltages, and the major desired property is therefore electric strength.

Vacuum-dried and oil-impregnated papers have a short-time electric strength of 300 to 900% of the untreated value. The electric strength-thickness relationship of such insulation is usually shown as $V = AT^n$, where A is the strength per unit thickness and T is the thickness. The exponent n is commonly determined at about 0.8 but for design purposes is better set at 0.5. This is because the values of both A and n vary with degree of impregnation, temperature, voltage gradient, electrode configuration, duration of voltage stress, and other factors. The effect of voltage stress duration is shown[8] in Fig. 12-4. This effect is roughly cumulative for repeated stressing. Because of the many variables involved in its performance, impregnated paper is used at voltages corresponding to only 25 to 35% of its "inherent" electric strength.

Paper is also impregnated with natural oils, synthetic ester fluids, silicones, and a variety of other materials. Though promising, these have not found wide use so far.

12-5. Altered cellulose and synthetic fiber papers have also found some use in insulation. The cellulose molecule has been chemically reacted to form cyanoethylated and acetylated papers,[9] which generally have greater thermal and moisture resistance than cellulose. Papers or "felts" are also made from polyester, acrylic, nylon, glass, and other fibers. These are generally of low electric strength as such and are therefore used as a base for resin-impregnated and laminated structures rather than by themselves.

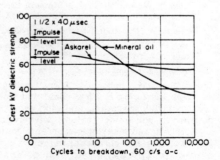

FIG. 12-4 Time dependence of electric strength in impregnated paper. (Dakin and Works, *Trans. AIEE*, 1952, volume F1-1, p. 321.)

ELECTRICAL ENGINEERING MATERIALS REFERENCE GUIDE

Nomex paper (E. I. du Pont de Nemours & Co.), made from a synthetic, aromatic poly-amide fiber, is used for insulation as such. This paper has capability for use at operating temperatures of 180°C or over. This paper has a dielectric constant of 2.6, an electric-loss factor of 0.011 at 1000 Hz, and an electric strength of 600 V/mil, all at 5-mil thickness. Mechanical properties are equal or superior to those of the best rope papers.

12-6. Wood. Wood, once extensively used as electrical insulation, today has been largely replaced by synthetic materials for general use. The dielectric constant of wood var-ies in the range 2.5 to 7.7; the dry resistivity is in the order of 10^{10} to 10^{13} $\Omega \cdot$ cm. Dry wood will withstand a potential gradient of 10 kV/in in service. These properties vary over a wide range, depending on the type of wood, season of cutting, grain direction, and especially the water content. The chief remaining use of wood in electrical apparatus is in operating rods and tie rods, where it is used in resin-impregnated form. Here its mechanical stiffness and impact resistance are of value.

By far the chief use of wood for insulation is in poles and other structural insulating members for electrical transmission lines.[10] Wood contributes to the impulse strength of such a structure at the rate of 10 to 130 kV/ft when dry. In green wood or wood wet by even a short rain, this value falls to 50 to 70% of the above for longer lengths and may approach zero in very short, wet lengths. For design purposes, these values apply with sufficient accuracy to common hard or soft woods.[11] Other electrical virtues of wood in such use lie in its arc-quenching properties, which can prevent formation of a continued power-follow arc after a flashover caused by lightning, and in the fact that the impulse strength of wood is not seriously impaired even by repeated flashover. The arc-quenching properties of wood depend on the fact that the voltage across a flashover arc in wood or on its surface does not drop to zero, as it would in air or over porcelain, but falls only to a minimum arc gradient averaging 1.1 kV rms/in. Thus, if the gradient imposed by line voltage on the wood is less than 0.5 kV rms/in, there is very low probability that a flashover caused by a surge will be sustained as an arc.[12]

12-7. References on Paper, Fiber, and Wood

1. Dakin, T. W.: The Absolute Dielectric Constant of Cellulose Fibers; *Ann. Rept., NRC Conf. on Electrical Insulation,* 1950.

2. Sakamoto, T., and Yoshida, Y.: Research on Dielectric Properties of Impregnated Paper as Composite Dielectrics; *E. T. J. (Japan),* p. 3, March 1956.

3. Race, H. H., Hemphill, R. J., and Endicott, H. S.: Important Properties of Electrical Insulating Papers; *Gen. Electr. Rev.,* vol. 43, p. 492, 1940.

4. Dieterle, W.: Acetylated Paper as an Electrical Insulating Material; *Bull. Swiss Electrotech. Inst.,* vol. 22, p. 3, 1955.

5. Flowers, L. C.: Moisture Evolution Rates Calculated for Cellulose Paper Undergoing Thermal Decomposition; *Insulation,* p. 23, September 1964.

6. Bullwinkel, E. P.: *NAS-NRC Publ.* 1238, *Conf. on Electrical Insulation,* p. 101, 1964.

7. Ford, J. G., Lockie, A. M., and Leonard, M. G.: INSULDUR—A New and Improved Heat-stabilized Insulation; *AIEE Conf. Paper* 60–936, June 1960.

8. Dakin, T. W., and Works, C. N.: Impulse Dielectric Strength Characteristics of Liquid-Impregnated Pressboard; *Trans. AIEE Paper* 52–228, 1952.

9. Dieterle, W.: Insulation Features of Acetylated Paper; *Insulation,* vol. 8, p. 19, June 1962.

10. Dean, P. S.: Insulation Tests for the Design and Uprating of Wood-Pole Transmission Lines, *IEEE Trans.,* vol. PAS-85, p. 1258, 1966.

11. Lusignan, J. T., and Miller, C. F.: *Trans. AIEE,* vol. 59, pp. 534–540, 1940.

12. Darveniza, M., Limbourn, G. J., and Prentice, S. A.: *AIEE* 31 TP 66–94.

12-8. Insulating Fabrics. Base materials for insulating fabrics include natural fibers such as cellulose, cotton, and silk; synethetic organic fibers of, for example, cellulose derivatives, polyamides (nylon), polyethylene terephthalates (for example, Dacron), poly-acrylonitrile, and many others; and inorganic fibers, chiefly glass and asbestos. Nonwoven fabrics of natural fibers overlap in properties with the papers, fiberboards, and asbestos composites. Nonwoven synthetic organic fibers are usually bonded into a fabric by use of a bonding resin or by fusion. They find electrical use chiefly as a base for resin-impregnated insulation.

PAPER, FIBER, WOOD, AND INSULATING FABRICS

Unimpregnated woven fabrics find some limited use in electrical insulation. The electric strength of such fabrics generally does not exceed the breakdown strength of an equivalent air gap and may indeed be less. Their chief use, therefore, is to confer mechanical strength, abrasion resistance, and mechanical spacing of conductors in low-voltage applications. The effects of ambient moisture and temperature on the properties of cellulosic fabrics are similar to those observed in cellulose papers (see Par. **12-2**). Often the properties of such fabrics are upgraded by impregnation of a varnish after application. Better results are generally obtained if the fabric is impregnated prior to application.

Nonwoven synthetic organic mats may also be composed of two types of fibrous material: fibrils, which compose the bulk of the material, and fibrids, which are interspersed with and have a lower fusion point than the fibrils. On hot calendering the fibrids fuse to bond the mat. Depending on the proportion of fibrids present and the calendering process, rather impervious nonwoven fabrics of fairly high electric strength may be obtained. One such product is Nomex (refer to Par. **12-5**).

Table 12-1 gives some of the properties of more common organic fibers used in insulation. It should be understood that the long-time service-temperature rating applicable to a fabric material cannot be assessed by its chemical type alone. It is also a function of the particular filament or fiber, the weave, and the treatment given in manufacture. Service temperature also varies widely with the nature of the application. It must therefore be separately determined for a specific case. The overload temperatures given in the table are intended only as a quick guide to danger areas in possible misapplication. Service temperatures of a given fabric or composites made therefrom may be lower or might even be higher.

12-9. Impregnated fabrics and those coated with elastomers are much more widely used for insulation than unimpregnated ones. The base fabric is generally woven. At least some degree of flexibility is required of all such products, ranging from smooth, rather stiff and elastic materials (for example, for slot cells) to soft and rubbery tapes. Stretchability of such materials to allow their conforming to irregular surfaces is obtained by using base fabrics with a bias weave.

TABLE 12-1   Properties of Some Polymer Fibers of Use in Insulation

Material	Examples of trade names	Tenacity dry, g/ denier	Density	Moisture absorption, %	Max, overload temp., °C*	Flammability
Silk	——	4–5	1.25	11	130°D	Burns
Cotton	——	3–5	1.54	7	110°D	Burns
Viscose rayon	——	1.5–2.4	1.53	12–13	110°D	Burns
Acetate rayon	——	1.3–1.5	1.32	3–6	120°D	Slow burn
Polyamide	Nylon 6	6.8–8.6	1.14	4	100°D	Self-exting.
Polyvinyl chloride	Vinyon HH, PCU	0.7–3.8	1.34	0.1–0.5	75°S	Self-exting.
Polyvinylidene chloride–vinyl chloride	Saran	1.5–2.0	1.70	0.1	100°D	Self-exting.
Polyacrylonitrile	Orlon, Creslan, acrilan	2.2–2.6	1.14	1.5	135°S	Slow burn
Acrylonitrile–vinyl chloride	Dynel, Vinyon N	2.5–3.3	1.30	0.3–0.4	130°S	Self-exting.
Polyethylene terephthalate	Dacron, Terylene	4.4–4.5	1.38	0.4	150°D	Slow burn
Polyethylene, LD	——	1–3	0.925	Nil	75°S	Slow burn
Polyetheylene, HD	Velon LP	5–8 (4–6)	0.955	Nil	100°D	Slow burn
Polypropylene	——	4.5–7.0	0.905	Nil	100°D	Slow burn
Polytetrafluoroethylene	Teflon	1.6–1.7	2.3	Nil	180°S	Incombustible
Aromatic polyamide	Nomex	5.5	1.38	8	200°D	Self-exting.

*These temperatures give a rough guide to the conditions a material can withstand for approximately 100 h. The suffixes indicate the chief reason for the limit: D = degrades; S = softens. These are not service temperatures. See text.

References:

1. Cook, J. G. "Handbook of Textile Fibres"; Watford, Herts, England, Merrow Publishing Co., 1960.
2. Hill, R. "Fibres from Synthetic Polymers"; Amsterdam, Elsevier Publishing Co., 1953.

TABLE 12-2 Varnish-Impregnated and -Coated Fabrics

Material	Thicknesses, in	Tensile strength MD, lb/in of width	Electric strength, V/ mil (short time)	Max.-use temp., °C	Application
1. Black varnished cambric tape	0.008–0.012	40–45	1,400–1,450	105	Flexible cable tape
2. Black bias tape and cloth	0.007–0.015	40	1,100–1,150	——	Wrapper insulation on armatures, cable joints, irregular shapes
3. Black varnished canvas	0.016–0.031	50–100	380–500	105	Coil supports, padding for end windings, bonding cushion
4. Tan varnished silk tape	0.005	16	1,600	105	Layer and wrapper insulation
5. Rag paper–cambric varnished composite	0.005–0.015	——	650–750	105	Slot cells and other applications requiring high toughness
6. Epoxy-varnished glass–polyester film	0.013–0.017	400	1,400	130	Slot and phase insulation
7. Black varnished-asbestos fabric	0.015–0.050	——	——	——	Padding
8. Epoxy-varnished polyester mat tape	0.0025–0.007	14–68	1,600–2,370	——	Dry-type transformer layer insulation
9. Class F varnished glass cloth	0.003–0.015	60–200	1,900	155	Layer and wrapper insulation
10. Self-sealing varnished glass–polyester tape	0.012	40	800	155	Encapsulation of wrapped coils
11. Silicone-varnished glass cloth	0.0035–0.013	70–250	450–1,500	180	Flexible layer, wrapper and tape
12. Silicone-rubber-coated glass cloth	0.007–0.010	125	900–1,000	180	Layer insulation and cable tape
13. Silicone-rubber-coated bias-weave glass cloth	0.007	100	585	180	Taping irregular surfaces and shapes

PAPER, FIBER, WOOD, AND INSULATING FABRICS

Electric strength is the chief electrical property demanded of impregnated fabrics, since other electrical properties, such as loss tangent, are relatively less important in the applications in which these are chiefly used. The electric strength of such fabrics is conferred upon them by the impregnation. It is a function of the completeness of impregnation primarily and secondarily of the type of resin used. The type of resin is of primary importance in determining whether electric strength is maintained after mechanical or thermal stress.

The breakdown voltage of such materials varies with thickness according to the usual logarithmic expression $V_B = AT^n$, where A is the electric strength of unit thickness and T is the thickness, expressed in the same units. The exponent n in this expression varies from about 0.66 to nearly 1.0 for impregnated fabrics. The values for both A and n for a given material vary with the test (or use) conditions approximately as follows: With increasing duration of voltage stress, A decreases sharply, often by a factor of 2 or more (or ultimately to the corona-starting voltage gradient), whereas n may go up or down; with increasing electrode area, A decreases, and n usually increases. The value of A also usually decreases with rising test temperature to a degree depending on the material. Methods for measuring electric strength and a discussion of the significance of the results are given in ASTM Standards, Pt. 29, Method D149.

12-10. Thermal endurance of varnish-impregnated fabrics may be somewhat higher than normally warranted by the base fabric used, since an impregnating resin which is itself of appropriate thermal stability will protect the fabric to some degree. In general, cellulosic fabric-based materials are limited to operating temperatures not to exceed 105°C and synthetic fiber-base materials to a maximum of 155°C (with the exception of aromatic polyimide and polytetrafluoroethylene fabrics). Above the latter temperatures glass and asbestos fabrics find use.

The maximum-use temperature is commonly determined by following the change in electric strength of an impregnated fabric with exposure time to a series of temperatures indefinitely higher than the intended service temperature. Such testing is described by ASTM Method D1830, in which curved electrodes imposing a 2% elongation on the test-sample outer surface are used for electric-strength test. This takes account of the fact that use conditions of impregnated-fabric insulation may be such as to stress the material mechanically. The logarithm of the life of the tested material to, for example, 50% decrease in electric strength may then be plotted against the reciprocal of the absolute (Kelvin) test temperature and the points extrapolated to yield the service temperature for a desired life.

Other tests on impregnated and coated fabrics for electrical insulation are described in ASTM Methods D295, D1000, D2148, D902, D350, D1458, D69. Specifications are set forth in ASTM D373, D372, D1459, D1931, and various NEMA, AIEE, and military specifications, such as MIL-1-17205C.

It should be remembered that the service voltage stresses that any material of this class may be expected to withstand in extended service are only a small fraction of the breakdown stresses determined by electric-strength tests.

12-11. Uses of impregnated and coated fabrics fall into several categories: Tapes for wrapping of conductors, cables, leads, bus bars, connections, and splices. These are made both with straight-weave substrate and with bias cut, the latter having more stretch for application to irregular surfaces. They may be varnish-impregnated, rubber- or silicone-rubber-coated, or supplied with a fusible coating to seal on heating or with a contact adhesive.

Cloth for coil wrapping or slot insulation: Often this is supplied as combinations of varnished fabric with fish paper, synthetic polymer films, or mica composites.

Sleeving for lead insulation: Generally this is a continuous braid tubular fabric impregnated with a very flexible resin or rubber.

Padding, used for spacing of coil ends and the like: This is often a resin-impregnated nonwoven fabric such as a felt or asbestoes. Electrical requirements are low.

An illustrative selection of materials typical of those available in the class of coated and impregnated fabrics (together with appropriate properties) is given in Table 12-2.

Chapter **13**

INORGANIC INSULATING MATERIALS

DOUGLAS M. MATTOX

13-1	Asbestos	13-1
13-5	Glass	13-1
13-12	Molded Mica	13-3
13-13	Porcelain	13-6

13-18	Steatite Ceramics	13-8
13-22	Inorganic Insulating Systems	13-9
13-24	Sheet Insulation	13-14
13-29	Ferroelectrics	13-15

13-1. Asbestos Sources. *Warning:* The widespread use of asbestos and products containing asbestos has been significantly curtailed because asbestos has been implicated in pulmonary disease following long term, high concentration exposure. Substitute fibers are continuously being introduced into the market. Where asbestos is used, it must be manufactured with approved employee protection and used in approved and protected applications. Consult the Occupations Safety and Health Administration.

Asbestos is a fibrous crystalline material, and chrysotile is the principal source for commercial use. It is a hydrated silicate of magnesium whose composition formula is $3MgO:2SiO_2:2H_2O$. However, the composition does vary from approximately 37 to 44% silicon oxide, 39 to 44% magnesium oxide, 12 to 15% H_2O, and up to 6% iron oxide. It is a fibrous mineral having a specific gravity of 2.2 to 2.6 to 4.0 on a Mohs scale. Individual fibers are very strong, with tensile strengths up to 400,000 lb/in². The mineral loses its water of crystallization at approximately 400°C and becomes weaker. At about 700°C all the water has been removed, and at 775°C the crystal structure changes. The melting temperature of asbestos is reported to be 1525°C. The principal sources are Canada, United States, Turkey, Rhodesia, and China. Most asbestos has magnetite and other iron oxides as impurities. Processes have been developed for removing the impurities in order to obtain good electrical insulation properties.

13-2. Fabrication of Asbestos. Asbestos fibers are extremely fine, ranging from 850,000 to 1,400,000/lin in. Under a high-powered microscope they appear to be very smooth, as if polished. Generally, fiber lengths up to 1 in are found in the commercial varieties. See warning in Par. **13-1**.

13-3. Asbestos Insulation. Asbestos is not a good insulator unless it is totally dry. It absorbs moisture too readily for many electrical applications. It is dried and impregnated with varnishes or inorganic binders like phosphate solutions. Refining processes have been developed which remove water-soluble electrolytes and iron oxide. The fibers are then dispersed and formed into thin sheets, as in papermaking. These products are known as Quinterra and Novabestos. The materials are then generally used in a treated manner, either by impregnation or in laminates. See warning in Par. **13-1**.

13-4. Untreated Asbestos. These materials were often used for low-voltage insulation or for barriers to provide separation where heat-resisting or arc-resisting properties are required. Asbestos cloth was used for electrical safety blankets. Magnet coils, mill and railway motors, and transformer coils were often insulated with asbestos-woven and paper tape. See warning in Par. **13-1**.

13-5. Glass: Constitution. Glass is an amorphous substance; it often consists of silicates, and in some cases borates, phosphates, and so forth are used. Most glass is made by fusing together some form of silica such as sand, an alkali oxide such as is found in potash

13-1

or soda ash, and some lime or lead oxide. It is a hard and brittle material when cold, but on heating sufficiently it softens, becoming plastic and, finally, very fluid. It generally breaks with a conchoidal fracture. It may be blown, pressed, cast, and cut to a great variety of medium-sized shapes. Colors are imparted by the addition of metallic oxides which absorb light of the appropriate wavelength. The number of formulas for glassmaking is infinite, but in general the chief constituent is SiO_2 (silica), which is usually 50 to 90% of the total context.

Glass generally possesses high resistivity and dielectric strength at ambient temperatures. However, there are glasses which can be made semiconducting by addition of oxides which can be multivalent, such as vanadium oxide (V_2O_5). The resistivity of glasses typically ranges from 10^{11} to 10^{18} $\Omega \cdot$cm at room temperature, while for a high-quality pure fused silica the resistivity may be as high as 10^{19} $\Omega \cdot$cm. The temperature coefficient is negative and very large. For most glasses, the conductivity is ionic, and the alkaline metal cations do most of the conducting. At high temperatures glasses become more conducting and eventually, when molten, have moderate conductivities. The conductivity is sufficient to permit resistive heating and melting. Fig. 13-1 shows dc volume resistivities as a function of temperature for some commercial glasses. The mechanical properties of glass in general are rather poor, with tensile strengths very low, in the range of 2000 to 12,000 lb/in². However, the compressive strengths are much higher, varying from 20,000 to 50,000 lb/in². The specific gravity of ordinary glass varies from 2.2 to 2.8; however, flint glass, which contains lead silicate, can have a density as high as 5.9. Coefficients of thermal expansion vary considerably with composition and range from 0.8×10^{-6} to 13×10^{-6}/°C. Specific heat is approximately 0.2. Thermal conductivity generally is approximately 0.0025 cgs unit and increases with temperature. The dielectric constant generally varies from 3.8 for fused silica to as high as 10 for lead silicate glasses having high lead content.

13-6. Dielectric Strength of Glass. The dielectric strength of glass, as with all solids, varies with the test conditions. For typical use conditions the dielectric strength is governed by the thermal breakdown process, in which the heat dissipated in the conductivity process

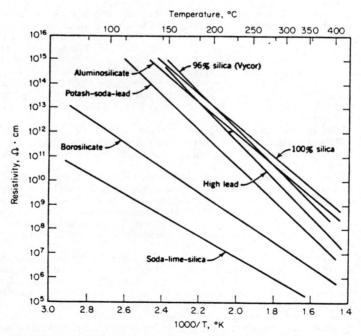

FIG. 13-1 Resistivity as a function of temperature for commercial glasses.

INORGANIC INSULATING MATERIALS

increases the temperature of the glass, leading to an increased conductivity, which in turns leads to further temperature increase of the glass. This typical process eventually leads to a runaway condition in which the conductivity of the glass increases to such a point that the glass is no longer an insulating material. Dielectric strength for glasses ranges from 50 to 3000 kV/cm. The impulse strength of glass is generally close to the dc or 60-Hz peak value. One advantage of glass, especially in the form of glass flake, is that its corona-withstand voltage, or equivalently its life under corona-discharge conditions, is very much higher than that of organic materials. However, the corona resistance of mica flakes is even higher.

13-7. Typical Commercial Glasses. · Data supplied by the Corning Glass Works are included in Table 13 and show the approximate analysis, designation, and product number of typical glasses (see *Engineering Materials Handbook;* McGraw-Hill, New York.

13-8. Electrical and Mechanical Properties of Commercial Glasses. Electrical and mechanical properties of commercial glasses are included in Table 13-1 for the glasses listed in Table 13-2. The changes of dielectric properties with temperature are shown in Figs. 13-2 and 13-3, based on data supplied by the Corning Glass Works.

13-9. Glass Fiber and Ribbon. Glass in thin sheets or ribbon drawn in thicknesses down to 0.0025 cm is now used in capacitors. Also, processes have been developed for producing glass fibers of 0.0005 cm diameter, which are then spun into yarns and woven. The fibers are made by drawing molten glass through small orifices. These fibers are made into glass cloth and tape and used either untreated or treated with organic resins as a bonding agent. There are a wide variety of glass fibers available, and new ones are continually being developed. Often the glass cloth is combined with mica into a lamination for coil and slot insulation.

13-10. Fused Silica. Quartz is a form of silica occurring in hexagonal crystals which are commonly colorless and transparent. The density is about 2.65. Fused silica is made by melting sand in an electric furnace and shaping it on exit. It is also made by the flame pyrolysis of $SiCl_4$. It is an insulating material of excellent properties and is very heat-resistant. Crystalline quartz is an anisotropic material and on heating has a coefficient of thermal expansion on its main axis of $7.8 \times 10^{-6}/°C$ and along the transverse axis of $14 \times 10^{-6}/°C$. In the temperature range from 1100 to 1400°C, fused quartz undergoes gradual devitrification (usually from the surface). The vitreous nature, acid resistivity, and very low coefficient of expansion of fused quartz give it exceptional properties as an electrical insulator. Fused silica has a density of 2.0 to 2.2, continuous-use temperatures to 1200°C, specific heat of 0.168, coefficient of thermal expansion of $0.5 \times 10^{-6}/°C$, dielectric constant of 3.8, and resistivity of 10^{14} to 10^{17} $\Omega \cdot$ cm at room temperature.

13-11. Glass–Ceramics. This class of materials is produced by introducing a nucleating agent into the glass batch, processing the batch with routine glass-making procedures, and following this by a special heat-treating schedule which develops a largely crystalline body. The composition determines the crystalline phases, and those phases determine the properties. This class of materials can be made to have very high mechanical strength (around 100,000 lb/in^2 flexural) or excellent thermal shock resistance (thermal expansion coefficients can be obtained from 10×10^{-6} to $-1 \times 10^{-6}/°C$). These materials are available in different forms from several major glass manufacturers. An example of the extent to which the properties can be controlled is illustrated by the development of machinable glass–ceramics. The Corning trademark for this material is Macor. Glass–ceramic insulators are available in which the molten glass is cast around metal hardware and then crystallized so that the two materials have similar coefficients of expansion. These have been used as transformer bushings.

Properties of Crystallized Glasses. Electrical and mechanical properties of some commercial crystallized glasses are shown in Table 13-3 from data supplied by the Corning Glass Works.

13-12. Molded Mica and Glass. This is a composition made up of ground or crushed mica with a fusible vitreous binder of soft glass or lead borate, in the correct proportions of two parts mica to one part soft glass. The mixture is heated to a temperature sufficient to soften the glass, and then the product is molded, under hydraulic pressure, while hot. It is a vitreous material which may be machined, sawed, drilled, and tapped. Low-melting metals may be cast around the composite, or metal inserts may be used. The material has been used for vacuum tube faces, insulated pipe joints, etc. It may be used to temperatures of about 250°C. A trade name is Mycalex.

TABLE 13-1 Approximate Compositions of Commercial Glasses
(Corning Glass Works)

Designation	Corning Glass no.	Percent								
		SiO_2	Na_2O	K_2O	CaO	MgO	BaO	PbO	B_2O_3	Al_2O_3
Silica glass, fused silica	—	99.5+	—	—						—
90% silica glass*	7900, 7910, 7911	96.3	<0.2	<0.2	—	—	—	—	2.9	0.4
Soda-lime, plate glass	—	71–73	12–14	—	0–12	1–4	—	—	—	0.5–1.5
Soda-lime, electric lamp bulbs	0080	73.6	16	0.6	5.2	3.6	—	—	—	1
Lead silicate, electrical	0010	63	7.6	6	0.3	0.2	—	21	0.2	0.6
Lead silicate, high lead	0120	56.5	5.4	8.6	—	—	—	29.5		
Borosilicate, low expansion	7740	80.5	3.8	0.4	—	—	—	—	12.9	2.2
Borosilicate, low electrical loss†	7070	70.6	2.1	0.4	0.1	0.2	—	—	25.2	1.1
Borosilicate, tungsten sealing	7050	67.3	4.6	1.0	—	0.2	—	—	24.6	1.7
Aluminosilicate	1710, 1720	57	1.0	—	5.5	12	—	—	4	20.5

*Vycor brand.
†Pyrex brand.

TABLE 13-2 Mechanical and Electrical Properties of Glass Ceramics

Properties	Pyroceram 9606	Pyroceram 9608	Machinable glass-ceramics
Physical			
Specific gravity	2.61	2.50	2.67
Thermal conductivity, $(Btu \cdot ft)/(ft^2 \cdot h \cdot °F)$	1.95	1.13	1.07
Coefficient of thermal expansion, per °F, 77–570°F	3.2×10^{-6}	$0.2–1.1 \times 10^{-6}$	3.94×10^{-6}
Thermal shock resistance	Good	Good	Excellent
Specific heat, $Btu/(lb \cdot °F)$	0.185	0.190	NA*
Mechanical			
Modulus of rupture, lb/in^2	20,000	16,000–23,000	23,000
Electrical			
Log volume resistivity at 500°C, $\Omega \cdot cm$	7.1	5.2	6.1
Loss tangent at 25°C (10 kHz)	0.008	0.02	0.0054
Dielectric constant at 25°C (10 kHz)	5.6	6.8	6.6
Dielectric strength, V/mil at 25°C	250–350	NA*	1000–2000

*Data not available.

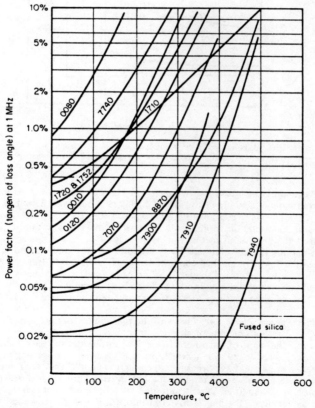

FIG. 13-2 Power factor as a function of temperature for commercial glasses. *(Data courtesy of Corning Glass Works.)*

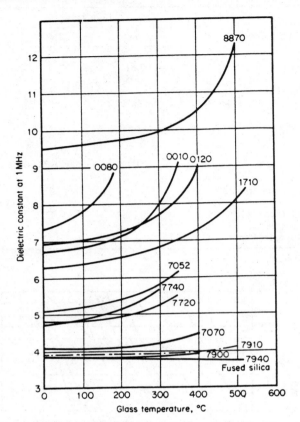

FIG. 13-3 Dielectric constant as a function of temperature for commercial glasses. *(Data courtesy of Corning Glass Works.)*

13-13. Porcelain: Constitution. A typical porcelain used for electrical insulation consists of approximately 28% china clay, 10% ball clay, 35% feldspar, 25% flint, and 2% talc. The feldspar, which has a composition $K_2O:Al_2O_3:6SiO_2$, serves as the fluxing material which assists in dissolving the more insoluble constituents, clay and quartz. The porcelain is usually fused at 1200°C. During the process there is about 12% shrinkage. The raw materials used in making the porcelain are finely ground and intimately mixed in a liquid suspension. The mixed material is then converted to a plastic state by filtration under pressure; it is next molded into the desired shape, dried, dipped in a glaze, and finally fired at a slowly rising temperature, which fuses the flux in the body of the material and also converts the glaze into a smooth glass coating. There are three different processes for making porcelain insulators.

13-14. Wet-Process Porcelain. Pieces of the plastic material about the consistency of putty are worked into a convenient size and shape and placed on plaster of paris molds which determine the shape of the inner surface. The surface not in contact with the mold is worked into the desired shape by machine forming ("jiggering") or pressing. After the pieces have been partly dried and have become stiff enough to handle, they are removed from the mold. The surfaces which were not in contact with the mold are finished to accurate dimensions. The pieces are then dried, glazed, and fired. Another method of making wet-processed porcelain is by extruding the plastic material into tubes or bars. In some cases, finishing operations are performed on lathes where the desired shapes are obtained by turning. This

TABLE 13-3 Electrical and Mechanical Properties of Commercial Glasses

	Volume resistivity, $\Omega \cdot cm$		Dielectric properties at 1 MHz and 20°C		Specific gravity	Young's modulus, lb/in^2	Coefficient of expansion, °C	Viscosity data, °C		
	250°C	350°C	Power factor	Dielectric constant				Strain point*	Annealing point	Softening point
Silica glass, fused silica	10^{12}	10^9	0.0002	3.78	2.20	10×10^6	5.5×10^{-7}	1070	1140	1667
99% silica glass (7900)	10^9	10^8	0.0005	3.8	2.18	9.7×10^6	8×10^{-7}	820	910	1500
99% silica glass (7911)	10^{11}	10^9	0.0002	3.8	2.18	9.7×10^6	8×10^{-7}	820	910	1500
Soda lime, plate glass	$10^1 - 10^7$	10^6	0.004–0.010	7.0–7.6	2.46–2.49	10×10^6	87×10^{-7}	515	550	735
Soda lime, electric lamp bulbs	10^7	10^6	0.009	7.2	2.47	9.8×10^6	92×10^{-7}	478	510	696
Lead silicate, electrical	10^8	10^7	0.0016	6.6	2.85	9.0×10^6	93×10^{-7}	397	428	626
Lead silicate, high lead	10^{10}	10^8	0.0012	6.7	3.05	8.6×10^6	89×10^{-7}	395	435	630
Borosilicate, low expansion	10^8	10^6	0.0016	4.6	2.23	9.8×10^6	32×10^{-7}	515	555	820
Borosilicate, low electrical loss	10^{11}	10^9	0.0006	4.0	2.13	6.8×10^6	32×10^{-7}	455	490	——
Borosilicate, tungsten sealing	10^8	10^7	0.0033	4.9	2.25	——	46×10^{-7}	461	496	703
Aluminosilicate	10^{11}	10^9	0.0037	6.3	2.63	12.7×10^6	42×10^{-7}	672	712	915

*Safe operating temperature 10 to 20°C below strain point, on the assumption that temperature differences within the body are not severe. Mechanical strength retained up to this temperature. (Data courtesy of Corning Glass Works.)

13-8   ELECTRICAL ENGINEERING MATERIALS REFERENCE GUIDE

process is being tried extensively in making high-voltage porcelain. Disks for suspension insulators are made by hot pressing. The mold is plaster of paris into which the soft clay is placed. The heated plunger presses the material into the mold and also forms the other surface.

13-15. Casting Process for Porcelain. Porcelains of high electric strength and of complicated form can be made by pouring water-based clay, sand, and feldspar suspensions into multipart plaster of paris forms. The cast pieces are removed from the molds after they have stiffened sufficiently to permit handling. Plaster of paris is particularly useful because it absorbs water and accelerates the drying. The molded shapes are dried, finished, glazed, and fired. The large high-voltage insulators commonly used are made by casting.

13-16. Dry-Process Porcelain. The preparation of the body consists in taking the pressed material and drying, crushing, dampening, and pulverizing it. The result of pulverization of the dampened mass is production of granules. These are pressed in a die, where they adhere to each other. The molded shapes are dried, finished to dimensions, glazed, and fired. The general result is a nonuniform porous mass which is not capable of withstanding high voltages but is suitable for low-voltage applications and for use in dry locations.

13-17. Porcelain Glazes. The impervious character of porcelain is not obtained by the glaze but is the result of the vitrification of the primary body. The glaze is a smooth glass coating which provides a finish and prevents the accumulation of dirt. In order to avoid cracks, it is important to employ a glaze which has the same coefficient of thermal expansion as the underlying porcelain.

13-18. Steatite Ceramics. These are products formed by firing steatite, or talc, which is a magnesium silicate in the form of $Mg_3(OH)_2:Si_4O_{10}$. The natural material is a soft creamy white and has a soapy feel. The material is machined to dimension and then fired, resulting in a very hard product with small dimensional changes. The water absorption in the naturally fired stone is high. These ceramics are used for low-tension and high-frequency insulators and for use at elevated temperatures for the thermal shock resistances desired. Trade names of steatite insulators include Lava, Lava Rock, Lavite, and AlSiMag.

13-19. Titania Bodies. These are formed from rutile crystals of titanium dioxide. The mineral rutile (TiO_2) has a dielectric constant of 183 parallel to the crystalline axis and 83 perpendicular to the crystalline axis. Titania bodies are used for high-frequency low-loss insulators and dielectrics.

13-20. Other Ceramic Insulators. Many other minerals are used to produce ceramics having special characteristics such as low thermal coefficient of expansion, low loss, and so forth. Some of these materials are alumina (Al_2O_3), forsterite (Mg_2SiO_4), mullite $(Al_6Si_2O_{13})$, sillimanite $(AlSiO_5)$, spodumene $(Li_2Al_2Si_4O_{12})$, zircon $(ZrSiO_4)$, and wollastonite $(CaSiO_3)$. Spark plug porcelains are generally high-alumina materials, the remainder consisting of oxides of silicon, zirconium, magnesium, and manganese. A comprehensive listing of mechanical, thermal, and electrical properties is given in Table 13-4. Fig. 13-4 shows the thermal resistivity of a variety of ceramics over a wide temperature range.

The science of ceramics has also led to the production of essentially pure single-phase polycrystalline bodies. These are now available from several suppliers as the most commonly used oxides (for example, Al_2O_3, MgO, spinel, etc.).

Single-crystal ceramics are also available as continuously drawn tapes and tubes. The outstanding properties of single crystals thus become available from low-cost processing.

It should be noted that the properties of ceramic materials generally exhibit a strong dependence on impurity level, crystalline imperfection, stoichiometry, and microstructure. In addition, the method of measurement is important, particularly at elevated temperatures. Thus, the values given in the literature by different investigators may be inconsistent, although an estimate of consistency may be obtained from density or purity values, which should be given.

13-21. References on Inorganic Insulating Materials

1. Norton, F. H.: *Refractories,* 3d ed.; McGraw-Hill, New York, 1949.

2. Green, A. T., and Stewart, G. H. (eds.): *Ceramics, A Symposium;* British Ceramics Society, London, 1953.

3. Ryschkewitsch, E.: *Oxydkeramik der Einstuffsystemme;* Springer-Verlag, Berlin, 1949.

4. Lebean, P., and Trombe, F. (ed.): *Les Hautes Temperatures et leur Utilization en Chimie,* vol. 2; Masson et Cie, Paris, 1950.

INORGANIC INSULATING MATERIALS

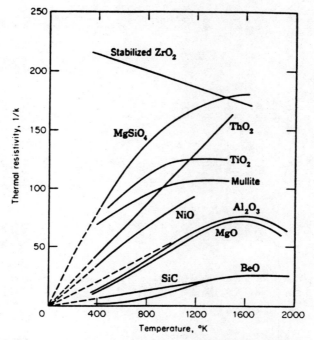

FIG. 13-4 Thermal resistivity as a function of temperature for ceramics.

5. Campbell, E. I. (ed.): *High-Temperature Technology;* John Wiley, New York, 1956.
6. Dodd, A. D.: *Dictionary of Ceramics;* George Newnes, London, 1964.
7. Kingery, W. D., Bowen, H. K., and Uhlmann, D. R.: *Introduction to Ceramics,* 2d ed.; Wiley, New York, 1976.
8. Popper, P.: *Speical Ceramics;* Academic, New York, 1960.
9. Kingery, W. D.: *Property Measurements at High Temperature;* Wiley, New York, 1959.
10. Ryshkewitsch, E.: *Oxide Ceramics;* Academic, New York, 1960.
11. Clark, F. M.: *Insulating Materials for Design and Engineering Practice;* Wiley, New York, 1962.
12. *Standard Handbook for Electrical Engineers,* 9th ed.; McGraw-Hill, New York, 1957.
13. Doremus, R. H.: *Glass Science;* Wiley, New York, 1973.
14. McMillan, P. W.: *Glass Ceramics,* 2d ed.; Academic, London, 1979.

13-22. Inorganic Insulating Systems. A relatively new concept in electrical insulation is complete systems composed only of inorganic materials for insulating electrical devices. Such materials as mica, glass, asbestos, cements, and ceramics have been used for years, but mostly in conjunction with organic resins or insulation. With the advent of metallic phosphates, alkaline silicates, and glasses as replacements for the organic resins, a complete insulation system has evolved. Some of the materials most commonly used in inorganic insulation systems are described below.

A new composite insulation system was developed for the U.S. Air Force to insulate the rotor field coils in a high-temperature generator for 300°C service. The heart of the insulation is a transferable ceramic tape, approximately 2 to 3 mils thick (5.1 to 7.6 × 10^{-2} mm), and composed of a mixture of glass and alumina. The most important innovation of this tape is a fugitive organic fiber backing, which imparts mechanical strength to the tape, allows easy wrapping of the conducting wire, and burns off completely during subsequent

TABLE 13-4 Mechanical, Thermal, and Electrical Properties of Refractories

	Composition	Porosity, vol. %	Fusion temp., °C	Max. normal use temp., °C	Density bulk (b), true (t), g/cm^3	Specific heat capacity, cal/(g·°C), 20–1000°C	Linear-expansion coeff., 10^{-6} in/(in·°C), 20–1000°C
Special refractories							
Sapphire crystal	99.9 Al$_2$O$_3$	0	2030	1950	3.97(t)	0.26	8.6
Sintered alumina	99.8 Al$_2$O$_3$	3–7	2030	1900	3.97(t)	0.26	8.6
Sintered beryllia	99.8 BeO	3–7	2570	1900	3.03(t)	0.50	8.9
Hot-pressed boron nitride	98 BN, 1.5 B$_2$O$_3$	3–7	2730	1700	2.25(t)	0.39	0.77–7.51
Hot-pressed boron carbide	99.5 B$_4$C	2–5	2450	1900	2.52(t)	0.36	4.5
Sintered calcia	99.8 CaO	5–10	2600	2000	3.32(t)	0.23	13.0
Graphite	99.9 C	20–30	3700	2600	2.22(t)	0.34	1.5–2.5
Sintered magnesia	99.8 MgO	3–7	2800	1900	3.58(t)	0.25	13.5
Molybdenum	99.8 Mo	0	2625	2200	10.2(t)	0.065	5.45
Sintered molybdenum silicide	99.8 MoSi$_2$	0–10	2030	1700(air)	6.2(t)	0.11	9.2
Sintered mullite	72 Al$_2$O$_3$, 28 SiO$_2$	3–10	1810	1750	3.03(t)	0.25	5.3
Sintered forsterite	99.5 Mg$_2$SiO$_4$	4–12	1885	1750	3.22(t)	0.23	10.6
Sintered spinel	99.8 Mg Al$_2$O$_4$	3–10	2135	1850	3.58(t)	0.25	8.8
Platinum	99.9 Pt	0	1774	1550	21.45(t)	0.035	10.1
Platinum-20 rhodium	80 Pt-20 Rh	0	1900	1650	18.74(t)	0.048	10.3
Dense silicon carbide	98 SiC, 1–2 Si, <1C	2–5	>2700	1600(air)	3.22(t)	0.20	4.0
Tantalum	99.8 Ta	0	3000	2000	16.6(t)	0.036	6.5
Sintered titanium carbide	98 TiC, <1C, ≤1O	3–10	3140	2500	4.25(t)	0.18	7.4
Sintered titania	99.5 TiO$_2$	3–7	1840	1600	4.24(t)	0.20	8.7
Tungsten	99.8 W	0	3410	3000	19.3(t)	0.034	4.0
Sintered thoria	99.8 ThO$_2$	3–7	3050	2500	10.00(t)	0.06	9.0
Sintered yttria	99.8 Y$_2$O$_3$	2–5	2410	2000	4.50(t)	0.13	9.3
Sintered urania	99.8 UO$_2$	3–10	2800	2200	10.96(t)	0.06	10.0
Sintered stabilized zirconia	92 ZrO$_2$, 4 HfO$_2$, 4 CaO	3–10	2550	2200	5.6(t)	0.14	10.0
Sintered zircon	99.5 ZrSiO$_4$	5–15	2420	1800	4.7(t)	0.16	4.2

Commercial refractories

Silica glass	99.8 SiO_2	0	1710	1100	2.20(t)	0.18	0.5
Vycor glass	96 SiO_2, $4B_2O_3$	0	——	950	2.18(t)	0.19	0.7
Pyrex glass	81 SiO_2, 13 B_2O_3, 2 Al_2O_3, 4 M_2O	0	——	650	2.23(t)	0.20	3.2
Mullite porcelain	70 Al_2O_3, 27 SiO_2, 3 MO + M_2O	2–10	1750	1400	2.8(b)	0.25	5.5
High-alumina porcelain	90–95 Al_2O_3, 4–7 SiO_2, 1–4 MO + M_2O	2–5	1800	1500	3.75(b)	0.26	7.8
Steatite porcelain	35 MgO, 60 SiO_2, 5 Al_2O_3	2–5	1450	1200	2.7(b)	0.26	10.2
Superduty fireclay brick	40–45 Al_2O_3, 55–50 SiO_2	10–20	1750	1650	2.1(b)	0.26	5.3
Magnesite brick	83–93 MgO, 2–7 FeO	10–20	2100	1750	3.2(b)	0.25	14.0
Bonded silicon carbide	90 SiC, 6–9 SiO_2, 2–4 Al_2O_3	20	2200	1450	2.95(b)	0.20	4.5
2000°F I.F.B.	Fireclay or diatomaceous	80–85	1600	1090	0.52(g)	0.25	5.3
2600°F I.F.B.	Fireclay	72–77	1750	1430	0.75(b)	0.25	5.3
3000°F I.F.B.	Aluminous fireclay	60–65	1800	1650	1.04(b)	0.25	5.3

	Thermal conductivity, $(cal \cdot cm)/(s \cdot cm^2 \cdot °C)$		Modulus of rupture (MR) or tensile strength (TS), lb/in^2		Modulus of elasticity, $10^6 \, lb/in^2$	Thermal stress resistance	Electrical resistivity, $\Omega \cdot cm$		Notes
	At 100°C	At 1000°C	At 20°C	At 1000°C			20°C	1000°C	
Special refractories									
Sapphire crystal	0.072	0.019	40,000–150,000 (MR)	30,000–100,000 (MR)	55	Very good	$>10^{14}$	10^8	1
Sintered alumina	0.069	0.014	30,000 (MR)	22,000 (MR)	53	Good	$>10^{14}$	5×10^7	2
Sintered beryllia	0.500	0.046	20,000 (MR)	10,000 (MR)	45	Excellent	$>10^{14}$	10^8	3
Hot-pressed boron nitride	0.04–0.07	0.03–0.06	7,000–15,000 (MR)	1,000–2,150 (MR)	12	Good	10^{10}	10^4	4
Hot-pressed boron carbide	0.07	0.05	50,000 (MR)	40,000 (MR)	42	Good	0.5	——	5
Sintered calcia	0.033	0.017	——	——	——	Fair–poor	$>10^{14}$	10^6	6
Graphite	0.30	0.10	3,500 (TS)	4,000 (TS)	1.3	Excellent	10^{-3}	10^{-3}	7
Sintered magnesia	0.082	0.016	14,000 (MR)	12,000 (MR)	30.5	Fair–poor	$>10^{14}$	10^7	8
Molybdenum	0.35	0.28	90,000–250,000 (TS)	30,000 (TS)	45	Excellent	5.2×10^{-6}	24×10^{-6}	9
Sintered molybdenum silicide	0.075	0.03	100,000 (TS)	40,000 (TS)	59	Good	2.2×10^{-6}	——	10
Sintered mullite	0.013	0.008	12,000 (MR)	7,000 (MR)	21	Good	$>10^{14}$	——	11
Sintered forsterite	0.010	0.005	10,000 (MR)		——	Fair–poor	$>10^{14}$	10^6	12
Sintered spinel	0.033	0.013	12,300 (MR)	11,000 (MR)	34.5	Fair	$>10^{14}$	10^6	13
Platinum	0.166	0.22	24,000 (TS)	8,000 (TS)	21.3	Excellent	11.4×10^{-6}	45×10^{-6}	14

TABLE 13-4 Mechanical, Thermal, and Electrical Properties of Refractories (*Continued*)

Platinum-20 rhodium	—	—	70,000 (TS)	30,000 (TS)	28	Excellent	20.8×10^{-6}	32×10^{-6}	15
Dense silicon carbide	0.133	0.05	24,000 (MR)	24,000 (MR)	68	Excellent	10	4	16
Tantalum	0.130	0.12	50,000–180,000 (TS)	25,000 (TS)	27	Excellent	12.4×10^{-6}	54×10^{-6}	17
Sintered titanium carbide	0.08	0.02	160,000 (TS)	140,000 (TS)	45	Very good	1×10^{-4}	—	18
Sintered titania	0.015	0.008	8,000 (MR)	6,000 (MR)	—	Excellent	5.48×10^{-6}	25×10^{-6}	19
Tungsten	0.40	0.30	100,000–500,000 (TS)	60,000 (TS)	60	Fair–poor	$>10^{14}$	10^4	20
Sintered thoria	0.022	0.007	12,000 (MR)	7,000 (MR)	21	Fair–poor	$>10^{14}$	10^5	21
Sintered yttria	—	—	—	—	Fair–poor			22	22
Sintered urania	0.020	0.007	12,000 (MR)	18,000 (MR)	25	Fair–poor	—	—	23
Sintered stabilized zirconia	0.005	0.005	20,000 (MR)	15,000 (MR)	22	Fair–good	10^8	500	24
Sintered zircon	0.015	0.008	12,000 (MR)	6,000 (MR)	30	Good	$>10^{14}$	10^5	25
Commercial refractories									
Silica glass	0.004	0.012	15,500 (MR)	—	10.5	Excellent	$>10^{14}$	10^6	26
Vycor glass	0.004	—	10,000 (MR)	—	10.5	Excellent	$>10^{14}$	—	27
Pyrex glass	0.004	—	10,000 (MR)	—	10	Very good	$>10^{14}$	—	28
Mullite porcelain	0.007	0.006	10,000 (MR)	6,000 (MR)	10	Good	$>10^{14}$	10^4	29
High-alumina porcelain	0.05	0.015	50,000 (MR)	—	53	Very good	$>10^{14}$	10^4	30

Steatite porcelain	0.008	0.006	20,000 (MR)	——	10	Fair–poor	$>10^{14}$	10^5	31
Superduty fireclay brick	0.004	0.005	750 (MR)	700 (MR)	14		10^8	10^4	
						Poor			32
Magnesite brick	0.040	0.009	4,000 (MR)	4,000 (MR)	25	Fair	10^8	10^5	33
Bonded silicon carbide	0.080	0.030	2,000 (MR)	2,000 (MR)	50		10^5	10^3	
						Excellent			34
2000°F I.F.B.	——	40 (MR)	——	——		Fair–good			
2600°F I.F.B.	0.0005	0.0011	170 (MR)	——	——	Fair–good	——	35	36
3000°F I.F.B.	0.0007	0.0014	290 (MR)	——	——	Fair–good	——	——	37

Notes: (1) Single crystals available; superior high temperature strength. (2) Best mechanical properties of sintered oxides. (3) Highest thermal conductivity of oxides; powder highly toxic. (4) Properties strongly directional; oxidizes in air above red heat. (5) Very hard, oxidizes readily at red heat. (6) Hydrates in damp air; unstable in strongly reducing atmosphere. (7) Strength increased to max. 7,000 lb/in (TS) at 2500°C; oxidizes air; reacts $H_2 > 2500$°C. (8) Dissociated in vacuo and reducing atmospheres > 1700°C. (9) Vapor pressure 2 μm at 2150°C. (10) Brittle; forms protective SiO_2 coating in air. (11) Good load-bearing capacity up to fusion temperature. (12) Good for some electrical applications and for relatively high expansion coeff. (13) Few advantages over alumina. (14) Tensile strength 2,000 lb/in^2 at 1600°C; 5×10^{-5} %/h creep at 500 lb/in^2, 750°C. (15) Tensile strength 8,000 lb/in^2 at 1600°C; 1×10^{-1} %/h creep at 500 lb/in^2, 750°C. (16) Excellent load resistance at high temperatures. (17) Good machinability; embrittled in hydrogen. (18) Oxidizes rapidly at temperatures above red heat. (19) Loses oxygen in nonoxidizing atm.; samples weak unless fine-grained. (20) Single-crystal tensile strength of 3,400 lb/in^2 at 2250°C reported. (21) Excellent chemical and mechanical stability. (22) Other rare earth oxides similar. (23) Gains oxygen and shatters on heating in air. (24) Good deformation resistance at very high temperatures; good slag resistance. (25) Dissociates into $ZrO_2 + SiO_2$ at high temperatures. (26) Devitrifies at temperatures above 1100°C. (27) Can be used in place of fused silica in many applications. (28) Most useful laboratory glass. (29) Good for laboratory ware. (30) Best porcelain available. (31) Mainly for electrical applications. (32) Not much used in laboratory. (33) Good resistance to basic slags. (34) Good thermal conductivity and load-bearing capacity. (35) Light weight, good thermal insulation, easy to shape and use. (36) Light weight, good thermal insulation, easy to shape and use. (37) Light weight, good thermal insulation, easy to shape and use.

Source: Kingery, W. D.: *Property Measurements at High Temperature;* Wiley, New York, 1959.

13-14　　ELECTRICAL ENGINEERING MATERIALS REFERENCE GUIDE

firing. These advantages are totally absent in standard ceramic tapes. The final consolidation of the rotor coil into a monolithic structure is done by using an inorganic encapsulation compound obtained commercially.

13-23. Conductor Insulation. The electrical insulation on a conductor is primarily used to direct the current flow in the conductor in the desired direction. Some examples are as follows:

a. *Fibrous insulation:* The covering of conductors with fibrous glass and fibrous crystal is the most usual insulating process. The glass fibers have the lower space factor (3 to 7 mils) and lower temperature resistance (400°C).

b. *Enamel and refractory glass:* A combination of finely divided, low-melting lead glass and a refractory such as silica or alumina is dispersed in an organic wire enamel. This enamel composition is applied to a conductor and cured, and the conductor is used in an electrical device. The conductor, which is held in place by an inorganic encapsulant, is then fired to about 650 to 700°C to burn out the organic enamel and fuse the glass–refractory combination. The insulated conductor has a conductor-to-conductor electric strength of 450 V and a temperature use limit of 500°C.

c. *Glass coatings:* Glass coatings fused onto a conductor have provided electrical insulation capabilities up to 400°C and 400 V electric strength between conductors. The coating produces very small cracks on bending and may fracture and separate on small radius bends. A very thin organic coating such as a polytetraflouroethylene (Teflon) coating may be applied over the glass to improve handling characteristics.

d. *Chemical coatings:* Anodized aluminum wire has found some uses at temperatures up to 500°C. The outer layer of aluminum on the conductor has been converted electrochemically to aluminum oxide, an electrical insulator. The conductor-to-conductor electric strength averages 400 V, despite the fact that microscopic cracks are found when the conductor is wound into electrical devices.

13-24. Flexible Sheet Insulation. Flexible sheet insulation is primarily used on curved or uneven surfaces to prevent grounding of electrical conductors to metal components and to prevent electrical breakdown between voltage differentials.

a. *Mica:* The only positive barrier to electrical breakdown is provided by mica. Mica splittings or paper can be combined with an inorganic binder to provide a semiflexible sheet material having an electrical strength of 400 V/min. The temperature limit of use is 550°C.

b. *Synthetic mica paper:* A highly flexible synthetic mica paper is physically strong and thermally stable to 980°C. The electric strength is 140 V/mil at room temperature, with little loss up to 600°C. A reversible dehydration amounting to 4.5% by weight occurs at 110°C and may cause delamination of the paper.

c. *Other fibrous papers:* A number of other fibrous papers can be used as thin electrical insulating sheets, but all have low (about 35 to 50 V/mil) electric strength. Electrical-grade glass cloth has high initial strength but loses most of its physical strength after exposure to 450°C. The new S-type glass cloth retains more strength at elevated temperatures and can be used to 500°C. Synthetic aluminum silicate fiber mats are electrically useful to 750°C. Asbestos in paper or tape form has fair to good physical strength, especially when combined with organic or glass fibers. Dehydration at about 500°C substantially weakens it, but electrical properties are maintained. A great number of other fibers have appeared on the market in paper, mat, and fabric form for use in specialized applications.

13-25. Rigid Sheet Insulation. Rigid sheet is generally used where both electrical insulation and structural strength are required.

a. *Mica sheet:* Both mica flakes and mica paper are made into rigid sheets using glass and metal phosphate as a bond. Electric strengths up to 2500 V/mil can be obtained with flexural strengths up to 20,000 lb/in². Some grades can be used up to 800°C.

INORGANIC INSULATING MATERIALS 13-15

b. *Asbestos sheet:* A number of grades of asbestos boards are made using cement or alkaline silicate bonds. The length of the asbestos fiber, from almost a powder to 2 in, greatly influences the structural strength of the sheet. The low-density, short-fiber sheets have flexural strengths of about 500 lb/in^2, while the long-fiber sheets can achieve a strength of 32,000 lb/in^2. All are somewhat porous, thus giving a low electrical strength of 35 to 60 V/mil. The asbestos fiber generally used is chrysotile, which dehydrates and loses strength rapidly above 500°C, so its use temperature is limited to this value. See warning in Par. **13-1**.

c. *Ceramic sheet:* A number of ceramics can be made into sheets for insulation and structural purposes.

d. *Glass-mica sheet:* Glass-bonded mica sheet material is made from both natural and synthetic mica. The natural mica composition is usable to 450°C and the synthetic mica to 620°C. Maximum electric strength is 500 V/mil, and flexural strength is 15,000 lb/in^2.

13-26. Hybrid Sheet Insulation. These systems include those where a flexible sheet or tape is formed and subsequently fired. The tape process involves the manufacture of a flexible green tape of ceramic powder intimately dispersed in a plasticizer of organic resin. During the subsequent sintering operation, the organics are removed at some low temperature (for example, 400°C), and the powder is sintered into a monolithic body at the sintering temperature to form a rigid shape. This technique has commonly been applied to the production of electronic substrates.

A second technique is the tape transfer technique, whereby a ceramic powder is organically bonded with an adhesive binder to a flexible organic tape. The organic tape is separated from the ceramic film, which is then pressure-bonded to the conductor. Subsequent thermal treatment removes the organic and sinters the ceramic powder.

13-27. Encapsulants. An encapsulant is primarily used to provide structural integrity and environmental protection to an electrical device. The inorganic encapsulants are cements made from various inorganics, usually metal oxides as fillers and a bonding material. The bonding material is generally a hydraulic or water-setting compound or a chemical-setting compound such as an alkaline silicate or a metal phosphate. All cements are porous and thus have low electric strength (35 to 60 V/mil).

a. *Alkaline silicate cements:* These hard, strong, acid-resistant cements provide good electrical insulation to 500°C.

b. *Metal phosphate cements:* These alkali-resistant cements are the best for both structural and electrical properties. They can be used at temperatures up to 1000°C.

c. *Hydraulic, or water-setting, cements:* Plaster of paris and portland cement are examples of this type of cement. The presence of water gives them poorer electrical properties.
Dehydration of the cement above 100°C considerably reduces their already weakened physical strength.

d. *Other cements:* One of the most important of these cements is a magnesium oxysulfate-bonded magnesium oxide. This finds extensive use around heating elements because of its high thermal conductivity. Use temperature is 1000°C. Several other cements, made for specialized electrical applications, utilize both hydraulic and chemical setting.

13-28. References on Insulating Systems
1. Goldsmith, A., et al.: *Handbook of Thermophysical Properties of Solid Materials,* vol. 3; Macmillan, New York, 1961.
2. Hughes Aircraft Co., Electronic Properties Information Center: data sheets on electrical insulating materials.
3. Kueser, P. E., et al.: Electrical Conductors and Electrical Insulation Materials Topical Report, NASA-CR-54092.
4. Vondracek, C. H.: Inorganic Potting Compounds for High Temperatures; *Mater. Des. Eng.,* p. 100, December 1964.

13-29. Ferroelectrics and Ferroelectricity. A ferroelectric material possesses a reversible polarization as shown by a dielectric hysteresis loop (Fig. 13-5). It lacks a center of

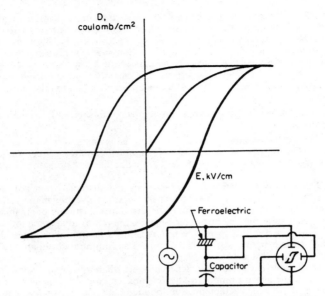

FIG. 13-5 Hystersis loop for a ferroelectric material showing a reversible polarization.

symmetry in the structure and possesses a polar axis. Furthermore, the hysteresis loop disappears above a certain temperature (Curie point), and a transition occurs in the crystal structure to a form of higher symmetry. A more workable definition would be that:

a. Upon applying an electric field E, a displacement field D is generated containing frequency components not present in E.

b. A linear combination of two electric fields does not produce a corresponding linear combination of the two displacement fields.

A number of ferroelectric materials (such as rochelle salt, alkali-metal dihydrogen phosphates and arsenates, and also a group represented by guanidine aluminum sulfate hexahydrate) have been discovered but have not proved useful in practical application because of such difficulties as chemical instability and the incompatibility of the material's Curie temperature and the temperature of practical working devices. The only group that has found widespread application is that possessing the perovskite structure, named for the mineral perovskite, $CaTiO_3$. Members of this group include solid solutions or mixtures of titanates, zirconates, niobates, tantalates, and stannates. The most widely used member of this group is $BaTiO_3$, which has been successfully used in capacitor, transducer, memory, and thermistor applications.

The polar axis caused by the small ionic displacement from symmetry is believed to contribute to the polarization of the structure. At the Curie temperature of $BaTiO_3$ (120°C), the thermal energy of the ions is sufficient to overcome the electrostatic forces holding the ions at these positions. When this occurs, the structure changes from tetragonal to cubic, and the ferroelectric effect disappears. The Curie temperature can be increased by substituting lead at a rate of approximately 4°C per mole percent of lead substitution. A strontium substitution will lower the Curie temperature at a rate of approximately 3°C per mole percent of strontium. Other additions can affect the Curie temperature of $BaTiO_3$.

13-30. Ferroelectric Capacitors. An advantage of ferroelectric material is its large dielectric constant, which permits the use of physically small capacitors. Several other characteristics that should be taken into consideration are the material's temperature coefficient, voltage or bias sensitivity, and breakdown characteristics. The temperature variance of the dielectric constant of several typical types of dielectrics with high dielectric constants is shown in Fig. 13-6. In general, the higher the dielectric constant, the more pronounced the effect of temperature. The curves may be shifted or flattened by changes in composition or process treatment.

The voltage or bias sensitivity shown in Fig. 13-7 indicates that the materials with higher dielectric constant are more sensitive to increasing bias fields. The K_β/K_O ratio is the incremental change in the dielectric constant, where K_β is the value of dielectric constant with dc bias voltage and K_O is that value without the dc bias voltage. In ferroelectrics a high electric field applied to a device can cause voltage breakdown. The general rule is: the higher the dielectric constant, the lower the dielectric strength. A figure of 2.5 kV/mm at room temperature for high-dielectric-constant materials (4000 to 6000) is a nominally accepted value, and this can be increased to 15 kV/mm for low-dielectric-constant ($\sim$30) dielectrics. These values would have to be derated for elevated temperatures.

Capacitors with very high dielectric constants, in the order of 100,000 and higher, have been developed by treating $BaTiO_3$ in a reducing atmosphere. These capacitors are semiconductor in nature and rather lossy and are therefore limited to very low-voltage applications such as those in transistor circuits. Other ferroelectrics and their properties are shown in Table 13-5.

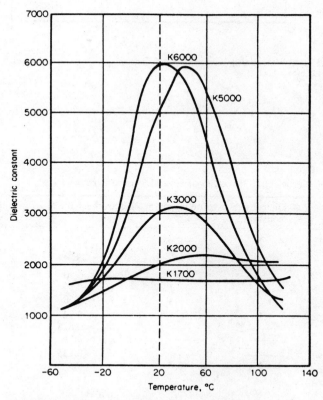

FIG. 13-6 Dielectrics with a high dielectric constant as a function of temperature.

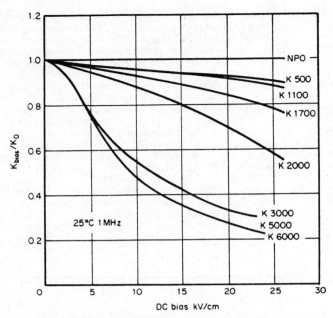

FIG. 13-7 High-dielectric-constant materials are more sensitive to increasing bias fields.

TABLE 13-5 Piezoelectric Properties of Several Typical Ferroelectrics

Material	Dielectric constant	Curie temp., °C	Electromechanical coupling, k	Strain constant, C/N, d
Rochelle salts	500	$-18, +24$	$k_{14} = 0.78$	$d_{14} = 870$ $d_{25} = -53$ $d_{36} = 12$
BaTiO$_3$	1,500	120	$k_{31} = 0.20$ $d_{33} = 0.50$ $k_{15} = 0.50$	$d_{31} = -56$ $k_{33} = 150$
Lead zirconate titanate	1,500 —— ——	350 —— ——	$k_{31} = 0.32$ $k_{33} = 0.675$ $k_{radial} = 0.54$	$d_{31} = -140$ $d_{33} = 320$
BaTiO$_3$ (88%) PbTiO$_3$ (12%)	—— 850	—— 150	$k_{31} = 0.125$ $k_{33} = 0.365$	$d_{31} = 30$ $d_{33} = 90$
PbNbO$_2$ (80%) BaNbO$_3$ (20%)	400	425	$k_{radial} = 0.20$	$d_{31} = 25$
PbNbO$_3$ (60%) SrNbO$_3$ (40%)	755	310	$k_{radial} = 0.26$	$d_{31} = -53$

13-31. References on Ferroelectrics

1. Kanzig, W.: Ferroelectrics and Anti-ferroelectrics, in *Solid State Physics,* vol. 4, Seitz, F., and Turnbull, D., (eds.); Academic, New York, 1957.

2. Schicke, H. M.: *Essentials of Dielectromagnetic Engineering;* Wiley, New York, 1961.

3. Burfoot, J. C.: *Ferroelectrics: An Introduction to Physical Principles,* Van Nostrand, New York, 1962.

4. IRE Standards on Piezoelectric Crystals; *Proc. IRE,* vol. 49, no. 7, July 1961.

5. Saburi, O.: Processing Techniques and Applications of PTC Thermistors; *IEEE Trans. Component Parts,* vol. CP-10, no. 2, June 1963.

6. Ichikawa, Y., and Carlson, W. G.: Yttrium-Doped Ferroelectric Solid Solutions with PTC of Resistance; *Am. Ceram. Soc. Bull.,* vol. 42, no. 5, May 1963.

7. Hench, L. L., and Dove, D. B.: *Physics of Electronic Ceramics;* Marcel Dekker, New York, 1971.

Chapter 14

INSULATING VARNISHES

E. J. CROOP

14-1 Definition   14-1
14-2 Functions of Insulating Varnishes   14-2
14-3 Manufacture of Varnishes   14-2
14-4 Resins Used in Varnish   14-2
14-5 Drying Oils   14-4
14-6 Solvents and Thinners   14-4

14-7 Dryers   14-5
14-8 Applications   14-5
14-9 Testing   14-6
14-10 Thermal Endurance   14-6
14-11 Methods of Applying   14-6
14-12 Specification for Shellac   14-7
14-13 References   14-7

14-1. Definition. Varnish is usually defined as a liquid composition, normally a solution of resinous matter in an oil or volatile solvent, which, after application to a surface, dries by either evaporation or chemical action to form a hard, lustrous coating which is more or less resistant to air, moisture, and various chemical agents. Varnishes are sometimes classified by color as "clear" or "black" and/or by functions, as coil-bonding varnish, mica-sticking varnish, or coil-impregnating varnish. Varnish does not contain pigments like paint. The dried film is usually transparent or translucent. When pigments are added to varnish, the mixture is called an "enamel."

Varnishes have been traditionally divided into three classes:

a. *Oil or oleoresinous varnishes,* which are essentially solutions of natural or synthetic resins or asphalts, in drying oils (especially linseed, tung, or soybean oil).

b. *Spirit varnishes,* which are solutions of natural or synthetic resins or asphalts in volatile solvents such as alcohol, acetone, and turpentine. They do not contain a drying oil.

c. *Synthetic, heat-reactive varnishes,* which are solutions of synthetic resins in organic volatile solvents. They are thermoset and they dry by polymerization due to heat alone, not by surface oxidation of oil or oleoresinous materials. Heat-reactive varnishes are in widespread use because of their superior properties of thermal stability, chemical resistance, and physical toughness.

Waterborne varnishes are a recent development brought about by efforts to reduce air pollution. Part or all of the volatile organic solvents are replaced in the manufacture of the varnish with water. Their drying is accomplished in a manner similar to the solvent-containing varnishes.

Other attempts to reduce air pollution by the photochemically reactive solvents in conventional varnishes involve replacing such solvents in varnishes with nonphotochemically reactive solvents. These have been termed "Rule 66 solvents," to meet the requirements of the Los Angeles County (California) Regulation 66 defining amount and type of solvent emission permitted from manufacturing plants. This regulation has been adopted by many other localities in the United States.

Another present trend is toward the use of solvent-free varnishes. They are composed of liquid synthetic resins, but contain no solvent or water. This results in more thorough impregnation of coils and windings, and eliminates air pollution. They are particularly suited for use in the "trickle process."

14-1

14-2 ELECTRICAL ENGINEERING MATERIALS REFERENCE GUIDE

Drying of varnishes takes place by (1) evaporation of the solvent, (2) oxidation of the oils, (3) polymerization of the resin and oil. It is usual to classify varnishes as either air-drying or baking, depending on the method of drying.

14-2. Functions of Insulating Varnishes. The chief functions of insulating varnishes are the protection of fibrous insulation (such as fabrics, glass, asbestos, and wood) from mechanical damage, chemical contamination, and moisture penetration; improvement of insulating properties; bonding the layers of laminated materials or conductors; imparting a surface finish; and increasing heat transfer. For properties, see Tables 14-1 and 14-2.

14-3. Manufacture of Varnishes. The details of actual processes used in making varnishes are in large degree held as trade secrets by the various manufacturers, although such processes are relatively simple and easily acquired by those practicing the art.

The oil varnishes are made by treatment and bodying of drying oils at elevated temperatures, then by cooking in varnish kettles a mixture of the oils and resins to attain specific chemical properties, with subsequent addition of required solvents and dryers. The spirit varnishes are made by dissolving natural or synthetic resins in appropriate solvents, usually with some heating and agitation. Recently synthetic resins have been developed which may be used as varnishes without oils or solvents, drying either in air or by baking (polyesters, epoxies, etc.).

The raw materials used in manufacture of insulating varnishes consist of drying oils, resins, solvents, and dryers. They are briefly described in the following paragraphs.

14-4. Resins Used in Varnish. A resin is a solid, semisolid, or liquid organic compound, noncrystalline in nature, with color varying from water-white to yellow or brown (usually), transparent or translucent, and soluble in such organics as ether and alcohol but insoluble in water. Chemically resins differ widely, but all are rich in carbon and hydrogen and also contain some oxygen. Some of the more thermally stable synthetic resins may also contain some nitrogen, sulfur, etc. Resins may be divided into four groups: animal, vegetable, mineral, and synthetic.

Animal Resins. Shellac is the most important animal resin. It is a purified resin made from lac, which consists of excretions from certain insects found on trees in India and the Far East. The crude lac is known as "stick lac"; it is crushed and sifted to produce so-called "grain lac," which is then melted and strained. When cold, it is marketed in various shapes, such as flakes, buttons, and cakes, known respectively as "shellac," "button lac," and "garnet lac." Shellac is extensively used in making spirit varnishes. It has a resistivity on the order of 10^{15} to 10^{16} cm and a dielectric constant of 2.9 to 3.7 approximately. For information concerning the testing of shellac see ASTM Designations D29-81 and D411-73.

Vegetable Resins are obtained from various trees. These resins have been called "copals." The types that are found in fossilized deposits are called "hard copal." Their melting points vary from 75 to 450°C. The types known as "soft copal" come from living trees of many species and have melting points from 75 to 100°C (see Table 14-3). While relatively inexpensive, these resins are being largely displaced by synthetic resins.

Mineral Resins. Asphalts are sometimes called mineral resins. They are soluble in carbon disulfide and benzene, are soft at 70°C, and melt at about 100°C. Natural and mineral tars and pitches are in this group. Gilsonite is one of the best for black varnishes. Some of the mineral resins are still in use today, but they are being replaced by the superior, high-performance synthetic resins.

Synthetic Resins. These are formed directly from organic compounds. They are largely replacing the natural resins because of their uniformity of composition and (in comparison with older varnishes) their superior performance in imparting chemical resistance, thermal stability, and mechanical durability. The types having widest application in oleoresinous varnishes are the modified or unmodified phenolic resins and the oil-modified glycerol phthalate resins or alkyds. Synthetic resins are also used in spirit varnishes and in some cases are sufficiently fluid for use alone, drying in air or by baking. Some of the more important commercial resins and trade names are: (*a*) alkyd (Amberlac, Beckosol, Dulux, Durez, Glyptal, Plaskon, Rezyl); (*b*) ethyl cellulose; (*c*) phenolic (Amberal, Bakelite, Durez, Harvel, Super Beckacite); (*d*) urea (Beckamine, Beetle, Uformite); (*e*) vinyl (Vinylite, Formvar, Geon, Exon); (*f*) melamine formaldehyde (Melmac, Uformite, Super Beckamine); (*g*) silicone; (*h*) epoxy (Epon, Araldite, Epotuf, Epirez, DER); (*i*) urethane (Spenkel, Mondur, Nacconate); (*j*) polyester (Alkanex, Isonel); (*k*) aromatic polymer (ML); (*l*) acrylic (Elvacite, Acryloid).

TABLE 14-1 Properties of Typical Organic Insulating Varnishes*

Number	Type	Viscosity, Demmler #1, s	Solids, %	Thermal class, °C†	Curing temp., °C	Curing time, h‡	Dielectric strength, V/mil§	Solvent
B-222-N	Clear, air-drying, impregnating, coating, modified phenolic	75–110	54	————	30	3	1650	Mineral spirits
B-225	Clear, air-drying, impregnating, coating, oleoresinous	30–60	50	————	30	3–4	1400	Mineral spirits
B-6-614	Red, air-drying sealing coating, alkyd	100–300	55	————	30	4	1400	Xylol
B-142-1	Clear, thermoset, baking impregnating, coating, modified polyester	45–85	48	190	135–160	1–4	3900	Xylol
B-280-2	Clear, thermoset, baking impregnating, coating, epoxy ester	135–180	50	160	150–170	1–2	2300	Xylol
B-185-4	Clear, thermoset, baking impregnating, modified polyester	30–60	48	195	135–160	6–12	3050	Xylol
B-529 (OMVAR)	Clear, thermoset, baking polyaromatic insulating varnish	125–200	46	200	175–185	6–8	3000	Xylol
B-540-1	Clear, thermoset, baking impregnating coating, modified polyester, low-temperature cure	25–50	45	195	125–140	2	4000	Xylol
B-7-373	Clear, solventless, thermoset, baking polyester insulating varnish	900–1200 CPS	100	180	150	4	Power Factor 4–5% @ 150°C	None
B-535	Clear, water-reducible, thermoset	450–700	60	205	125–150	2–4	3700	Water
B-549-1	Clear, water-reducible, thermoset, baking, modified polyester insulating varnish	250–450	60	195	160–175	2–4	4000	Water

*Supplier: Insulating Materials Division, Westinghouse Electric Corporation.
†Per ASTM-D2307, twisted MW-35 wire pairs, 20,000 h.
‡Curing time refers to internal mass curing in a deep layer and not to film drying time.
§Dielectric strength for internal curing is determined with ¼-in electrodes on varnish which is baked on 2½-mil bond paper. The air-drying varnish dielectric strength is determined by test method from MIL-I-24092.
NOTE: Most of the above varnishes are available in "Rule 66" solvents and can be tailored to meet specific needs.

ELECTRICAL ENGINEERING MATERIALS REFERENCE GUIDE

TABLE 14-2 Properties of Silicone Insulating Varnishes

Number	Type	Specific gravity at 25°C	Solids, %	Viscosity at 25°C, cP	Cure temp., °C	Cure time, h	Electric strength, V/mil	Solvent
994	Internal-curing coating varnish	1.00–1.02	49	80–150	220	2	2500	Xylene
997	Internal-curing impregnating	1.01	49	110	200	3	2000	Xylene
996	Internal-curing impregnating	1.00–1.02	49	100–200	150	3	1600	Xylene

Source: Dow Corning Corporation Technical Data Sheets.

TABLE 14-3 Typical Commercial Vegetable Resins

Name	Type and source	Usefulness
Amber	Fossil resin, Germany	Oil varnishes
Kauri	Fossil copal, New Zealand	Oil varnishes
Pale East India	Damar, East Indies	Oil and spirit varnishes
Black East India	Damar, East Indies	Oil varnishes
Congo copal	Fossil resin, Africa	Oil varnishes
Manila	Soft copal, East Indies, Philippines	Oil and spirit varnishes
Mastic	India, North Africa	Spirit varnishes
Sandarac	Africa, Australia	Spirit varnishes
Pine resin gum "thus"	Southern United States, France, India	Oil varnishes, etc.

Some of the synthetic resins may be coreacted with each other to achieve a resin with final properties that are roughly intermediate between those of the original coreactants; that is, phenolic and alkyd. See *Modern Plastics Encyclopedia* for complete listings and descriptions of commercial resins.

14-5. Drying oils used in varnish manufacture are classified according to their drying properties. They are naturally occurring, largely triglycerides of long-chain saturated and unsaturated fatty acids. The drying ability is a function of the amount of unsaturation and the number of conjugated double bonds. The drying mechanism is considered a combination of oxidation and polymerization at the double bonds to form an insoluble and infusible gel. The drying oils include linseed oil and tung oil (china wood oil). The semidrying oils include cottonseed and soybean oil. The nondrying oils are castor, olive, and rosin oils.

Linseed oil is the drying oil used in by far the greatest quantity. It is derived from flaxseed and has a density of 0.932 to 0.936 at 15°C. It dries largely by oxidation to a hard gum and is a good insulator. The action is hastened by using bodied linseed oil or by adding dryers or by baking. For specifications, see ASTM D234-82 and D260-80. Tung oil is obtained from the nuts of the oriental tung tree, which is now being grown in the southeastern United States. The density is 0.938 to 0.942 at 15°C. Drying is by polymerization as well as oxidation. See ASTM Specifications D12-80. Rosin oil is also used for impregnating paper insulation in power cables because of its penetrating power and high electric strength, although it is giving way to the use of isopropyl biphenyl. The chlorinated hydrocarbons, such as Askarel or Interteen, have been banned for this purpose by the Environmental Protection Agency (EPA).

The market price of drying oils fluctuates widely and is sometimes a factor in deciding which type of varnish is used at any particular time. Great care is required to ensure that varnish changeovers do not disrupt manufacturing procedures or compatibility with other components of the insulation system.

14-6. Solvents and thinners are organic liquids used to reduce the viscosity of varnish and facilitate its application. They dissolve the varnish ingredients, are present in a balanced

INSULATING VARNISHES

ratio of nonvolatile solids to solvents, then disappear by evaporation after the wet film is applied. Those commonly used in varnish manufacture and application are listed in Table 14-4.

14-7. Dryers. The drying properties of certain oils (see Par. **14-5**) can be increased by the addition of metallic catalysts, which on heating with the oil (as in a varnish) promote faster polymerization or directly liberate oxygen and thus promote drying. These materials are usually metal–organic compounds of cobalt, lead, or manganese dioxide. However, there are objections from the electrical point of view to the use of any of these metallic salts, even in small quantities. Fortunately, they are effective in concentrations of less than 1% addition by weight. The unnecessary use of dryers should be avoided.

14-8. Types of Insulating Varnishes for Various Applications. Catalogs of insulating-varnish manufacturers list the following general types:

a. Clear baking varnish for armatures, field coils, and instruments.

b. Black baking varnish for armatures, field coils, and transformers when higher electric strength and resistance to moisture, acids, and alkalies are wanted. Less oil resistance than that of the clear.

TABLE 14-4 Typical Solvents and Thinners Used in Varnishes

Solvent	ASTM specification	Specific gravity	Flash point of closed cup, °F	Toxicity
Acetone	D329-81	0.791–0.799	2	Slight
Amyl acetate	D318-58	0.860–0.865	77	Slight
Amyl alcohol	D319-80	0.812–0.820	97	Slight
Benzine, petroleum spirits		0.69–0.77	70	Slight
High-flash naphtha		0.87–0.89	113–151	Considerable
Ethyl acetate	D302-80	0.883–0.888	25	Slight
Ethyl alcohol		0.790	57	Slight
Ethylene glycol monoethyl ether	D3128-72	0.927–0.933	107	In question
Methyl alcohol, methanol	D1152-72	0.796	32	Considerable
Methylated spirits, denatured alcohol		0.790	57	Slight
MEK, methyl ethyl ketone		0.806	24	Medium
MIBK (methyl isobutyl ketone)	D1153-72	0.805	75	Medium
Petroleum spirits, mineral spirits	D235-77	0.762–0.814	100	Slight
Toluene	D-362-71	0.860–0.870	41	Considerable
Triacetin (plasticizer)		1.16–1.17	270	Slight
Turpentine	D-13-65	0.860–0.875	90	Slight
VM&P naphtha		0.750–0.765	53	Slight
White spirit, mineral turpentine			78	Slight
Xylol, xylene	D364-77, D846	0.859	83	Considerable

NOTES: For an extensive listing of solvents and thinners with properties, see T. H. Durrans, *Solvents,* 1957. ASTM has specifications for a number of solvents.

Hazardous or chlorinated solvents formerly used, such as benzene, the cellosolves, carbon tetrachloride, and trichloroethylene, have been severely restricted in applications by the Occupational Safety and Health Administration (OSHA).

ELECTRICAL ENGINEERING MATERIALS REFERENCE GUIDE

c. Clear or black baking varnish (internal-curing type) in which the varnish thermosets throughout the depth of the coil during the baking operation and assists in bonding the components of the coil.

d. Clear or black air-drying varnish used where baking is not convenient.

e. Clear red or black finishing varnish, usually containing shellac or synthetic resins, used for producing a harder surface over baking varnishes, for protection against oil, moisture, dirt, and metal dust, and for improving appearance.

f. Sticking varnish used in cementing cloth, paper, mica, etc.

g. Core-plate varnish (air drying, baking, and flashing) for insulating armature and transformer laminations. The air drying is not suitable for oil-immersed operation.

h. Epoxy resin varnish (baking) for all coil impregnation, internal curing, where superior durability and chemical and moisture resistance are required.

i. Silicone resin varnish (air drying and baking) for motor stators and rotors, transformers, and coils used under high-temperature and high-humidity conditions.

j. Polyester resin varnishes (baking) for motor stators and rotors, transformers, and coils, for high-temperature service not so severe as to require silicones.

k. Phenolic varnishes (baking) for hermetic motor coils and bonding of formwound coils.

14-9. Methods of Testing Varnish. The following specifications and methods of testing have been developed by ASTM (in referring to these standards, always consult the latest annual issue of the ASTM Index).

a. Methods of Testing Varnishes Used for Electrical Insulation (D115-82).

b. Testing Electrical Insulating Varnish for 180°C and Higher (D1346-68).

c. Sampling and Analysis of Shellac (D29-81).

d. Methods of Testing Shellac Used for Electrical Insulation (D411-73).

e. Specifications for Orange Shellac and Other Indian Lacs for Electrical Insulation (D784-61).

f. Specifications for Shellac Varnishes (D360-73).

g. Methods of Testing Silicone Insulating Varnishes (D1346-68).

h. Thermal Endurance of Flexible Insulating Varnish (D1932-67).

i. Thermal Degradation of Electrical Insulating Varnish by Helical Coil Method (D3145-75).

j. Thermal Aging Characteristics of Electrical Insulating Varnishes Applied over Film Coated Magnet Wire (D3251-73).

Other test methods for varnish cover density, viscosity, flash point, time of drying, electric strength as liquid and when solid at two temperatures and after water immersion, heat flexibility, oilproof test, draining or working viscosity, nonvolatile and volatile matter. Other properties which should be determined depend on the application but may include some of the following: impregnating properties, internal drying of saturating varnish, hardness, elasticity, toughness, bonding strength, flex life, resistance to chemicals and corrosive actions, and appearance.

14-10. Thermal Endurance of Varnishes. It is known that the thermal stability of varnishes determined by simple test methods as well as by field experience varies considerably. ASTM D1932-67 and the former IEEE 57 test methods are generally used to determine thermal stability in the laboratory, but experience with present tests does not show that any are reliable for rating varnishes for a particular service, although they do give comparative data. Operating data are the best guide for the selection of a varnish.

14-11. Methods of Applying Varnishes. Two general methods are in use for treating coils, windings, and insulating parts with insulating varnishes: (1) by vacuum impregnation, and (2) by hot dipping. Finishing varnishes are usually applied by brush or spray. Mica sticking varnishes are applied by brush or sometimes by machine (as by passing a roller which dips in the varnish). There has been a revival of the trickle process whereby a varnish is allowed to drip or trickle onto a heated rotating component such as a motor armature. This provides more thorough impregnation, especially if the varnish contains very little solvent, or preferably is solventless. Synthetic varnishes are frequently used for impregnation by dipping and require baking to develop their properties fully. Baking ovens of the continuous type are frequently used where the class of production is suited to them. Infrared lamps are used with considerable success for baking varnish-treated coils; generally they reduce the time of bake.

INSULATING VARNISHES

14-12. Specification for Shellac Varnish. See ASTM Standard Specifications for Shellac Varnishes (D360).

14-13. References on Insulating Varnishes

1. Parker, D. H., *Principles of Surface Coating Technology,* Interscience Publishers, New York, 1965.

2. Gardner, H. A., *Physical Examination of Paints, Varnishes, and Colors,* 12th ed., Washington, D.C., 1962.

3. Moses, G. L., Lee, R., and Hillen, R. J., *Insulating Engineering Fundamentals,* Lake Publishing Co., Lake Forest, Ill., 1958, pp. 22–25, and 74–78.

4. Durrans, T. H., *Solvents,* D. Van Nostrand Company, Inc., Princeton, N.J., 1957.

5. Chatfield, H. W., *Paint & Varnish Manufacture,* George Newnes, Ltd., London, 1955.

6. Gordon, P. L., and Gordon, R., *Paint and Varnish Manual,* Interscience Publishers, Inc., New York, 1955.

7. Bidlack, V. C., and Fasig, E. W., *Paint and Varnish Production Manual,* John Wiley & Sons, Inc., New York, 1951.

8. Moses, G. L., *Electrical Insulation: Its Application to Shipboard Electrical Equipment,* McGraw-Hill Book Company, New York, 1951.

9. Von Fischer, W., *Paint and Varnish Technology,* Reinhold Publishing Corporation, New York, 1948.

10. ASTM Publ. 310, *Solvents.*

11. *Paint, Varnish, Lacquer, and Related Materials; Methods of Inspection, Sampling and Testing,* Federal Standard 141-B.

12. Commercial Resins Chart, *Modern Plastics Encyclopedia,* 1974.

Chapter **15**

COATING POWDERS

E. J. CROOP

15-1 General 15-1
15-2 Coating Processes 15-1
15-3 Applications 15-1

15-4 Advantages 15-1
15-5 Coating Materials 15-1
15-6 References 15-2

15-1. General. If one attempts to insulate a sharp edge or corner (as in the bare slots of a small motor stator) with a varnish or a paint, it is impossible to insulate the edges to withstand more than a few volts, because surface tension causes the liquid to draw away from the sharp edges. This is especially severe as the coated component is heated in an oven to cure the coating. The heat causes a decrease in viscosity, which makes the uncured coating flow more readily away from corners and edges in response to surface tension. This problem has been overcome by the development of solid coating powders, formulated to have a thixotropic character when molten, which are applied to the preheated object to be coated.

15-2. Coating Processes. In one method, called the fluidized-bed method, the solid coating powder, which may be thermoplastic or thermosetting, is suspended in air by an air stream blown in from the bottom of a container through a fritted glass filter or other means which distributes the air uniformly. The mixture behaves like a liquid in which air acts as the solvent or suspending medium. The object to be coated, usually a metallic material of relatively high heat capacity, is preheated to a temperature above the fusion point of the powdered coating material, then dipped in the fluidized bed, and withdrawn. The heat stored in the object to be coated is great enough to cause the solid particles to melt, coalesce, and flow to form a smooth continuous coating. If the powder is properly formulated and the treating conditions are properly controlled, the coating does not pull away from the edges and uniform high electric strengths are obtained.

Another technique is to apply the powders to the preheated or even cold object by electrostatic spraying. Automatic machines are used in applying coating powders to insulate the slots of small motor stators by this method. Flocking techniques are also used to apply such powders. More recently, electrostatically charged fluidized beds, electrostatically charged cloud chambers (pherostatic coaters), and electrogasdynamic powder-spraying tunnels have been used in the commercial application of coating powders.

15-3. Applications. Typical applications are to industrial machinery and component parts, electrical equipment (such as transformers and switchgear), vehicles, (tractors and other farm implements), and consumer items (such as air-conditioner housings, equipment cabinets, shelving, refrigerators, lighting fixtures, bicycles, fences, and lawn furniture). In most cases, coating powders are also being used to replace conventional paint, enamel, and lacquer as a decorative finish where durability is a prime requirement.

15-4. Advantages. A chief advantage of a coating-powder-cured film is that it yields a high-quality finish in only one coat which may be as thick as desired—usually many times as thick as a single coat of paint or varnish. Any powder overspray is easily and 100% recoverable. Powder-coating processes are ecologically compatible, since there is no solvent to dispose of.

15-5. Coating Materials. The technique is applicable to a wide variety of materials, including epoxy, polyamide, and cellulose derivatives (see Table 15-1). Epoxy coatings are

TABLE 15-1 Materials Currently Used in Coating Powders

Material	Minimum coating temperature, °C
Epoxy	120–150
Polyethylene	200
Polyvinyl chloride	185
Polypropylene	250
Cellulose acetate butyrate	260
Nylon-6	300
Penton (chlorinated polyether)	290
Teflon TFE	400
Teflon FEP	340
Polychlorotrifluoroethylene, PCTFE	250

used in low-voltage automotive, aircraft, and some fractional-horsepower appliance motors to replace paper and polyester film slot cells. The advantages of these coatings are lower cost and higher temperature capability over the previously used insulating materials.

15-6. References on Coating Powders

1. German patent No. 933,019, *Process and Apparatus for the Preparation of Protective Coatings from Pulverulent, Synthetic, Thermoplastic Materials,* Erwin Gemmer (Knapsack-Griesheim Aktiengesellschaft), Sept. 15, 1955.

2. Stott, L. F., "Fluidized Bed Method of Coating," *Org. Finish.,* **17**:16–17 (June 1956).

3. "Now: Fluidized Coatings," *Chem. Eng.,* **63**:236–237 (January 1956).

4. Newman, I. A., and Bockhoff, F. J., "Fluidized Plastic Coating for Corrosion Resistance," *Prod. Eng.,* **28**:140–143 (January 1957).

5. Checkel, R. L., "Fluidized Polymer Deposition," *Mod. Plast.,* **36**:125–132 (October 1958).

6. U.S. patent 2,844,489, *Fluidized Bed Coating Process,* Erwin Gemmer (Knapsack-Griesheim Atkiengesellschaft), July 22, 1958.

7. Thielking, R. H., and McClenahan, D. L., "Fluidized Coating—A Method of Slot Insulation," *Conf. Paper CP59-471,* presented at the winter general meeting, New York, Feb. 4, 1959.

8. Elbling, I. N., and Saunders, H. E., "Electrostatic Powder Coatings Attract the Electrical Market," *Plast. Eng.,* October 1973, pp. 60–63.

9. Elbling, I. N., and Saunders, H. E., "Determining Acceptability of Powder Coatings for Production Use," *S.M.E. Technical Paper FC73-537,* presented at the 3rd North American Powder Coating Conference, Cincinnati, Ohio.

10. Gemmer, E., "Das Wiebelsintern-Entwicklungen and neuere Erkenntnisse," *Kunststoffe,* **47**:510–512 (1957).

11. Armstrong, C. W., "Fluidized-bed Processing Equipment," conference paper presented at the regional technical conference, Society of Plastics Engineers, Fort Wayne, Ind., May 22, 1959.

12. Parent, J. D., Yagol, N., and Steiner, C. S., "Fluidizing Processes," *Chem. Eng. Prog.,* **43**(8):429–436 (1947).

13. U.S. patent 2,711,387, *Treating Subdivided Solids,* G. L. Matheson and H. J. Hall, June 21, 1955.

14. Elbling, I. N., "Powdered Insulating Finishes," *Official Digest,* **31**(419):1625–1639 (1959).

15. Hate, F. L., "Powder Coating," *Resin News,* vol. 4, no. 12, 1964.

16. Sprackling, J. M., "Why Use Epoxy Powder Coatings?" *Resin News,* vol. 4, no. 12, (1964).

Chapter 16

IMPREGNATING AND FILLING COMPOUNDS

J. D. B. SMITH

16-1	Nature and Purpose 16-1		16-8	Polybutadienes 16-5
16-2	Bitumens, Asphalt, Pitch, and Waxes		16-9	Solventless Resins 16-5
	16-1		16-11	Low-Temperature Resins 16-7
16-3	Resins 16-2		16-12	References 16-8
16-7	Urethanes 16-5			

16-1. Nature and Purpose. Two different classes of materials are dealt with here, namely: (1) the older impregnating and filling compounds, primarily the bitumens and waxes, which are generally melted in place and which remain permanently heat-softening, and (2) the newer, important synthetic products of modern polymer chemistry—the solvent-reactive or so-called "solventless" resins. The latter, when properly applied (usually with resort to vacuum and pressure), can then be induced to react (polymerize) in situ and thus provide solid, more nearly void-free insulation. Where appreciable voltage stresses will be encountered in service, the impregnation and sealing of all forms of porous insulating materials (windings, wire coverings, joints, etc.) are essential to securing satisfactory overall insulation. In using either class of materials, the purpose is to provide, as nearly as possible, a void-free insulating structure to seal out moisture, chemicals, and other electrically destructive contamination throughout the useful life of the equipment. A wide variety of commercially available materials is delineated in Ref. 1. To make an intelligent selection among the myriad materials which are available today, attention should be paid to many things other than basic electrical and physical properties. Some of these considerations involve such esoteric electrical properties as long-time resistance to the destructive effects of corona. Closely allied is the property of voltage endurance (that is, ability of the insulation system to withstand required voltage stress throughout the life expectancy of the electrical equipment).

No less important are considerations involving the thermal endurance of electrical insulation systems, which frequently are affected by a combination of thermal and oxidative degradation of the organic materials that make up the impregnant, whether solid or liquid. Of vital importance in dealing with these considerations is the concept of *functional evaluation.* All the affected insulating materials, combined with the active elements in an operating system, should be functionally tested to destruction, under specified (usually accelerated) conditions. Thus the materials interactions which occur can reveal major flaws and shortcomings of an operating insulation system. For further information on these considerations, see Moses,[2] Clark,[3] and Sillars.[4]

16-2. Bitumens, Asphalt, Pitch, and Waxes. Bitumen and asphalt[5] are impure hydrocarbon materials obtained from mineral sources, presumably of vegetable origin. Materials such as gilsonite, grahamite, wurtzilite, and impsonite are mined in solid form; others are derived from petroleum distillation. The term "pitch" is normally retained for a solid black material derived from wood or coal distillation.

Wax[6] is defined as any of a class of natural or synthetic substances composed of carbon, hydrogen, and oxygen and consisting chiefly of esters other than those of glycerin or of free fatty acids. In the natural wax class are included montan wax, beeswax, spermaceti, Chinese wax, and carnauba wax. The mineral waxes such as ozokerite differ chemically from the

ELECTRICAL ENGINEERING MATERIALS REFERENCE GUIDE

true waxes in that they contain no oxygen. These waxes when melted are good impregnating materials, being waterproof but usually not oilproof. Most of them are vulnerable to oxidation at elevated temperatures, developing increased acidity and a deterioration in electrical properties. The synthetic and modified ester-type waxes appear not to have had any significant use as impregnating or filling compounds for insulation applications, their main application being in cosmetic and pharmaceutical products and in polishes.

For further information on the properties and uses of bitumens, asphalt, and waxes as impregnating and filling compounds, the reader is referred to Secs. 4-361 to 4-362 of the *Standard Handbook for Electrical Engineers,* 11th ed., by Donald G. Fink and H. Wayne Beaty, McGraw-Hill, 1978.

16-3. Solventless Impregnating and Filling Resins. The so-called solventless resins are generally mixtures of reactive liquid monomers or solutions of viscous or solid polymers in liquid reactive monomers or diluents. They are characterized by being essentially 100% reactive. A wide variety of commercial resins are available, which provide, when cured, thermoset (nonremeltable) products which range from rigid to flexible, depending on selection. A full range of viscosities are available. The low-viscosity materials are generally used for impregnation, the higher-viscosity products for embedment, encapsulation, and filling.

During cure, polymerization occurs, with volume shrinkage and increase in density as the polymer converts from a liquid of increasing viscosity, through a soft gel state, finally to a solid. For this reason, if the process is not carefully controlled, cracking of the substrate can occur and mechanical stresses are set up in the embedded structure. A wide variety of mineral fillers (for example, silica, alumina, calcium carbonate, and various clays) is frequently incorporated in the solventless resin to produce, among other changes, reduced shrinkage, reduced coefficient of thermal expansion, increased thermal conductivity, and change in electrical and physical properties of the cured product. Several widely differing classes of solventless resins are available; namely, the polyesters, epoxies, silicones, acrylics, urethanes, and 1,2-polybutadienes (see Table 16-1). For information on the physical and electrical properties of solventless resins, as well as the numerous manufacturing and design considerations involved in their electrical application, see Clark[3] and Harper.[7]

16-4. Polyester Resins.[8,9] These resins frequently consist of a styrene or vinyl toluene solution of unsaturated reactive polymer, the latter having the ester group as the recurring functional unit. They are thermosetting, usually utilizing peroxide catalysts to initiate "vinyl" polymerization, and can be formulated to provide a wide degree of flexibility in the cured product. While they are low in cost, perhaps their greatest shortcoming is a high degree of shrinkage on curing (as high as 10%), which can create cracking problems. This phenomenon can usually be overcome by appropriate design modifications. Typical properties are shown in Table 16-2.

16-5. Epoxy Resins.[10,11] Epoxy resins are probably the most popular choice of impregnating and potting resins used at the present time. Their superior electrical properties, combined with outstanding resistance to moisture, chemicals, and solvents, make them very attractive materials for insulation applications, for example, in the manufacture of turbine-generator coils, motors, and transformers.

Since the discovery of the original family of bisphenol A epoxy resins in the late thirties, several new types of epoxy resins have become available commercially. These include cycloaliphatic epoxies,[12] epoxy-novolacs,[13] and glycidyl esters.[14] In general, each of these "families" of epoxy resins is available as low-viscosity liquids or as high-molecular-weight solids, thereby offering a significant degree of versatility in resin-processing and -fabricating techniques.

The curing agents (or hardeners) employed with these epoxy resins are usually di- or polyamines, polyamides, di- or polycarboxylic acids or anhydrides, Lewis acid-amine complexes (such as BF_3:400), and various organometallic compounds.[15] Some epoxies will cure at room temperature; others require elevated-temperature cure. Frequently, accelerators are added to the curing agent (for example, with anhydrides) to speed up the rate of cure.[16,17,18] Shrinkage during polymerization is substantially lower (usually about 4% in an unfilled resin) than that found with impregnants such as polyesters and acrylics. Excellent bonding to substrate is usually found even at elevated temperatures. For most applications, addition of flexibilizers is usually recommended. Typical properties of cured epoxy resins are shown in Table 16-2.

TABLE 16-1 Different Types of Resins Used as Solventless Impregnants

Resin type	Reactive diluent	Advantages	Disadvantages
Epoxy	Glycols, mono- or diepoxy compounds, carboxylic acids, anhydrides	Good bond strength, low shrinkage	Addition of flexibilizers or accelerators is usually required
Polyester	Vinyl monomer (usually styrene or vinyl toluene)	Low viscosity and fast gel times	High shrinkage ($\sim$10%), susceptibility to hydrolysis
Acrylic	Acrylic monomer	Low viscosity and fast gel times	High shrinkage ($\sim$10%), rigidity
Urethane	Polyol	Good flexibility on cure	Limited thermal capability, susceptibility to hydrolysis
1,2-Polybutadiene	Vinyl monomer (usually styrene or vinyl toluene)	Excellent electrical characteristics at elevated temperatures	Sluggish cure rates requiring high-temperature cure ($>$200°C)
Silicones		High-temperature capability	High cost

TABLE 16-2 Properties of Typical Polyester and Epoxy Resins*

Proeprty	Polyester resins		Epoxy resins	
	Rigid	Flexible	Unfilled	Silica-filled
Specific gravity	1.10–1.46	1.10–1.20	1.11–1.23	1.6–2.0
Tensile strength, lb/in^2	6000–10,000	800–1800	4000–13,000	5000–8000
Elongation, %	5	40–310		
Modulus of elasticity in tension, $10^5 \, lb/in^2$	3.0–6.4		4.5	
Compressive strength, lb/in^2	13,000–36,500		15,000–18,000	17,000–28,000
Flexural strength, lb/in^2	8500–18,300		14,000–21,000	8000–14,000
Impact strength, Izod, ft · lb/in notch (½- by ½-in notched bar)	0.2–0.4	7.0	0.2–0.6	0.3–0.45
Hardness	M70–M115 (Rockwell)	84–94 (Shore A)	M80–M100 (Rockwell)	M85–M120 (Rockwell)
Thermal conductivity, $10^4 \, cal/(S)(cm^2)(°C)(cm)$	4		4–5	10–20
Thermal expansion, $10^{-5}/°C$	5.5–10		4.5–6.5	2.0–4.0
Resistance to heat (continuous), °F	250	250	250–600	250–600
Heat-distortion temperature, °F	140–400		115–550	160–550
Volume resistivity (at 50% RH and 23°C), $\Omega \cdot cm$	10^{14}		$10^{13}–10^{17}$	$10^{13}–10^{16}$
Electric strength, ⅛ in thick, V/mil				
Short-time	380–500	250–400	400–500	400–500
Step-by-step	280–420	170	380	
Dielectric constant, 60-Hz	3.0–4.4	4.4–8.1	3.5–5.0	3.2–4.5
Dissipation (power) factor, 60-Hz	0.003–0.028	0.026–0.31	0.002–0.010	0.008–0.03
Arc resistance, S	125	135	45–120	150–300
Water absorption (24-h, ⅛ in thick), %	0.15–0.60	0.50–2.5	0.08–0.13	0.04–0.10

*Taken from Harper, *Electronic Packaging with Resins,* McGraw-Hill, New York, 1961.

IMPREGNATING AND FILLING COMPOUNDS

16-6. Acrylic Resins. Acrylic-based impregnating resins are usually comprised of a difunctional acrylate (for example, tetraethylene glycol diacrylate) or an unsaturated alkyd blended with acrylic monomers, such as methyl methacrylate. The cured materials are usually strong, rigid, and resistant to sharp blows. Physical and electrical properties are not affected by weathering, so that one of the major applications for acrylics is in outdoor electrical equipment. Good bonding to substrate is usually found. Typical electrical and physical properties of acrylic impregnating resins are shown in Table 16-3.

TABLE 16-3 Typical Properties of Acrylic Impregnating Resins

Properties	Unfilled	Filled
Physical		
Hardness (Rockwell)	M40–M70	M85–M105
Modulus of elasticity, lb/in^2 $\times$ 10^5	4.5	4.5
Flexural strength, lb/in^2 $\times$ 10^2	10–17	13–17
Elongation, %	1.0–5.0	3.0–10.0
Thermal conductivity, (cal)(cm)/ (s)(cm^2)(°C) $\times$ 10^{-5}	4–6	4–6
Thermal expansion, in/(in)(°C) $\times$ 10^{-5}	8.0	5.0–9.0
Electrical		
Dissipation factor		
60 Hz	0.04–0.06	0.04–0.06
10^3 Hz	0.03–0.05	0.03–0.05
Dielectric constant		
60 Hz	3.0–3.7	3.5–4.5
10^3 Hz	3.0–3.5	3.0–3.5
Dielectric strength, short-time, V/mil	450–500	450–550
Volume resistivity, $\Omega \cdot$ dm	>10^{15}	>10^{14}

16-7. Urethanes. A whole family of thermosetting impregnating resins can be produced by reacting isocyanate compounds with other organic compounds. When they are reacted with polyols, polyurethanes (more commonly known as urethanes) are formed. The polyols are usually glycols, polyesters, and polyethers. Because of the versatility of urethane chemistry, a wide range of impregnating resins with significantly differing properties can be formulated. Through careful choice of materials, varying flexibilities, hardnesses, flexural strengths, and tensile moduli can be achieved. In general, urethanes display very good insulation properties.[19,20]

Typical physical and electrical properties of urethane impregnating resins are shown in Table 16-4.

16-8. Polybutadienes. Polybutadiene resins are high-vinyl-content materials derived from butadiene. There are two main types: 1,4-polybutadiene, which usually exists as a solid elastomeric material, and 1,2-polybutadiene, which can exist as a liquid or semiliquid at ambient[21,22] but will change to a very hard and rigid material after curing.

High 1,2-content polybutadiene impregnating resins usually contain a significant amount of vinyl monomer (for example, styrene or vinyl toluene) and a free radical initiator (for example, organic peroxide). The cured products can vary from pliable to rigid. Because they are 100% hydrocarbon thermoset polymers, they display excellent dielectric properties over a broad range of frequencies. Also, outstanding chemical resistance, high heat distortion, and low water-absorption properties are shown by these materials. Typical properties found with some of these polybutadiene impregnating resins are shown in Table 16-5.

16-9. Solventless Silicone Potting, Encapsulating, and Filling Resins. Silicone polymers are characterized by the presence of the recurring siloxane group made up of alternating silicon and oxygen atoms, to which organic groups are attached through silicon. These resins, therefore, partake somewhat of the nature of inorganic materials and are very stable to heat (for review, see Rochow[23] and Meals and Lewis[24]). An extremely versatile line of thermally stable solventless silicone polymer systems is available. The cured products vary

ELECTRICAL ENGINEERING MATERIALS REFERENCE GUIDE

TABLE 16-4 Typical Properties of Urethane Impregnating Resins

Properties	Unfilled[20]	Filled[1]
Physical		
Hardness (Shore)	55A–70D	
Tensile strength, lb/in^2	410–4000	100–600
Elongation, %	60–200	
Specific gravity, g/cm^3	1.06–1.12	1.24–1.26
Coefficient of linear thermal expansion, in/(in)(°C) $\times 10^{-5}$	20–24	10–20
Thermal conductivity, (cal)(cm)/ (s)(cm^2)(°C) $\times 10^{-4}$	3.4–4.2	
Electrical		
Volume resistivity, $\Omega \cdot$ cm	2.7×10^{14} -3.0×10^{15}	2×10^{11}
Dielectric constant		
60 Hz		6.7–7.5
10^3 Hz	2.9–4.3	6.7–7.5
Dissipation factor		
60 Hz		0.015–0.017
10^3 Hz	0.03–0.13	0.050–0.060
Dielectric strength, V/mil	490–650	450–500

in consistency from the so-called dielectric gels, through rubberlike products, to rigid analogs of polyesters and epoxies. Properties of a few of these are listed in Table 16-6.

16-10. New Heat-Resistant Solventless Impregnating, Encapsulation, and Potting Resins. In recent years, several new types of high-performance thermoset solventless resins have become readily available based on modified polyimide and epoxy resin chemistry. Mitsubishi Electric Company has pioneered the polyimide area of solventless resin chemistry with two high-temperature resin systems, MDT and BT. Thermally stable triazine

TABLE 16-5 Properties of Polybutadiene Impregnating Resins*

Properties	Unfilled†[22]	Filled[21]
Physical		
Hardness	70 (Shore D)	E70–E95 (Rockwell)
Specific gravity (25°C)	0.905	2.05
Flexural strength, lb/in^2	3,400	11,000–14,000
Flexural modulus, lb/in^2 $\times 10^5$	1.04	13–14
Coefficient of linear expansion, 10^{-6}/°F	56.8	
Thermal conductivity, (cal)(cm)/ (s)(cm^2)(°C) $\times 10^{-4}$	——	5.2
Electrical		
Dissipation factor (23°C)		
60 Hz	0.0003	
10^3 Hz	0.0024	0.004
Dielectric constant (23°C)		
60 Hz	2.95	3.7
10^3 Hz	2.77	3.7
Dielectric strength, short-time, V/mil	564	500–1,000 +
Volume resistivity, $\Omega \cdot$ cm	10^{15}–10^{17}	1.7–7.1 ($\times 10^{15}$)

*High 1,2-vinyl-content resins.
†Containing 20% w/w styrene monomer.

IMPREGNATING AND FILLING COMPOUNDS

TABLE 16-6 Properties of Some Solventless Silicone Insulating Resins

Properties	Rigid unfilled resin	Rigid filled resin		Rubbery (RTV) resin	Dielectric gel
		Silica	Zircon		
Physical					
Specific gravity	1.11	1.70	3.3	1.12	0.97
Coefficient of thermal expansion $10^{-5}/°C$	12.5 (lin.)	8.0 (lin.)	1.3 (lin.)	77.0 (vol.)	96.0 (vol.)
Thermal conductivity, $\times 10^4$	3.6	20.0	27.0	5.0	7.0
Electric strength, V/mil, ASTM D149	350–0.125 in (¼-in elec.)	350–0.125 in (¼-in elec.)	240–0.125 in (¼-in elec.)	400–0.063 in (¼-in elec.)	1000–0.10 in (½-in ball elec.)
Electrical					
Volume resistivity, $\Omega \cdot$ cm	5×10^{15}	4×10^{15}	4×10^{15}	0.5×10^{14}	5×10^{14}
Dielectric constant (ASTM D150) (23°C), 400 Hz	2.82	3.62	7.30	3.14	3.00
Dissipation factor (ASTM D150) (23°C), 400 Hz	0.002	0.007	0.008	0.01	0.00008

units are used as the modifier in these systems to prepare long-chain polyimide-type structures.[25] BT resin can be used to prepare impregnated tapes, insulating prepreg boards, adhesives, molding compounds and casting resins. Significant improvements in the thermal, mechanical, and electrical properties over analogous epoxy and polyester resin systems are claimed.

The Toshiba Corporation of Japan is marketing a similar heat-stable product identified as TC resin.[26]

The Hitachi Company of Japan has developed a heat-resistant solventless resin system based on hybrid epoxy/diisocyanate chemistry. The heterocyclization reaction of diisocyanates with epoxies can be programmed stoichiometrically to undergo controlled heterocyclization reactions producing thermally stable isocyanurate and oxazolidone ring structures.[27] Table 16-7 shows cured resin data at 200°C for one of these hybrid epoxy-diisocyanate resins (Hitachi IO) compared to epoxy and silicone resins at the same temperature.[27]

16-11. Low-Temperature Curable Impregnating and Potting Resins. New low-temperature anaerobic- and ultraviolet-curable impregnating and potting solventless resins are now readily available.[28] The anaerobic-curable resins are unique in the fact that they are air-stable but will cure very rapidly under anaerobic conditions (i.e., in the absence of oxygen). Most of these resin systems appear to be based on acrylic or modified-acrylic chemistry and would be expected to exhibit the characteristics of acrylic resins.

An interesting extension of the anaerobic resin chemistry is the development of low-viscosity solventless resins for vacuum-pressure impregnation (VPI) applications.[29] These impregnating resins, which have been developed for insulating motor and generator coils, have the property of being stable (i.e., not curing) when stored in air but will undergo effective cure at room temperature when exposed to a pressurized or flowing inert gas such as nitrogen.

TABLE 16-7 Comparative Data for the Hitachi "IO" Resin, Silicone and Epoxy Resins[27]

Cured resin properties at 200 C	Resin type		
	IO	Silicone	Epoxy
Dissipation factor, %	0.3	5.5	3.0
Dielectric constant	3.0	2.7	4.0
Tensile strength, lb/in^2	8,700	29	2,900
Tensile shearing strength, lb/in^2	2,175	——	15
Volume Resistivity, $\Omega \cdot$ in	10^{12}	10^9	10^9

16-8 ELECTRICAL ENGINEERING MATERIALS REFERENCE GUIDE

16-12. References

1. *Insulation/Circuits, Desk Manual,* vol. 28, no. 7, prepared by the Lake Publishing Corporation, Libertyville, Ill., 1982.

2. Moses, G. L., et al., *Insulation Engineering Fundamentals,* Lake Publishing Corporation, Libertyville, Ill., 1958.

3. Clark, F. M., *Insulating Materials for Design and Engineering Practice,* John Wiley & Sons, New York, 1962.

4. Sillars, R. W., *Electrical Insulating Materials and Their Application,* Chap. 6, IEE Monograph Series 14, Peter Peregrinus, Ltd., 1973.

5. Neppe, S. L., "The Chemistry and Rheology of Asphatl Bitumen," *Pet. Refiner,* **31**(2):137–142 (1952).

6. Warth, A. H., *The Chemistry and Technology of Waxes,* 2d ed., Reinhold, New York, 1956.

7. Harper, C. A., *Electronic Packaging with Resins,* McGraw-Hill, New York, 1961.

8. Beonig, H. V., *Unsaturated Polyesters, Structure, and Properties,* American Elsevier, New York, 1964.

9. Doyle, E. N., *The Development and Use of Polyester Products,* McGraw-Hill, New York, 1969.

10. Lee, A., and Neville, K., *Handbook of Epoxy Resins,* McGraw-Hill, New York, 1967.

11. May, C. A., and Tanaka, Y., *Epoxy Resins; Chemistry and Technology,* Marcel Dekker, New York, 1973.

12. Bruins, Paul F. (ed.), *Epoxy Resins Technology,* Chap. 2 by J. J. Stevens, Jr., Interscience Publishers, John Wiley & Sons, New York, 1968.

13. Dow Chemical, "Experimental Resin XD-7855," *Epoxy Resin News,* April 1973.

14. Celanese Resins, *Tech. Bull. 1068.*

15. Perez, R. J., Ref. 12, Chap. 4.

16. Lee, H. (ed.), *Advances in Chemistry Series,* 92, Chap. 4, "Epoxy Resins," by A. M. Partansky, American Chemical Society Publication, New York, 1970.

17. Bowen, D. O., and Whiteside, R. C., Jr., Ref. 16, Chap. 5.

18. Smith, J. D. B., "Epoxy Resins", Chap. 3, Advances in Chemistry Series, 114, American Chemical Society Publication, 1979, edited by R. S. Bauer.

19. Conap, Inc., *Tech. Bull. T-117,* 1976.

20. N. L. Industries, Inc., *Tech. Bull. PB-100.*

21. Firestone, *Tech. Bull. 0372.*

22. Richardson Company, *Tech. Bull. RIC-101.*

23. Rochow, E. G., *An Introduction to the Chemistry of the Silicones,* 2d ed., John Wiley & Sons, New York, 1951.

24. Meals, R. N., and Lewis, F. M., *Silicones,* Reinhold, New York, 1959.

25. Motoori, S., Kinbara, H., Gaku, M., and Ayano, S., "BT Resin for Electrical Insulation Material," *Conference Proceedings No. 81CH1717-8,* 15th Electrical/Electronics Insulation Conference, Chicago, Ill., Oct. 19–22, 1981, pp. 168–171.

26. Suzuki, S., Ito, T., Wada, M., Gonoi, S. and Yajima, N., "New Class H Solvent-Free Impregnating Resin," *Toshiba Review,* no. 120, March–April 1979.

27. Kinjo, N., Koyama, T., Narahara, T. and Mukai, J., "New Heat Resistant Solventless Varnish IO Resin" (Part II), *Conference Proceedings No. 79CH1510-7-E1,* 14th Electrical/Electronics Insulation Conference, Boston, Mass., Oct. 8–11, 1979, pp. 129–132.

28. Loctite Corporation, *Engineering Adhesive Bulletin,* no. 3, 1983.

29. Smith, J. D. B., and Bich, G. J., "Anaerobically Curable Vacuum Pressure Impregnating Resins for Motor and Generator Insulation," Reference 25, pp. 46–50.

Chapter 17

STRUCTURAL MATERIALS

By ROBERT W. BOHL

Definitions and Properties of Structural
  Materials   17-1
Structural Iron and Steel   17-6
Steel Strand and Rope   17-16

Corrosion of Iron and Steel   17-18
Nonferrous Metals and Alloys   17-21
Stone, Brick, Concrete, and Glass Brick
  17-30

Definitions of Properties of Structural Materials

17-1. Stress is the intensity at a point in a body of the internal forces or components of force that act on a given plane through the point. Stress is expressed in force per unit of area (pounds per square inch, kilograms per square millimeter, etc.). There are three kinds of stress: tensile, compressive, and shearing. Flexure involves a combination of tensile and compressive stress. Torsion involves shearing stress. It is customary to compute stress on the basis of the original dimensions of the cross section of the body, though "true stress" in tension or compression is sometimes calculated from the area of the time a given stress exists, rather than from the original area.

17-2. Strain is a measure of the change, due to force, in the size or shape of a body referred to its original size or shape. Strain is a nondimensional quantity but is frequently expressed in inches per inch, etc. Under tensile or compressive stress, strain is measured along the dimension under consideration. Shear strain is defined as the tangent of the angular change between two lines originally perpendicular to each other.

17-3. A stress-strain diagram is a diagram plotted with values of stress as ordinates and values of strain as abscissas. Diagrams plotted with values of applied load, moment, or torque as ordinates and with values of deformation, deflection, or angle of twist as abscissas are sometimes referred to as stress-strain diagrams but are more correctly called "load-deformation diagrams." Six stress-strain diagrams are shown in Fig. 17-1, where curve I is typical of normalized high-carbon steel; curve II is typical of low-carbon ductile steels, which have a yield point shown at Y; and curve III is typical of some of the nonferrous alloys; curve IV represents a heat-treated alloy steel; curve V is typical for a gray cast iron;

and curve VI shows approximately the type of curve obtained for timber. The stress-strain diagram for some materials is affected by the rate of application of the load, by cycles of previous loading, and again by the time during which the load is held constant at specified values; for precise testing, these conditions should be stated definitely in order that the complete significance of any particular diagram may be clearly understood.

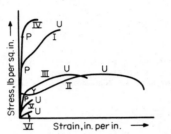

FIG. 17-1 Typical stress-strain diagrams for tensile stress.

17-4. The modulus of elasticity is the ratio of stress to corresponding strain below the proportional limit. For many materials the stress-strain diagram is approximately a straight line below a more or less well-defined stress known as the "proportional limit." As there are three kinds of stress, there are three moduli of elasticity for a material, that is, the modulus in tension, the modulus in compression, and the modulus in shear. The value in tension is practically the same, for most ductile metals, as the modulus in compression; the modulus in shear is only about 0.36 to 0.42 of the modulus in tension. The modulus is expressed in pounds per square inch (or kilograms per square millimeter) and measures the elastic *stiffness* (the ability to resist elastic deformation under stress) of the material.

17-5. Elastic Strength. To the user and the designer of machines or structures one significant value to be determined is a *limiting stress below which the permanent distortion of the material is so small that the structural damage is negligible and above which it is not negligible.* The amount of plastic distortion which may be regarded as negligible varies widely for different materials and for different structural or machine parts. In connection with this limiting stress for elastic action a number of technical terms are in use; some of them are

a. *Elastic Limit.* The greatest stress which a material is capable of withstanding without a permanent deformation remaining upon release of stress. Determination of the elastic limit involves repeated application and release of a series of increasing loads until a set is observed upon release of load. Since the elastic limit of many materials is fairly close to the proportional limit, the latter is sometimes accepted as equivalent to the elastic limit for certain materials. There is, however, no fundamental relation between elastic limit and proportional limit. Obviously the value of the elastic limit determined will be affected by the sensitivity of apparatus used.

b. *Proportional Limit.* The greatest stress which a material is capable of withstanding without a deviation from proportionality of stress to strain. The statement that the stresses are proportional to strains below the proportional limit is known as *Hooke's law.* Proportional limits for the metals in Fig. 17-1 are located at the points P; however, the numerical values of the proportional limit are influenced by methods and instruments used in testing and the scales used for plotting diagrams.

c. *Yield point* is the lowest stress at which marked increase in strain of the material occurs *without increase in load.* It is indicated at point Y on the stress-strain curve II in Fig. 17-1. If the stress-strain curve shows no abrupt or sudden yielding of this nature, then there is no yield point; for example, curve I in Fig. 17-1 exhibits no yield point. Iron and low-carbon steels have yield points, but most metals do not, including iron and low-carbon steels immediately after they have been plastically deformed at ordinary temperatures.

d. *Yield strength* is the stress at which a material exhibits a specified limiting permanent set. Its determination involves the selection of an amount of permanent set that is considered the maximum amount of plastic yielding which the material can exhibit, in the particular service condition for which the material is intended, without appreciable structural damage. A set of 0.2% has been used for several ductile metals, and values of yield strength for various metals are for 0.2% set unless otherwise stated. On the stress-strain diagram for the material (see Fig. 17-2) this arbitrary set is laid off as q along the strain axis, and the line mn drawn parallel to OA, the straight portion of the diagram. Since the stress-strain diagram for release of load is approximately parallel to OA, the intersection r may be regarded as determining the stress at the yield strength. The yield strength is generally used to determine the elastic strength for materials whose stress-strain curve in the region pr is a smooth curve of gradual curvature. See discussion in ASTM Designation E6.

17-6. Ultimate strength (tensile strength or compressive strength) is the maximum stress which a material will sustain when slowly loaded to rupture. Ultimate strength is computed from the maximum load carried during a test and the original cross-sectional area of the specimen. For materials that fail in compression with a shattering fracture, the compressive strength has a definite value, but for materials that do not fracture, the compressive strength is an arbitrary value depending on the degree of distortion which is regarded as indicating complete failure of the material. In tensile tests of many materials, especially those having appreciable ductility, failure does not occur at the stress corresponding to the ultimate strength. For such materials, localized deformation, or necking, occurs and the nominal stress decreases because of the rapidly decreasing cross-sectional area until failure occurs.

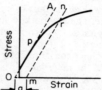

FIG. 17-2 Yield strength of a material having no well-defined yield point.

17-7. Shearing strength is the maximum shearing stress which a material is capable of developing. The general remarks in Par. **17-6** regarding methods of failure are also applicable to failures in shear. Owing to experimental difficulties of obtaining true shearing strength, the values of modulus of rupture in torsion are usually reported as indicative of the shearing strength.

17-8. Modulus of rupture in flexure (or torsion) is the term applied to the computed stress, in the extreme fiber of a specimen tested to failure under flexure (or torsion), when computed by the arbitrary application of the formula for stress with disregard of the fact that the stresses exceed the proportional limit. Hence the modulus of rupture does not give the true stress in the member but is useful only as a basis of comparison of relative strengths of materials.

17-9. Ductility is that property of a material which enables it to acquire large permanent deformation and at the same time develop relatively large stresses (as drawing into a wire). Though ductility is a highly desirable property required by almost all specifications for metals, the quantitative amount needed for structural applications is not entirely clear but probably does not exceed about 3% elongation after the structure is fabricated (see *Proc. ASTM,* vol. 40, p. 551). The commonly used measures of ductility are

a. *Elongation* is the ratio of the increase of length of a specimen, after rupture under tensile stress, to the original gage length; it is usually expressed in percent. The percentage of elongation for any given material depends upon the gage length, which should always be specified.

b. *Reduction of area or contraction of area* is the ratio of the difference between the original and the fractured cross section to the original cross-sectional area; it is usually expressed in percent.

c. *Bend test* measures the angle through which a given specimen of material can be bent, at a specified temperature, without cracking. In some cases the maximum angle through which the specimen can be bent around a certain diameter or the number of bendings back and forth through a stated angle are measured. In other cases the elongation in a given gage length across the crack on the tension side of the bend specimen is measured. See ASTM Standard E16.

17-10. Plasticity permits a material to assume permanent deformations under loads without recovery of the strain when the loads are removed. Plasticity permits shaping of metal parts by plastic deformation; plastic materials deform instead of fracturing under load.

ELECTRICAL ENGINEERING MATERIALS REFERENCE GUIDE

17-11. *Brittleness* is defined as the ability of a material to fracture under stress with little or no plastic deformation. Brittleness implies a lack of plasticity.

17-12. *Resilience* is the amount of strain energy (or work) which may be recovered from a stressed body when the loads causing the stresses are removed. Within the elastic limit the work done in deforming the bar is completely recovered upon removal of the loads; the total amount of work done in stressing a unit volume of the material to the elastic limit is called the *modulus of resilience.*

17-13. *Toughness* is the ability to withstand large stresses accompanied by large strains before fracture. The toughness is usually measured by the total work done in stressing a unit volume of the material to complete fracture and may be interpreted as the total area under the stress-strain curve (Fig. 17-1). Ductility differs from toughness in that it deals only with the ability of the material to deform, whereas toughness is measured by the energy-absorbing capacity of the material.

17-14. *Impact Resistance.* The ability of a material to resist impact or energy loads without permanent distortion is measured by the modulus of resilience. The ultimate resistance to impact before fracture is measured by the toughness of the material. For members with abrupt changes of section (holes, keyways, fillets, etc.), the resistance to a rapidly applied load depends greatly on the "notch sensitivity" (the resistance to the formation and spread of a crack); above certain critical velocities of loading and below certain critical temperatures, the impact strength is greatly reduced. Relative notch sensitivity under repeated loads is not the same as that in a single-blow notched-bar test. Impact values are influenced by speed of straining, shape and size of specimen, and type of testing machine.

Charpy or *Izod* impact bend tests measure the energy required to fracture small notched specimens (1 cm square) under a single blow. These tests are used as an indication of toughness, a property that is very sensitive to the composition and thermal-mechanical history of the material. Tests should be carried out over a range of temperatures to determine the temperature at which the alloy fails by brittle rather than ductile failure. (See ASTM Standard E23.)

17-15. *Fracture Mechanics.* Three primary factors have been identified that control the susceptibility of a structure to brittle failure: material toughness (affected by composition and metallurgical structure as well as temperature, strain rate, and constraints to plastic yielding); flaw size (internal discontinuities such as porosity or small cracks from welding, fatigue, fabrication, etc.); and stress level (applied or residual). Fracture mechanics attempts to interrelate these variables in order to predict the occurrence of brittle fracture on a quantitative design basis rather than depend upon qualitative relationships between experience and results of impact tests such as Izod and Charpy. Fracture mechanics has had excellent success when applied to high-strength materials. The material parameter defined, called the "fracture toughness K_C," can be measured experimentally and utilized to specify safe loading conditions in the presence of a given size and geometry of flaw. See "Fracture Toughness," ASTM Spec. Tech. Publ. 514, 1971.

17-16. *Hardness* is the resistance which a material offers to small, localized plastic deformations developed by specific operations such as scratching, abrasion, cutting, or penetration of the surface. Hardness does not imply brittleness, as a hard steel may be tough and ductile. The standard Brinell hardness test is made by pressing a hardened steel ball against a smooth, flat surface under certain standard conditions; the Brinell hardness number is the quotient of the applied load divided by the area of the surface of the impression. A different method of test is employed in the Shore scleroscope, in which a small, pointed hammer is allowed to fall from a definite height onto the material, and the hardness is measured by the height of the rebound, which is automatically indicated on a scale. The Rockwell hardness machine measures the depth of penetration in the metal produced by a definite load on a small indenter of spherical or conical shape. Vickers or Tukon hardness machines measure hardness on a microscopic scale. The dimensions of the impression of a lightly loaded diamond pyramid indenter on a polished surface are related to hardness number. (For further data see ASTM Standards E10, E18, and E92; also Metals Handbook, ASM, 8th ed., vol. 11, 1976, pp. 1–20.)

17-17. *Fatigue strength (fatigue limit)* is a limiting stress below which no evidence of failure by progressive fracture can be detected after the completion of a very large number of repetitions of a definite cycle of stress. The fatigue limits usually reported are those for completely reversed cycles of flexural stress in polished specimens. For stress cycles in which

STRUCTURAL MATERIALS   17-5

an alternating stress is superimposed on a steady stress the endurance limit (based on the maximum stress in the cycle) is somewhat higher. Most ferrous metals have well-defined limits, whereas the fatigue strength of many nonferrous metals is arbitrarily listed as the maximum stress that is just insufficient to cause fracture after some definite number of cycles of stress, which should always be stated. The fatigue strength of actual members containing notches (holes, fillets, surface scratches, etc.) is greatly reduced and depends entirely on the "stress-raising" effect of these discontinuities and the sensitivity of the material to the localized stresses at the notch.

17-18. Composition and Structure. Chemical analysis is employed to determine whether component elements are present within specified amounts and impurity elements are held below specified limits. Mechanical and physical properties, however, depend on the size, shape, composition, and distribution of the crystalline constituents that make up the structure of the alloy. Chemical analysis does not reveal these features of the structure. Metallographic techniques, which involve examination of carefully polished and etched surfaces by optical and electron microscopy or x-ray methods, are required to provide this vital information. Nondestructive testing (NDT) methods are useful in detecting the presence of flaws of various kinds in finished parts and structures. These techniques depend on the interference of the defect with some easily measured physical property, such as x-ray absorption, magnetic susceptibility, propagation of acoustical waves, or electrical conductivity. NDT techniques have particular application where defects are difficult to detect and quite likely to occur (as in welded structures) and where high-integrity performance requires 100% inspection.

17-19. Aging is a spontaneous change in properties of a metal with time after a heat-treatment or a cold-working operation. Aging tends to restore the material to an equilibrium condition and to remove the unstable condition induced by the prior operation and usually results in increased strength of the metal with corresponding loss of ductility. The fundamental action involved is generally one of precipitation of hardening elements from the solid solution, and the process can usually be hastened by slight increase in temperature. This is a very important strengthening mechanism in a variety of ferrous and nonferrous alloys, e.g., high-strength aluminum alloys.

17-20. Corrosion Resistance. There is no universal method of determining corrosion resistance, because different types of exposure ordinarily produce entirely dissimilar results on the same material. In general the subject of corrosion is rather complicated; in some cases corrosive attack appears to be chiefly chemical in its nature, while in others the attack is by electrolysis. Owing to the great diversity of materials exposed to corrosive influences in service and the wide range of service conditions, it is impracticable to formulate any universal measure of corrosion resistance. If the service life is likely to be determined by corrosion resistance, the degree of impairment which marks the end of usefulness will ordinarily be established by considerations of safety and reliability or perhaps of appearance. Corrosion testing is conducted in general by two methods: (a) normal exposure in service with periodic observations of corrosive action as it progresses under such conditions; (b) some type of artificially accelerated test, which may serve merely to obtain comparative results or, again, may simulate the conditions of service exposure. For specific information see H. H. Uhlig, "Corrosion and Corrosion Control," New York, John Wiley & Sons, Inc., 1963; ASTM STP290, Twenty Year Atmospheric Corrosion Investigation of Zinc-Coated and Uncoated Wire and Wire Products, 1961; and M. Fontana and N. Greene, *Corrosion Engineering,* McGraw-Hill, 1967.

17-21. Powder Metallurgy. Many alloys and metallic aggregates having unusual and very valuable properties are being produced commercially by mixing metal powders, pressing in dies to desired shapes, and sintering at high temperatures. Parts may be produced to close dimensional tolerance, and the process enables the mixing of dissimilar materials which will not normally alloy or which cannot be cast because of insolubility of the constituents. Wide use of powder metallurgy is made in producing copper-molybdenum alloys for contact electrodes for spot welding, extremely hard cemented tungsten carbide tips for use in metal cutting tools, and copper-base alloys containing either graphite particles or a controlled dispersion of porosity for bearings of the "oilless" or oil-retaining types. Silver-nickel and silver-molybdenum alloys (tungsten or graphite may be added) for contact materials having high conductivity but good resistance to fusing can be produced by the method. Powdered iron is being used to manufacture gears and small complex parts where the sav-

17-6 ELECTRICAL ENGINEERING MATERIALS REFERENCE GUIDE

ings in weight of metal and machining costs are able to offset the additional cost of metal and processing in the powdered form. Small Alnico magnets of involved shape which are exceedingly difficult to cast or machine can be produced efficiently from metallic powders and require little or no finishing. Solid mixtures of metals and nonmetals, such as asbestos, can be produced to meet special requirements. The size and shape of powder particles, pressing temperature and pressure, sintering temperature and time, all affect the final density, structure, and physical properties. For further details see W. D. Jones, *Fundamental Principles of Powder Metallurgy,* London, E. Arnold, 1960.

Structural Iron and Steel

17-22. Classification of Ferrous Materials. Iron and steel may be classified on the basis of composition, use, shape, method of manufacture, etc. Some of the more important ferrous alloys are described in the sections below.

17-23. Ingot iron is commercially pure iron, and contains a maximum of 0.15% total impurities. It is very soft and ductile and can undergo severe cold-forming operations. It has a wide variety of applications based on its formability. Its purity results in good corrosion resistance and electrical properties, and many applications are based on these features. The average tensile properties of Armco ingot iron plates are: tensile strength 320 MPa (46,000 lb/in^2); yield point 220 MPa (32,000 lb/in^2); elongation in 8 in, 30%; Young's modulus 200 GPa (29 $\times$ 10^6 lb/in^2) (see ASTM Designation A345).

17-24. Plain carbon steels are alloys of iron and carbon containing small amounts of manganese (up to 1.65%) and silicon (up to 0.50%) in addition to impurities of phosphorus and sulfur. Additions up to 0.30% copper may be made in order to improve corrosion resistance. The carbon content may range from 0.05 to 2%, although few alloys contain more than 1.0%, and the great bulk of steel tonnage contains from 0.08 to 0.20% and is used for structural applications. Medium-carbon steels contain around 0.40% carbon and are used for constructional purposes—tools, machine parts, etc. High-carbon steels have 0.75% carbon or more, and may be used for wear and abrasion-resistance applications such as tools, dies, and rails. Strength and hardness increase in proportion to the carbon content while ductility decreases. Phosphorus has a significant hardening effect in low-carbon steels, while the other components have relatively minor effects within the limits they are found. It is difficult to generalize the properties of steels, however, since they can be greatly modified by cold working or heat-treatment.

17-25. High-strength low-alloy steels are low-carbon steels (0.10 to 0.15%) to which alloying elements such as phosphorus, nickel, chromium, vanadium, and niobium have been added to obtain higher strength. This class of steel was developed primarily by the transportation industry to decrease vehicle weight, but the steels are widely used. Since thinner sections are used, corrosion resistance is more important, and copper is added for this purpose. See ASTM Designation A242.

17-26. Free-Machining Steels. Additions of manganese, phosphorus, and sulfur greatly improve the ease with which low-carbon steels are machined. The phosphorus hardens the ferrite, and the manganese and sulfur combine to form nonmetallic inclusions that help form and break up machining chips. The improvement in machinability is gained at some loss of mechanical properties, and these steels should be used for noncritical applications. Small amounts of lead also improve machining characteristics of steel by helping break up chips as well as providing a self-lubricating effect. Lead is more often added to higher-carbon steels where the effect on mechanical properties is less detrimental than that caused by sulfide inclusions. (See ASM *Metals Handbook,* 8th ed., vol. 3.)

17-27. Alloy Steels. When alloying ingredients (in addition to carbon) are added to iron to improve its mechanical properties, the product is known as an alloy steel. Heat-treatment is a necessary part of the manufacture and use of alloy steels; only through proper quenching and tempering can the full beneficial effects of the alloys be obtained. The chief advantages obtained from the addition of alloys to steel are (a) to increase the depth of hardening on quenching, thus making it possible to produce more uniform properties throughout thick sections with a minimum of distortion; (b) to form chemical compounds which when properly distributed develop desirable properties in the steel, that is, extreme hardness, corrosion or heat resistance, high strength without excessive brittleness.

STRUCTURAL MATERIALS

17-7

The most commonly used alloy steels have been classified by the American Iron and Steel Institute and the Society of Automotive Engineers and are identified by a nomenclature system that is partially descriptive of the composition. The system of steel designations is shown in Table 17-1. Approximate strengths of several alloy steels after specific heat-treatments are given in Table 17-3.

Mechanical properties of the alloy steels vary over a wide range depending upon size, composition, and thermomechanical treatment. (See also E. C. Bain and H. Paxton, *Alloying Elements in Steel*, 2d ed.; American Society for Metals, 1961.)

17-28. Cast Iron. Iron ore is reduced to the metallic form in a blast furnace yielding a product of molten iron saturated in carbon (about 4%). Most commonly, this "hot metal" is immediately processed to steel by a refining process without allowing it to solidify. Occasionally it is cast into bars; this product is called pig iron. Cast iron is made by remelting pig iron and/or scrap steel in a cupola or electric furnace, and casting it into molds to the desired shape of the finished part. Cast iron has a much higher carbon content than steel, usually between 2.5 and 3.75%.

17-29. Gray Cast Iron. In gray cast iron, the excess carbon beyond that soluble in iron is present as small flake-shaped particles of graphite. The flakes of graphite account for some of the unique properties of gray iron; in particular, its low tensile strength and ductility, its ability to absorb vibrational energy (damping capacity), and its excellent machinability. Cast iron is easy to cast because it has a lower melting point than steel, and the formation of the low-density graphite offsets solidification shrinkage so that minimal dimensional changes occur on freezing. Other elements in the composition of ordinary gray cast iron are important chiefly insofar as they affect the tendency of carbon to form as graphite rather than in chemical combination with the iron as iron carbide (Fe_3C). Silicon is most effective in promoting the formation of graphite. Slower cooling rates during freezing also favor the formation of graphite as well as increase the size of the flakes. Cooling rate also affects the mode of decomposition of the carbon retained in solution during freezing. Slow cooling favors complete precipitation as graphite, leaving a soft ferrite matrix, while fast cooling produces a stronger matrix containing Fe_3C (as pearlite). The tensile strength of gray cast iron typically ranges from 140 to 410 MPa (20,000 to 60,000 lb/in²). Corresponding compressive strengths are 575 to 1300 MPa (85,000 to 190,000 lb/in²). Young's modulus may range from 70 to 150 GPa (10×10^6 to 20×10^6 lb/in²), depending on the microstructure. See ASTM Specifications A48 and A126, ASM *Metals Handbook*, 9th ed., vol. 1, 1978.

17-30. White Cast Iron. Careful adjustment of composition and cooling rate can cause all the carbon in a cast iron to appear in the combined form as pearlite or free carbide. This structure is very hard and brittle and has few engineering applications beyond resistance to abrasion. This product does serve, however, as an intermediate product in the production of malleable cast iron described in the following paragraph.

17-31. Malleable Cast Iron. By annealing white cast iron at about 950°C, the combined carbon will decompose to graphite. This graphite grows in a spheroidal shape rather than the flakelike shape that forms during the freezing of gray cast iron. Because of this difference in graphite shape, malleable iron is much tougher and stronger. If the castings are slowly cooled from the malleabilizing temperature, the matrix can be converted to ferrite, with all the carbon appearing as graphite; this is a very tough product. Faster cooling will yield a pearlitic matrix with greater strength and hardness. It is also possible to quench and temper malleable iron for optimum combinations of strength and toughness. By careful control of composition, the malleabilizing cycle can be carried out in 8 to 20 h. See ASM *Metals Handbook*, 9th ed., vol. 1; ASTM Specification A48; "Malleable Iron Castings," Malleable Founders Society, Cleveland, 1960 (see Table 17-2).

17-32. Nodular cast iron has the same carbon content as gray iron; however, the addition of a few hundredths of 1% of either magnesium or cerium causes the uncombined carbon to form spheroidal particles during solidification instead of graphite flakes. Strength properties comparable with those of steel may be achieved in the pearlitic iron. The softer ferritic and pearlitic as-cast irons exhibit considerable ductility, 10% elongation or more. As the hardness and strength are increased by appropriate heat-treatment or the thickness of the casting decreased below approximately ¼ in, the ductility decreases. An austenitic form of nodular iron may be obtained by adding various amounts of silicon, nickel, manganese, and chromium. For many purposes nodular iron exhibits properties superior to those of either gray or malleable cast iron. For more complete information see ASM *Metals Handbook*, 9th ed., vol. 1.

TABLE 17-1 AISI-SAE System of Designations

Numerals and digits	Type of steel and nominal alloy content	Numerals and digits	Type of steel and nominal alloy content	Numerals and digits	Type of steel and nominal alloy content
Carbon steels:		**Nickel-chromium-molybdenum steels:**		**Chromium steels:**	
10XX(a)	Plain carbon (Mn 1.00% max)	43XX	Ni 1.82; Cr 0.50 and 0.80; Mo 0.25	50XXX	Cr 0.50
11XX	Resulfurized			51XXX	Cr 1.02
12XX	Resulfurized and rephosphorized	43BVXX	Ni 1.82; Cr 0.50; Mo 0.12 and 0.25; V 0.03 min	52XXX	Cr 1.45
15XX	Plain carbon (max Mn range—1.00 to 1.65%)	47XX	Ni 1.05; Cr 0.45; Mo 0.20 and 0.35	**Chromium-vanadium steels:**	
Manganese steels:		81XX	Ni 0.30; Cr 0.40; Mo 0.12	61XX	Cr 0.60, 0.80 and 0.95; V 0.10 and 0.15 min
13XX	Mn 1.75	86XX	Ni 0.55; Cr 0.50; Mo 0.20	**Tungsten-chromium steel:**	
Nickel Steels:		87XX	Ni 0.55; Cr 0.50; Mo 0.25		
23XX	Ni 3.50	88XX	Ni 0.55; Cr 0.50; Mo 0.35	72XX	W 1.75; Cr 0.75
25XX	Ni 5.00	93XX	Ni 3.25; Cr 1.20; Mo 0.12	**Silicon-manganese steels:**	
Nickel-chromium steels:		94XX	Ni 0.45; Cr 0.40; Mo 0.12		
		97XX	Ni 0.55; Cr 0.20; Mo 0.20	92XX	Si 1.40 and 2.00; Mn 0.65, 0.82 and 0.85; Cr 0.00 and 0.65
31XX	Ni 1.25; Cr 0.65 and 0.80	98XX	Ni 1.00; Cr 0.80; Mo 0.25		
32XX	Ni 1.75; Cr 1.07	**Nickel-molybdenum steels:**		**High-strength low-alloy steels:**	
33XX	Ni 3.50; Cr 1.50 and 1.57	46XX	Ni 0.85 and 1.82; Mo 0.20 and 0.25		
34XX	Ni 3.00; Cr 0.77			9XX	Various SAE grades
Molybdenum steels:		48XX	Ni 3.50; Mo 0.25	**Boron steels:**	
		Chromium steels:			
40XX	Mo 0.20 and 0.25	50XX	Cr 0.27, 0.40, 0.50 and 0.65	XXBXX	B denotes boron steel
44XX	Mo 0.40 and 0.52	51XX	Cr 0.80, 0.87, 0.92, 0.95, 1.00 and 1.05	**Leaded steels:**	
Chromium-molybdenum steels:					
41XX	Cr 0.50, 0.80 and 0.95; Mo 0.12, 0.20, 0.25 and 0.30			XXLXX	L denotes leaded steel

50XXX, 51XXX, 52XXX: } C 1.00 min

(a) XX in the last two digits of these designations indicates that the carbon content (in hundredths of a percent) is to be inserted.

STRUCTURAL MATERIALS

17-9

TABLE 17-2 Average Properties of Three Grades of Cupola Malleable Iron

Properties	No. 1	No. 2	No. 3
Tensile strength, MPa (lb/in^2)	330 (49,700)	285 (43,000)	285 (43,000)
Yield strength, MPa (lb/in^2)	375 (41,000)	220 (33,000)	205 (31,000)
Elongation in 2 in	8.1	7.0	6.5

17-33. Chilled cast iron is made by pouring cast iron into a metallic mold which cools it rapidly near the surfaces of the casting, thus forming a wear-resisting skin of harder material than the body of the metal. The rapid cooling decreases the proportion of graphite and increases the combined carbon, resulting in the formation of white cast iron.

17-34. Alloy cast iron contains specially added elements in sufficient amount to produce measurable modification of the physical properties. Silicon, manganese, sulfur, and phosphorus, in quantities normally obtained from raw materials, are not considered alloy additions. Up to about 4% silicon increases the strength of pure iron; greater content produces a matrix of dissolved silicon that is weak, hard, and brittle. Cast irons with 7 to 8% silicon are used for heat-resisting purposes and with 13 to 17% silicon form acid- and corrosion-resistant alloys, which, however, are extremely brittle. Manganese up to 1% has little effect on mechanical properties but tends to inhibit the harmful effects of sulfur. Nickel, chromium, molybdenum, vanadium, copper, and titanium are commonly used alloying elements. The methods of processing or of making the alloy additions to the iron influence the final properties of the metal; hence a specified chemical analysis is not sufficient to obtain required qualities. Heat-treatment is also employed on alloy irons to enhance the physical properties. (See *Cast Metals Handbook,* American Foundrymen's Association, 1957.)

17-35. Density of cast iron varies considerably depending upon the carbon content and the proportion of the carbon that is present as graphite. Using the density of pure iron, 7.86, as a reference, the density of cast iron may range from 7.60 for white cast iron to as low as 6.80 for gray cast iron.

17-36. Thermal Properties of Cast Iron. Thermal properties vary somewhat with the composition and the proportions of graphitic carbon. The average specific heat from 20 to 110°C is 0.119; thermal conductivity, 0.40 W/(cm^3)(°C); coefficient of linear expansion, 0.0000106/(°C) at 40°C.

17-37. Values of modulus of elasticity for ferrous metals may be assumed approximately as shown in Table 17-3. The values for all steels are fairly constant, whereas for cast irons the modulus increases somewhat with increased strength of material. Alloy steels have practically the same modulus as plain carbon steels unless large amounts, say 10%, of alloying material are added; for large percentages of alloying elements the modulus decreases slightly. The modulus of steels is not affected by heat-treatment.

17-38. Heat-Treatment of Steel. The properties of steels can be greatly modified by thermal treatments which change the internal crystalline structure of the alloy. *Hardening* of steel is based on the fact that iron undergoes a change in crystal structure when heated above its "critical" temperature. Above this critical transformation temperature, the structure is called austenite, a phase capable of dissolving carbon up to 2%. Below the critical temperature, the steel transforms to ferrite, in which carbon is insoluble and precipitates as an iron carbide compound, Fe_3C (sometimes called cementite). If a steel is cooled rapidly

TABLE 17-3 Approximate Modulus of Elasticity for Ferrous Metals

Metal	Modulus in tension-compression, GPa (lb/in$^2 \times 10^{-6}$)	Modulus in shear, GPa (lb/in$^2 \times 10^{-6}$)
All steels	206 (30.0)	83 (12.0)
Wrought iron	186 (27.0)	75 (10.8)
Malleable cast iron	158 (23.0)	63 (9.2)
Gray cast iron, ASTM No. 20	103 (15.0)	41 (6.0)
Gray cast iron, ASTM No. 60	138 (20.0)	55 (8.0)

from above the critical temperature, the carbon is unable to diffuse to form cementite and the austenite transforms instead to an extremely hard metastable constituent called martensite in which the carbon is held in supersaturation. The hardness of the martensite depends sensitively on the carbon content. Low-carbon steels (below about 0.20%) are seldom quenched, while steels above about 0.80% carbon are brittle and liable to crack on quenching. Plain carbon steels must be quenched at very fast rates in order to be hardened. Alloying elements can be added to decrease the necessary cooling rates to cause hardening; some alloy steels will harden when cooled in air from above the critical temperature. It should be noted, however, that it is the amount of carbon that primarily determines the properties of the alloy; the alloying elements serve to make the response to heat-treatment possible.

Normalizing is a treatment in which the steel is heated over the critical temperature and allowed to cool in still air. The purpose of normalizing is to homogenize the steel. The carbon in the steel will appear as a fine lamellar product of cementite and ferrite called pearlite.

Annealing is similar to normalizing, except the steel is very slowly cooled from above the critical. The carbides are now coarsely divided and the steel is in its softest state, as may be desired for cold-forming or machining operations.

Process annealing is a treatment carried out below the critical temperature designed to recrystallize the ferrite following a cold-working operation. Metals become hardened and embrittled by plastic deformation, but the original state can be restored if the alloy is heated high enough to cause new strain-free grains to nucleate and replace the prior strained structure. This treatment is commonly applied as a final processing for low-carbon steels where ductility and toughness are important, or as an intermediate treatment for such products as wire that are formed by cold working.

Stress-relief annealing is a thermal treatment carried out at a still lower temperature. No structural changes take place, but its purpose is to reduce residual stresses that may have been introduced by previous nonuniform deformation or heating.

Tempering is a treatment that always follows a hardening (quenching) treatment. After hardening, steels are extremely hard, but relatively weak owing to their brittleness. When reheated to temperatures below the critical, the martensitic structure is gradually converted to a ferrite-carbide aggregate that optimizes strength and toughness. Figure 17-3 illustrates the effect of tempering on the properties of a typical medium-carbon alloy steel. When steels are tempered at about 260°C, a particularly brittle configuration of precipitated carbides forms; steels should be tempered above or below this range. Another phenomenon causing embrittlement occurs in steels particularly containing chromium and manganese that are given a tempering cycle that includes holding at, or cooling through, temperatures around

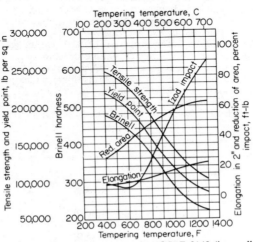

FIG. 17-3 Mechanical properties of SAE 3140 (in small sizes, ½ to 1½ in diameter or thickness). Quenched from 1475 to 1525°F, in oil. Tempered as indicated.

STRUCTURAL MATERIALS

567 to 621°C. Small molybdenum additions retard this effect, called "temper brittleness." It is believed to be caused by a segregation of trace impurity elements to the grain boundaries.

17-39. Mechanical Properties of Iron and Steel. Representative properties of selected ferrous alloys given various heat-treatments are given in Table 17-4.

17-40. Manganese Steels. Manganese is present in all steels as a scavenger for sulfur, an unavoidable impurity; otherwise the sulfur would form a low-melting constituent containing FeS and it would be impossible to hot-work the steel. The manganese content should be about five times the sulfur to provide protection against this "hot shortness." Beyond this amount, manganese increases the hardness of the steel and also has a strong effect on improving response to hardening treatment, but increases susceptibility to temper brittleness. Manganese can be specified up to 1.65% without the steel being classified as an alloy steel. The alloy containing 12 to 14% Mn and around 1% carbon is called Hadfield's manganese steel. This alloy can be quenched to retain the austenite phase and is quite tough in this condition. When deformed, this austenite transforms to martensite which confers exceptional wear and abrasion resistance. Applications for this unique steel include railroad switches, crushing and grinding equipment, dipper bucket teeth, etc. See ASTM Specification A128, *ASM Trans.*, vol. 49, 1957.

17-41. Vanadium Steels. In amounts up to 0.01%, vanadium has a powerful strengthening effect in microalloyed high-strength, low-alloy steels. In alloy steels, 0.1 to 0.2% vanadium is used as a deoxidizer and carbide-forming addition to promote fine-grained tough steels with deep hardening characteristics. Vanadium accentuates the benefits derived from other alloying elements such as manganese, chromium, or nickel, and it is used in a variety of quaternary alloys containing these elements. (See SAE 6120, Table 4-117.) Vanadium in amounts of 0.15 to 2.50% is an important element in a large number of tool steels. (For detailed data see "Vanadium Steels and Irons," New York, Vanadium Corporation of America, 1950; *Metallurgist Metals Tech.*, 1977, vol. 9, p. 375.)

17-42. Silicon Steels. Silicon is present in most constructional steels in amounts up to 0.35% as a deoxidizer to enhance production of sound ingot structures. Silicon increases the hardenability of steel slightly and also acts as a solid solution hardener with little loss of ductility in amounts up to 2.5%. Silicon in amounts of about 4.5% is a major ingredient in electrical steel sheets. Silicon improves the magnetic properties of iron, but even more importantly, these steels can be fabricated to produce controlled grain size and orientation. Since permeability is dependent upon crystal orientation, exceptionally small core losses are obtained by using grain-oriented silicon steel in motors, transformers, etc. (See ASM *Metals Handbook*, 9th ed., vol. 3.) Alloys containing 12 to 14% Si are exceptionally resistant to corrosion by acids. This alloy is too brittle to be rolled or forged, but it can be cast, and is widely used as drainpipe in laboratories and for containers of mineral acids.

17-43. Nickel Steels. Nickel is used as a ferrite strengthener, and improves the toughness of steel, especially at low temperatures. Nickel also improves the hardenability and is particularly effective when used in combination with chromium. Nickel acts similarly to copper in improving corrosion resistance to atmospheric exposure. Certain iron-nickel alloys have particularly interesting properties and are used for special applications: *Invar* (36% Ni) has a very low temperature coefficient of expansion; *Platinite* (46% Ni) has the same expansion coefficient as platinum, and the 39% Ni alloy has the same coefficient as low-expansion glasses. These alloys are useful as gages, seals, etc. *Permalloys* (45% Ni and 76% Ni) have exceptionally high permeability and are used in transformers, coils, relays, etc.

17-44. Chromium Steels. In constructional steels, chromium is used primarily as a hardener. It improves response to heat-treatment and also forms a series of complex carbide compounds that improve wear and high-temperature properties. For these purposes, the amount of chromium used is less than 2%. Alloys containing around 5% Cr retain high hardness at elevated temperatures, and have applications as die steels and high-temperature processing equipment. Alloys containing more than 11% Cr have exceptional resistance to atmospheric corrosion and form the basis of the stainless steels.

17-45. Stainless Steels. Iron-base alloys containing between 11 and 30% chromium form a tenacious and highly protective chrome oxide layer that gives these alloys excellent corrosion-resistant properties. There are a great number of alloys that are generally referred to as "stainless steels," and they fall into three general classifications.

Austenitic stainless steels contain usually 8 to 12% nickel, which stabilizes the austenitic

TABLE 17-4 Approximate Mechanical Properties of Iron and Steel

(Based on test data from various materials testing laboratories)

Metal	Strength in tension, lb/in^{2i}		Strength in compression, lb/in^{2i}		$Yield^a$ strength in shear, $lb/in^{2b,i}$	Endurance limit for reversed bending stress, lb/in^{2i}	Brinell hardness $No.^c$	Elongation in 2 in., %
	$Yield^a$	Ultimate	$Yield^a$	Ultimate				
Gray cast iron:								
ASTM 20	d	20,000	d	80,000	e	9,000	h	Less than 1
ASTM 35	d	35,000	d	125,000	e	15,000	h	Less than 1
ASTM 60	d	60,000	d	145,000	e	24,000	h	Less than 1
Gray cast iron with 1.15% nickel	d	50,000	d	156,000	e	20,000	h	Less than 1
Malleable cast iron	30,000	50,000	30,000	f	16,000	25,000	110	10
Commercial pure iron, annealed	19,000	42,000	19,000	f	12,000	26,000	69	48
Commercial wrought iron, as rolled	30,000	50,000	30,000	f	18,000	25,000	100	35
Structural steel, as rolled, and SAE 1020 steel, as rolled	35,000	60,000	35,000	f	21,000	30,000	120	35
SAE 1040 steel, water quenched, 1050°F temper	87,000	102,000	87,000	f	52,000	57,000	210	23
SAE 1095 steel, oil quenched, 850°F temper	97,000	188,000	97,000	f	58,000	98,000	380	10
SAE 2340 steel, oil quenched, 1200°F temper	91,000	112,000	91,000	f	54,000	67,000	248	24
Oil quenched, 400°F temper	174,000	282,000	174,000	f	96,000	112,000	488	8
SAE 3325 steel, oil quenched, 700°F temper	128,000	139,000	128,000	f	70,000	68,000	291	18
SAE 4140 steel, water quenched, 1100°F temper	116,000	140,000	116,000	f	63,000	64,000	250	16
SAE 5150 steel, oil quenched, 800°F temper	210,000	235,000	210,000	f	115,000	90,000	455	13
SAE 6120 steel, water quenched, 1100°F temper	130,000	164,000	130,000	f	72,000	92,000	350	16
SAE 9260 steel, oil quenched	100,000	158,000	100,000	f	60,000	62,000	240	16
AISI 8650, 1 in. diam., oil quenched 1000°F temper	158,000	170,000	158,000	f	h	h	350	14

Metal	Strength in tension, lb/in^{2i}		Strength in compression, lb/in^{2i}		Yield^a strength in shear, lb/in^{2b,i}	Endurance limit for reversed bending stress, lb/in^{2i}	Brinell hardness No.^{c}	Elongation in 2 in., %
	Yield^a	Ultimate	Yield^a	Ultimate				
AISI E 8740, 1 in. diam., quenched from 1525 F in oil tower, 1100°F temper	134,000	149,000	134,000	f	h	h	302	18.3
AISI E 9310, 1 in. diam., oil quenched from 1425 F, 300°F temper	118,000	145,000	118,000	f	h	h	302	15.5
AISI 9840, 1¼ in. diam., oil quenched from 1525 F, 800°F temper	205,000	220,000	205,000	f	h	h	430	12
18-8 stainless steel, 18% chromium, 8% nickel, water quenched	33,000	75,000	33,000	f	18,000	35,000	140	55
Steel casting, 0.35% C, 1.71% Mn, annealed	60,000	104,000	60,000	f	35,000	45,000	188	22
Steel casting, 0.25% C, 0.68% Mn, annealed	43,000	77,000	43,000	f	24,000	35,000	136	30
Cold-drawn steel rod, 0.20% C	60,000	80,000	60,000	f	36,000	38,000	150	18
Drawn wire, iron or soft steel	70,000	85,000	g	g	40,000	h	h	h
High-carbon steel wire	150,000	275,000	g	g	80,000	h	h	h

^{a}Yield strength taken as yield point, or at 0.2% nominal set.
^{b}Accurate data on ultimate strength in shear not available.
^{c}See Par. 17-16 for description of Brinell hardness test.
^{d}No well-defined yield strength.
^{f}For ductile metal the ultimate in compression is only slightly greater than yield strength.
^{g}Wire can offer resistance only in tension and in shear.
^{h}Data lacking.
^{i}1 lb/in^2 = 6.895 kPa.
NOTE: $t\cdot_C = (t\cdot_F - 32)/1.8$.

phase. These are the most popular of the stainless steels. With 18 to 20% chromium, they have the best corrosion resistance, and are very tough and can undergo severe forming operations. These alloys are susceptible to embrittlement when heated in the range of 593 to 816°C. At these temperatures, carbides precipitate at the austenite grain boundaries, causing a local depletion of the chromium content in the adjacent region, so that this region loses its corrosion resistance. Use of "extra low carbon" grades and grades containing stabilizing additions of strong carbide-forming elements such as niobium minimizes this problem. These alloys are also susceptible to stress corrosion in the presence of chloride environments.

Ferritic stainless steels are basically straight Fe-Cr alloys. Chromium in excess of 14% stabilizes the low-temperature ferrite phase all the way to the melting point. Since these alloys do not undergo a phase change, they cannot be hardened by heat-treatment. They are the least expensive of the stainless alloys.

Martensitic stainless steels contain around 12% Cr. They are austenitic at elevated temperatures, but ferritic at low; hence they can be hardened by heat-treatment. To obtain a significant response to heat-treatment, they have higher carbon contents than the other stainless alloys. Martensitic alloys are used for tools, machine parts, cutting instruments, and other applications requiring high strength. The austenitic alloys are nonmagnetic, but the ferritic and martensitic grades are ferromagnetic. Properties of some representative stainless steels are given in Table 17-5. See ASM *Metals Handbook,* 9th ed., vol. 3, 1980; C. Zapfee, *Stainless Steels,* ASM, 1949; *Enduro Stainless Steels,* Republic Steel, Cleveland.

17-46. Heat-resistant alloys are capable of continuous or intermittent service at temperatures in excess of 649°C. There are a great number of these alloys; they are best considered by class. *Iron-chromium alloys* contain between 10 and 30% chromium. The higher the chromium, the higher the service temperature at which they can operate. They are relatively low-strength alloys, and are used primarily for oxidation resistance. *Iron-chromium-nickel alloys* have chromium in excess of 18%, nickel in excess of 7%, and always more chromium

TABLE 17-5 Properties of Typical Stainless Steels

(From *Metals Handbook*)

Property	Austenitic		Martensitic	Ferritic
	AISI 309 Cr. 22–24% Ni. 12–15% annealed	AISI 321 Cr. 17–19% Ni. 8–11% annealed	AISI 410 Cr. 11.5–13.5% quenched and tempered at 1000°F	AISI 430 Cr. 14–18% annealed
Ultimate strength, lb/in^2	95,000	85,000	145,000	75,000
Yield strength, lb/in^2, 0.2% offset	40,000	30,000	115,000	40,000
Elongation in 2 in, %	45	55	20	30
Reduction of area, %	50–65	55–65	65	40–55
Hardness:				
Rockwell	B78–90	B75–90	C31	B79–90
Brinell	140–185	135–185	300	145–185
Density	7.9	7.9	7.7	7.7
Weight, kg/m^3	0.80	0.80	0.775	0.775
Thermal conductivity at 100°C, W/m$^2 \cdot$K	45.4	52.7	81.5	85.6
Coefficient of expansion per deg F (mean value from 32 to 1000°F)	9.6×10^{-6}	10.3×10^{-6}	7.2×10^{-6}	6.3×10^{-6}
Elastic modulus, lb/in^2	29×10^6	28×10^6	29×10^6	29×10^6
Scaling temp, deg F	2000	1650	1250	1500

NOTE: 1 lb/in^2 = 6.895 kPa; $t_{°C} = (t_{°F} - 32)/1.8$.

than nickel. They are austenitic alloys and have better strength and ductility than the straight Fe-Cr alloys. They can be used in both oxidizing and reducing environments and in sulfur-bearing atmospheres. Iron-nickel-chromium alloys have more than 10% Cr and more than 25% Ni. These are also austenitic alloys, and are capable of withstanding fluctuating temperatures in both oxidizing and reducing atmospheres. They are used extensively for furnace fixtures and components and parts subjected to nonuniform heating. They are also satisfactory for electric resistance-heating elements.

Nickel-base alloys contain about 50% Ni, and also contain some molybdenum. They are more expensive than iron-base alloys, but have better high-temperature mechanical properties.

Cobalt-base alloys contain about 50% cobalt and have especially good creep and stress-rupture properties. They are widely used for gas-turbine blades. Most of these alloys are available in both cast and wrought form; the castings usually have higher carbon contents and often small additions of silicon and/or manganese to improve casting properties. See ASM *Metals Handbook,* 9th ed., vol. 3; and publications of the International Nickel Company.

17-47. Creep Strength. Metals subjected to static loading at elevated temperatures continue to elongate (creep) with time. After an initial period of adjustment to a fairly constant velocity of flow the time rate of deformation under constant stress and temperature (expressed as percentage elongation per hour) is called the *creep rate.* Short-time tensile-test values are not reliable design criteria for metals used at elevated temperatures. The useful strength is limited to the stress that will not produce a damaging amount of deformation during the normal life of the structure. The *creep limit* for a material is the stress that will not produce more than a specified elongation (usually 1%) in a definite time interval (often taken as 10,000 or 100,000 h) at the given temperature. Determination of the creep limit involves long-time testing of a series of specimens to determine initial deformations and creep rates for various stresses. The data are plotted and arbitrarily extrapolated to obtain approximate total creep at future times. Some nonferrous materials such as lead and zinc creep at room temperatures (see Par. **17-84**), whereas, except for stresses nearly up to the ultimate, no appreciable creep has been observed for steels until temperatures above about 260°C (500°F) are exceeded. Variations in data reported on steels have led to the conclusion that creep characteristics are too sensitive an index of strength to permit exact duplication either in different laboratories or in duplicate tests in the same laboratory. Figure 17-4 shows the approximate variation of creep stress with temperature for several steels. For boilers, piping, etc., operating at temperatures above 538°C (1000°F), the maximum working stresses that can be used (without excessive creep) are so small as to make it difficult to produce economical and safe designs until better creep-resistant materials are available. Eleven grades of alloy steels for service at temperatures from 399 to 593°C (750 to 1000°F) are covered by ASTM Specifications A351, A335, and A193 for specific applications to castings, bolting materials, and seamless pipe. These range from ordinary carbon-molybdenum steels to 18% Cr, 8% Ni austenitic steels. In general, 0.4 to 1.5% molybdenum is used in steels for high-temperature service, since it is the only element thus far proved to be effective in increasing creep resistance when present in only small amounts. (For a comprehensive tabulation of creep data see "High Temperature Strength Data of Metals and Alloys," Punched Cards, ASTM, 1965; and annual reports of the Joint Research Committee on Effect of Temperature on the Properties of Metals in *Proc. ASTM.*)

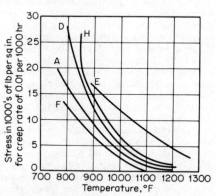

FIG. 17-4 Creep stress for a creep rate of 0.01% per 1000 h for several steels. *A.* Wrought 0.01 to 0.20% carbon steels. *D.* Wrought chromium-molybdenum bolt steel (0.40% carbon). *E.* Wrought 18% chromium, 8% nickel steel. *F.* Wrought carbon steels (carbon above 0.20%). *H.* Wrought 1.0 to 2.5% chromium. 0.50% molybdenum steel (0.20% carbon maximum).

17-16　　ELECTRICAL ENGINEERING MATERIALS REFERENCE GUIDE

17-48. General References on Iron and Steel

Bain, E., and Paxton, H.: *Alloying Elements in Steel;* Cleveland, ASM, 1961.

Campbell, J.E., et al.: *Application of Fracture Mechanics for Selection of Metallic Structural Materials,* Metals Park, Ohio, ASM, 1982.

Horger, O.J.: *Metals Engineering—Design,* ASME Handbook, 2d ed.; New York, McGraw-Hill, 1965.

Roberts, G. A., et al.: *Tool Steels,* 4th ed., Cleveland, Ohio, ASM, 1980.

American Society for Metals: *Sourcebook on Industrial Alloys and Engineering Data,* ASM, 1978.

Advances in the Technology of Stainless Steel and Related Alloys, ASTM STP 369, 1965.

Fracture Toughness, ASTM STP 514, 1971.

Making, Shaping, and Treating of Steel, 9th ed.; Pittsburgh, United States Steel Co., 1971.

Steel Strand and Rope

17-49. Iron and Steel Wire. Annealed wire of iron or very mild steel has a tensile strength in the range of 310 to 415 MPa (45,000 to 60,000 lb/in²); with increased carbon content, varying amounts of cold drawing, and various heat-treatments the tensile strength ranges all the way from the latter figures up to about 3450 MPa (500,000 lb/in²), but a figure of about 1725 MPa (250,000 lb/in²) represents the ordinary limit for wire for important structural purposes. For example, see the following paragraph on bridge wire. Wires of high carbon content can be tempered for special applications such as spring wire. The yield strength of cold-drawn steel wire is 65 to 80% of its ultimate strength. For examples showing the effects of drawing and carbon content on wire, see *Making, Shaping, and Treating of Steel,* U.S. Steel (Par. **17-48**).

17-50. Galvanized-Steel Bridge Wire. The manufacture of high-strength bridge wire like that used for the cables and hangers of suspension bridges such as the San Francisco–Oakland Bay Bridge, the Mackinac Bridge in Michigan, and the Narrows Bridge in New York is an excellent example of careful control of processing to produce a quality material. The wire is a high-carbon product containing 0.75 to 0.85% carbon with maximum limits placed on potentially harmful impurities. Rolling temperatures are carefully specified, and the wire is subjected to a special heat-treatment called "patenting." The steel is transformed in a controlled-temperature molten lead bath to ensure an optimum microstructure. This is followed by cold drawing to a minimum tensile strength of 1550 MPa (225,000 lb/in²) and a 4% elongation. The wire is given a heavy zinc coating to protect against corrosion. Joints or splices are made with cold-pressed sleeves which develop practically the full strength of the wire. Fatigue tests of galvanized bridge wire in reversed bending indicate that the endurance limit of the coated wire is only about 345 to 415 MPa (50,000 to 60,000 lb/in²).

17-51. Wire rope is made of wires twisted together in certain typical constructions and may be either flat or round. Flat ropes consist of a number of strands of alternately right and left lay, sewed together with soft iron to form a band or belt; they are sometimes of advantage in mine hoists. Round ropes are composed of a number of wire strands twisted around a hemp core or around a wire strand or wire rope. The standard wire rope is made of six strands twisted around a hemp core, but for special purposes, four, five, seven, eight, nine, or any reasonable number of strands may be used. The hemp is usually saturated with a lubricant, which should be free from acids or corrosive substances; this provides little additional strength, but acts as a cushion to preserve the shape of the rope and helps to lubricate the wires. The number of wires commonly used in the strands are 4, 7, 12, 19, 24, and 37, depending upon the service for which the ropes are intended. When extra flexibility is required, the strands of a rope sometimes consist of ropes, which in turn are made of strands around a hemp core. Ordinarily the wires are twisted into strands in the opposite direction to the twist of the strands in the rope. The make-up of standard hoisting rope is 6 × 19; extra-pliable hoisting rope is 8 × 19 or 6 × 37; transmission or haulage rope is 6 × 7; hawsers and mooring lines are 6 × 12 or 6 × 19 or 6 × 24 or 6 × 37, etc.; tiller or hand rope is 6 × 7; highway guard-rail strand is 3 × 7; galvanized mast-arm rope is 9 × 4 with a cotton center. The tensile strength of the wire ranges, in different grades, from 415 to 2415 MPa (60,000 to 350,000 lb/in²), depending on the material, diameter, and treatment. The maximum tensile efficiency of wire rope is 90%; the average is about 82.5%, being higher for 6 × 7 rope and lower for 6 × 37 construction. The apparent modulus of elasticity for steel cables in service may be assumed to be 62 to 83 × 10⁶ kPa (9 to 12 × 10⁶ lb/in²) of

TABLE 17-6 Diameters of Sheaves and Drums for Wire Rope

Rope	Sheave diameter ÷ rope diameter		
	Average	Minimum	Larger installations
6 × 7	72	42	96
6 × 19	45	30	90
6 × 37	27	18	
8 × 19	31	21	

TABLE 17-7 Nominal Strengths of Wire Rope
[6 × 19 Classification/Bright (Uncoated), Independent Wire Rope Core]

				Nominal strength*			
Nominal diameter		Approximate mass		Improved plow steel†		Extra improved plow steel†	
in	mm	lb/ft	kg/m	tons	metric tonnes	tons	metric tonnes
¼	6.4	0.12	0.17	2.94	2.67	3.40	3.08
⁵⁄₁₆	8	0.18	0.27	4.58	4.16	5.27	4.78
⅜	9.5	0.26	0.39	6.56	5.95	7.55	6.85
⁷⁄₁₆	11.5	0.35	0.52	8.89	8.07	10.2	9.25
½	13	0.46	0.68	11.5	10.4	13.3	12.1
⁹⁄₁₆	14.5	0.59	0.88	14.5	13.2	16.8	15.2
⅝	16	0.72	1.07	17.7	16.2	20.6	18.7
¾	19	1.04	1.55	25.6	23.2	29.4	26.7
⅞	22	1.42	2.11	34.6	31.4	39.8	36.1
1	26	1.85	2.75	44.9	40.7	51.7	46.9
1⅛	29	2.34	3.48	56.5	51.3	65.0	59.0
1¼	32	2.89	4.30	69.4	63.0	79.9	72.5
1⅜	35	3.5	5.21	83.5	75.7	96.0	87.1
1½	38	4.16	6.19	98.9	89.7	114	103
1⅝	42	4.88	7.26	115	104	132	120
1¾	45	5.67	8.44	133	121	153	139
1⅞	48	6.5	9.67	152	138	174	158
2	52	7.39	11.0	172	156	198	180
2⅛	54	8.35	12.4	192	174	221	200
2¼	57	9.36	13.9	215	195	247	224
2⅜	60	10.4	15.5	239	217	274	249
2½	64	11.6	17.3	262	238	302	274
2⅝	67	12.8	19.0	288	261	331	300
2¾	70	14.0	20.8	314	285	361	327

*To convert to kilonewtons (kN), multiply tons (nominal strength) by 8.896; 1 lb = 4.448 newtons (N).

†Available with galvanized wires at strengths 10% lower than listed, or at equivalent strengths on special request.

ELECTRICAL ENGINEERING MATERIALS REFERENCE GUIDE

cable section. Grades of wire rope are (from historic origins) referred to as traction, mild plow, plow, improved plow, and extra improved plow steel. The most common finish for steel wire is "bright" or uncoated, but various coatings, particularly zinc (galvanized), are used. See Wire Rope Users Manual, AISI, Washington, D.C., 2nd ed., 1981.

17-52. Diameter of Sheaves and Drums for Wire Rope. The average and minimum tread diameters, in accordance with the practice recommended by the American Steel & Wire Co., are shown in Table 17-6; also higher values should be used for larger hoisting installations. Diameters larger than those listed as minimum will give increased rope life.

17-53. Nominal Properties of Wire Rope. See Table 17-7.

Corrosion of Iron and Steel

17-54. Principles of Corrosion. Corrosion may take place by direct chemical attack or by electrochemical (galvanic) attack; the latter is by far the most common mechanism. When two dissimilar metals that are in electrical contact are connected by an electrolyte, an electromotive potential is developed, and a current flows. The magnitude of the current depends on the conductivity of the electrolyte, the presence of high-resistance "passivating" films on the electrode surfaces, the relative areas of electrodes, and the strength of the potential difference. The metal that serves as the anode undergoes oxidation and goes into solution (corrodes).

When different metals are ranked according to their tendency to go into solution, the galvanic series, or electromotive series, is obtained. Metals at the bottom will corrode when in contact with those at the top; the greater the separation, the greater the attack is likely to be. Table 17-8 is such a ranking, based on tests by the International Nickel Company, in which the electrolyte was seawater. The nature of the electrolyte may affect the order to some extent. It should also be recognized that very subtle differences in the nature of the metal may result in the formation of anode-cathode galvanic cells: slight differences in composition of the electrolyte at different locations on the metal surface; minor segregation of impurities in the metal; variations in the degree of cold deformation undergone by the metal; etc. It is possible for anode-cathode couples to exist very close to each other on a metal surface. The electrolyte is a solution of ions; a film of condensed moisture will serve.

17-55. Corrosion Prevention. An understanding of the mechanism of corrosion suggests possible ways of minimizing corrosion effects. Some of these include: (1) avoidance of metal combinations that are not compatible; (2) electrical insulation between dissimilar metals that have to be used together; (3) use of a sacrificial anode placed in contact with a structure to be protected (this is an expensive technique but can be justified in order to protect such structures as buried pipelines and ship hulls); (4) use of an impressed *emf* from an external power source to buck out the corrosion current (called "cathodic protection"); (5) avoiding the presence of an electrolyte—especially those with high conductivities; and (6) application of a protective coating to either the anode or the cathode. The problems of corrosion control are complex beyond these simple concepts, but since the use of protective coatings on iron and steel is extensive, this subject is treated in the following sections.

17-56. Protective coatings may be selected to be inert to the corrosive environment and isolate the base metal from exposure, or the coating may be selected to have reasonable resistance to attack but act sacrificially to protect the base metal. Protective coatings may be considered in four broad classes: paints, metal coatings, chemical coatings, and greases. Painting is commonly used for the protection of structural iron and steel but must be maintained by periodic renewal. Metal coatings take various ranks in protective effectiveness, depending on the metal used and its characteristics as a coating material. A wide variety of metals are used to coat steels: zinc, tin, copper, nickel, chromium, cobalt, lead, cadmium, and aluminum; coatings of gold and silver are also used for decorative purposes. Coatings may be applied by these principal methods: hot dipping, cementation, spraying, electroplating, and vapor deposition. The latter may involve simply evaporation and condensation of the deposited metal, or may include a chemical reaction between the vapor and the metal to be coated.

17-57. Zinc coatings are more widely used for the protection of structural iron and steel than coatings of any other type. The hot-dip process is the earliest type known and is very extensively used at the present time; two improvements, the Crapo process and the Herman, or "galvannealed," process, are used in galvanizing wire. The cementation, or sherardizing, process consists in heating the articles for several hours in a packing of zinc dust in a slowly

STRUCTURAL MATERIALS   17-19

TABLE 17-8   Galvanic Series of Alloys in Seawater

↑	Platinum
Noble or	Gold
cathodic	Graphite
	Titanium
	Silver
	⌈ Chlorimet 3 (62 Ni, 18 Cr, 18 Mo)
	⌊ Hastelloy C (62 Ni, 17 Cr, 15 Mo)
	⌈ 18-8 Mo stainless steel (passive)
	│ 18-8 stainless steel (passive)
	⌊ Chromium stainless steel 11–30% Cr (passive)
	⌈ Inconel (passive) (80 Ni, 13 Cr, 7 Fe)
	⌊ Nickel (passive)
	Silver solder
	⌈ Monel (70 Ni, 30 Cu)
	│ Cupronickels (60–90 Cu, 40–10 Ni)
	│ Bronzes (Cu-Sn)
	│ Copper
	⌊ Brasses (Cu-Zn)
	⌈ Chlorimet 2 (66 Ni, 32 Mo, 1 Fe)
	⌊ Hastelloy B (60 Ni, 30 Mo. 6 Fe, 1 Mn)
	⌈ Inconel (active)
	⌊ Nickel (active)
	Tin
	Lead
	Lead-tin solders
	⌈ 18-8 Mo stainless steel (active)
	⌊ 18-8 stainless steel (active)
	Ni-Resist (high Ni cast iron)
	Chromium stainless steel, 13% Cr (active)
	⌈ Cast iron
	⌊ Steel or iron
	2024 aluminum (4.5 Cu, 1.5 Mg, 0.6 Mn)
Active or	Cadmium
anodic	Commercially pure aluminum (1100)
	Zinc
↓	Magnesium and magnesium alloys

NOTE: Alloys will corrode in contact with those higher in the series. Brackets enclose alloys so similar that they can be used together safely.
SOURCE: Fontana and Green, *Corrosion Engineering;* New York, McGraw-Hill Book Company.

rotating container. Electroplating is also employed, and heavier coatings can be obtained than are usual with the hot-dip process, but adherence is difficult to obtain, and this process is not often used. See ASTM specifications for zinc-coated iron and steel products.

17-58. Aluminum coatings are applied by a cementation process which is commercially known as "calorizing." The articles to be coated are packed in a drum in a mixture of powdered aluminum, aluminum oxide, and a small amount of ammonium chloride. The articles are then slowly rotated and heated in an inert atmosphere, usually of hydrogen. Such coatings are very resistant to oxidation and sulfur attack at high temperatures. Aluminum coatings can also be applied by the hot-dipping method and then are heat-treated to improve the alloy bond. Aluminum can also be applied by spraying. Aluminum-coated steel is extensively used for oxidation protection, for example, for heat ducts and automobile mufflers. Aluminum-zinc coatings, applied by hot dipping, have been developed that combine the high temperature protection of aluminum with the sacrificial protection of zinc.

17-59. Tin Coatings. Almost all tin coatings are now applied by electrolytic deposition methods. The accurate control obtained by electrolytic deposition is important because of the high cost of tin. Unlike zinc, tin is electropositive to iron. The coating must remain intact; once penetrated, corrosion of the iron will be accelerated. If a zinc coating is penetrated, the zinc will still sacrificially protect the adjacent exposed iron. Tin has good corrosion resistance, is nontoxic, readily bonds to steel, is easily soldered, and is extensively utilized by the container industry for food and other substances.

17-60. Lead and Lead-Tin Coatings. The objective of lead coating of steel is to obtain an inexpensive corrosion-resistant coating. Lead alone will not alloy with iron; it is necessary to add some tin to the lead to obtain a smooth, continuous, adherent coating. Originally, about 25% tin (called "terne metal") was used, but the tin content has been reduced as the price of tin has increased. Since corrosion protection is less effective than with tin or zinc, and the surface is soft and easily scratched, terne-coated steel is not extensively used. Applications include uses where corrosion is not too critical or likely, such as gasoline tanks and roofing sheets, or where the lubricating properties of the soft lead surface helps forming operations.

17-61. Metal-spray coatings are applied by passing metal wire through a specially constructed spray gun which melts and atomizes the metal to be used as coating. The surface to be sprayed must be roughened to afford good adhesion of the deposited metal. Nearly all the commonly used protective metals can be applied by spraying, and the process is especially useful for coating large members or repairing coatings on articles already in place. Sprayed coatings can also be applied that will resist wear and can be used to build up worn parts such as armature shafts and bearing surfaces or to apply copper coatings to carbon brushes and resistors.

17-62. Chromium coatings can be applied by cementation or electroplating. In electroplating, the best results are secured by first plating on a base coating of nickel or nickel copper to receive the chromium. The great hardness of chromium gives it important applications for protection against wear or abrasion; it will also take and retain a high polish. Very thin coatings have a tendency to be inefficient as a result of the presence of minute pinholes.

17-63. Electroplated Coatings. Electroplating is employed in the application of coatings of nickel, brass, copper, chromium, cadmium, cobalt, lead, and zinc. Only cadmium, chromium, and zinc are electronegative to iron. The other metals mentioned are employed because of their own corrosion-resistant properties and because they afford surface finishes having certain desirable characteristics.

17-64. Protective paints are extensively employed to protect heavily exposed structures of iron and steel, such as bridges, tanks, and towers. The protection is not permanent but gradually wears away under weather exposure and must be periodically renewed. Various specially prepared paints are used for protecting the surface from dampness, oxidizing gases, and smoke. No one paint is suitable for all purposes but the choice depends on the nature of the corrosive influence present. Asphaltum and tar protect the surface by formation of an impervious film. A chemical protective action is exerted by paints containing linseed oil as the vehicle and red lead as the pigment; linseed oil absorbs oxygen from the atmosphere and forms a thick elastic covering, a formation hastened by adding salts of manganese or lead to the oil. All dryers, vehicles, and pigments used in paint must be inert to the steel; otherwise corrosion will be hastened instead of prevented. Graphite- and aluminum-flake pigments give very impermeable films but do not show the inhibitive action of red lead or zinc when the films are scratched. Aluminum has the advantage of reflecting both infrared and ultraviolet rays of the sun; hence it protects the vehicle from a source of deterioration and is used to paint gasoline-storage tanks to prevent excessive heating due to the sun's rays. A large number of new protective coatings have recently been developed from synthetic materials, such as silicones, artificial rubbers, and phenolic plastics. Many of these are tightly adhering compounds in the form of paints or varnishes which offer rather good protection against a wide variety of chemical attacks. The majority of these new coatings, however, are sensitive to abrasion, and many of them must be baked on to secure full effectiveness. (See Reports of Committee D1 in *ASTM Proc.;* also bibliography, Par. **17-70.**)

17-65. Chemical coatings which are corrosion-resistant include Parkerizing and Bower-Barff finish. *Parkerizing* consists in immersing the steel in a solution of manganese dihydrogen phosphate, then heating to about the boiling point. The pieces are allowed to remain in the bath until effervescence ceases. After oiling, the surface produced has the appearance of gunmetal. These phosphate coatings are used mainly as bases for finishing enamels and paints. *Bower-Barff finish* consists in heating the pieces in a closed retort to a temperature of 871°C. Alternate injections of superheated steam and carbon monoxide then reduce the oxides formed on the surface to Fe_3O_4. The operations may be repeated several times until a sufficient depth of oxide is obtained. Other types of chemical surface which develop increased surface hardness and resistance to corrosion are produced commercially by treating a steel with nitrogen or silicon for prolonged periods of time at high temperature.

STRUCTURAL MATERIALS

17-66. *Greases and oils of various grades* are used to protect the surface by applying a thin film to parts of machines, tools, bearings, and steels which are to be put in storage or are to be shipped. These slushing compounds are usually mineral oils, fats and waxes, lanolin, or greases; oil-soluble chromates are sometimes added to aid in preventing corrosion. Silicone oils and greases are particularly effective because of their imperviousness and repellency of moisture.

17-67. *Corrosion-resistant ferrous alloys* such as rustless or stainless iron and steel have come into use for both structural and ornamental purposes but on account of their chromium and nickel contents are relatively expensive in comparison with the ordinary structural steels (see Par. 17-45). Copper-bearing iron and steel, containing about 0.15 to 0.25% copper, are used extensively; the copper content tends to retard corrosion slightly but does not prevent it, and some protective coating is usually necessary. Some structural uses have been made of these steels without applying special protective coatings. A tightly adherent brown oxide surface film forms from weathering to serve as the future "protective coating."

17-68. Stainless-Clad Sheets. Plates and sheets of steel faced with stainless steels are widely used in chemical processing, paper mills, nuclear reactors, and transportation or storage of corrosive liquids, food products, etc. The stainless-steel layer usually constitutes 10 to 20% of the thickness of the plate, and either one or both sides may be coated, but the surfacing must be strongly bonded to the base metal. One method of production consists in casting two stainless-steel sheets in the center of a steel ingot with a separating material between. The composite "sandwich" ingot is hot-rolled to the desired size and then separated into two sheets each clad on one face with stainless steel. Another method of production is to bond stainless sheet to a mild-steel plate by means of a great number of resistance welds closely spaced all over the sheet. These methods provide a stainless-clad surface on either thick or thin plates without the added expense of a solid stainless-steel sheet.

17-69. Copperweld. A series of steel products, including wire, wire rope, bars, clamps, ground rods, and nails, that contain a copper-clad surface are made by the Copperweld process. The copper coating is intimately bonded to the steel by pouring a ring of molten copper about a heated steel billet fastened in the center of a refractory mold. The solidified composite ingot is then hot-rolled to bar stock and subsequently cold-drawn to the various wire sizes. The thickness of the copper coating on wire is 10 to 12½% of the wire radius and produces a high-strength steel wire with a resistance to corrosion similar to that of a solid copper wire. Their increased electrical conductivity over that of a solid steel wire or rod makes the Copperweld products suitable for high-strength conductors, ground rods, aerial cable messengers, etc.

17-70. *References to Technical Literature on Corrosion and Protective Coatings*

Uhlig, H. H. (ed.): *The Corrosion Handbook;* New York, John Wiley & Sons, Inc., 1948.

Evans, U. R.: *An Introduction to Metallic Corrosion,* 3d ed.; London, Arnold, 1981.

Metal Finishing Guidebook and Directory; Hackensack, N.J., Metals and Plastics Publications, Inc. Published annually.

Reports of Committee A5, Corrosion of Iron and Steek, *Ann. Proc. ASTM.*

Payne, H. F.: *Organic Coating Technology;* New York, John Wiley & Sons, Inc., 1961.

Fontana, M. G., and Greene, N. D.: *Corrosion Engineering;* New York, McGraw-Hill, 1967.

ASM *Metals Handbook,* 9th ed., vol. 5, 1982.

Nonferrous Metals and Alloys

17-71. Copper. Numerous commercial "coppers" are available. The standard product is "tough-pitch" copper, which contains about 0.04% oxygen. If it has been electrolytically refined, it is called "electrolytic tough pitch." This copper cannot be heated in reducing atmospheres because the oxygen will react with hydrogen and severely embrittle the alloy. Various deoxidized varieties are made. When deoxidized with phosphorus, there is some loss of electrical conductivity depending on the amount of residual phosphorus and the extent to which other impurities are reduced and redissolve in the metal. As a general principle, alloying elements that dissolve in copper reduce conductivity sharply; those that are insoluble have little effect.

Copper castings are improved by using special deoxidizers such as Boroflux and silico-calcium copper alloy. By the use of these deoxidizers the castings are improved structurally,

17-22   ELECTRICAL ENGINEERING MATERIALS REFERENCE GUIDE

and the electrical conductivity can be increased to about 80 to 90% of standard annealed copper. Boroflux is a mixture of boron suboxide, boric anhydride, magnesia, and magnesium; for data on its use, see publications of the General Electric Company.

Oxygen-free high-conductivity copper is deoxidized with carbon, and thus is free of residual oxide or deoxidizer. It is a more expensive product but does not suffer the potential embrittlement of tough pitch, and is capable of more severe cold-forming operations at the cost of a slight loss of electrical conductivity. Free-machining copper contains lead or tellurium that drops conductivity 3 to 5%. Since copper is a very difficult material to machine, this may be a small sacrifice for certain applications. Small amounts of silver improve resistance to elevated-temperature softening with no loss of physical or mechanical properties.

17-72. Brass.   Brasses are alloys of copper and zinc; commercial brasses contain from 5 to 45% zinc. A wide variety of properties are obtainable in the brasses. In general, the alloys have excellent corrosion resistance, good mechanical properties, colors ranging from red to gold to yellow to white, and are available in a wide variety of cast and wrought shapes. The alloy of 30% zinc has an optimum combination of strength and ductility. It is called "cartridge brass," since an early application was drawing of cartridge shells. It is the most commonly used brass alloy.

Muntz metal contains nominally 40% zinc, and is a two-phase alloy that is readily hot-worked in the high-temperature form and develops good strength when cooled. It is used for extruded shapes and for bolts, fasteners, and other high-strength applications. The properties of brass can be modified by small additions of numerous alloying elements; those commonly used include silicon, aluminum, manganese, iron, lead, tin, and nickel. The addition of 1% tin to cartridge brass results in an alloy called Admiralty brass which has very good corrosion resistance and is extensively used in heat exchangers.

Brasses, especially the high zinc-bearing alloys, are subject to a corrosion phenomenon called *dezincification*. It involves a selective loss of zinc from the surface and the formation of a spongy copper layer accompanied by deterioration of mechanical properties. It is more likely to occur with the high-zinc brasses in contact with water containing dissolved CO_2 at elevated temperatures. Like many other metals, the brasses are susceptible to stress-corrosion cracking—an embrittlement due to the combined action of stress and a selective corrosive agent. In the case of brass, the particular agent responsible for stress-corrosion cracking is ammonia and its compounds. Brass products that might be exposed to such environments should be stress-relief annealed before being placed in service. For details on compositions and properties of brasses, see the appropriate ASTM specifications and the publications of brass producers.

17-73. Bronze, or Copper-Tin, Alloys.   Bronze is an alloy consisting principally of copper and tin and sometimes small proportions of zinc, phosphorus, lead, manganese, silicon, aluminum, magnesium, etc. The useful range of composition is from 3 to 25% tin and 95 to 75% copper. Bronze castings have a tensile strength of 195 to 345 MPa (28,000 to 50,000 lb/in²), with a maximum at about 18% of tin content. The crushing strength ranges from about 290 MPa (42,000 lb/in²) for pure copper to 1035 MPa (150,000 lb/in²) with 25% tin content. *Cast bronzes* containing about 4 to 5% tin are the most ductile, elongating about 14% in 5 in. Gunmetal contains about 10% tin and is one of the strongest bronzes. Bell metal contains about 20% tin. *Copper-tin-zinc alloy* castings containing 75 to 85% copper, 17 to 5% zinc, and 8 to 10% tin have a tensile strength of 240 to 275 MPa (35,000 to 40,000 lb/in²), with 20 to 30% elongation. *Government bronze* contains 88% copper, 10% tin, and 2% zinc; it has a tensile strength of 205 to 240 MPa (30,000 to 35,000 lb/in²), yield strength of about 50% of the ultimate, and about 14 to 16% elongation in 2 in; the ductility is much increased by annealing for ½ h at 700 to 800°C, but the tensile strength is not materially affected. *Phosphor bronze* is made with phosphorus as a deoxidizer; for malleable products, such as wire, the tin should not exceed 4 or 5%, and the phosphorus should not exceed 0.1%. *United States Navy bronze* contains 85 to 90% copper, 6 to 11% tin, and less than 4% zinc, 0.06% iron, 0.2% lead, and 0.5% phosphorus; the minimum tensile strength is 310 MPa (45,000 lb/in²), and elongation at least 20% in 2 in. *Lead bronzes* are used for bearing metals for heavy duty; an ordinary composition is 80% copper, 10% tin, and 10% lead, with less than 1% phosphorus. *Steam or valve bronze* contains approximately 85% copper, 6.5% tin, 1.5% lead, and 4% zinc; the tensile strength is 235 MPa (34,000 lb/in²), minimum, and elongation 22% minimum in 2 in (ASTM Specification B61). The bronzes have a great many industrial applications where their combination of tensile properties and corrosion resistance is especially useful.

STRUCTURAL MATERIALS

17-74. Beryllium-copper alloys containing up to 2.75% beryllium can be produced in the form of sheet, rod, wire, and tube. The alloys are hardenable by a heat-treatment consisting of quenching from a dull red heat, followed by reheating to a low temperature to hasten the precipitation of the hardening constituents. Depending somewhat on the heat-treatment, the alloy of 2.0 to 2.25% beryllium has a tensile strength of 415 to 650 MPa (60,000 to 193,000 lb/in²), elongation 2.0 to 10.0% in 2 in, modulus of elasticity 125 $\times$ GPa (18 $\times$ 10⁶ lb/in²), and endurance limit of about 240 to 300 MPa (35,000 to 44,000 lb/in²). An outstanding quality of this alloy is its high endurance limit and corrosion resistance; it can be hardened by heat-treatment to give great wear resistance and has high electrical conductivity. Typical applications include nonsparking tools for use where serious fire or explosion hazards exist and many electrical accessories such as contact clips and springs or instrument and relay parts. Beryllium is a toxic substance, and care should be taken to avoid ingesting airborne particles during such operations as machining and grinding. For details on properties and uses of beryllium bronzes, see publications of Kawecki Berylco Industries, New York.

17-75. General References on Copper and Copper Alloys

Mendenhall, J. H.: *Understanding Copper Alloys,* Olin Corp, New York, 1977.

Copper and Its Alloys; London, Institute of Metals, 1970.

Bridgeport Copper Metals Handbook; Bridgeport, Conn., Bridgeport Brass Company, 1964.

Source Book on Copper and Copper Alloys, Am. Soc. for Metals, 1979.

Flinn, R. A.: *Copper, Brass, and Bronze Castings,* Cleveland, NonFerrous Founders Society, 1963.

17-76. Nickel is a brilliant metal which approaches silver in color. It is more malleable than soft steel and when rolled and annealed is somewhat stronger and almost as ductile. The tensile strength ranges from 415 MPa (60,000 lb/in²) for cast nickel to 795 MPa (115,000 lb/in²) for cold-rolled full-hard strip; yield strength 135 to 725 MPa (20,000 to 105,000 lb/in²); elongation in 2 in, 2% when full hard to about 50% when annealed; modulus in tension about 205 GPa (30 $\times$ 10⁶ lb/in²). Nickel takes a good polish and does not tarnish or corrode in dry air at ordinary temperatures. It has various industrial uses in sheets, pipes, tubes, rods, containers, and the like, where its corrosion resistance makes it especially suitable.

The greatest tonnage use of nickel is as an alloying element in steels, principally stainless and heat-resisting steels. There are also a variety of copper-nickel alloys whose main applications are based on their excellent corrosion resistance, for example, condenser tubes. Additions of aluminum and titanium to nickel-base alloys result in age-hardening characteristics, and they can be heat-treated to exceptionally high strengths that are retained to high temperatures. The International Nickel Company publishes an extensive list of bulletins describing the characteristics of nickel and nickel alloys.

17-77. Monel metal is a silvery-white alloy containing approximately 66 to 68% nickel, 2 to 4% iron, 2% manganese, and the remainder copper. It can be cast, forged, rolled, drawn, welded, and brazed and is easily machined. It melts at 1360°C and has a density of 8.80, coefficient of expansion of 14 $\times$ 10⁻⁶ per degree Celsius, thermal conductivity of 0.06 cgs unit, specific heat of 0.127 cal/(g)(°C), and modulus of 175 GPa (25 $\times$ 10⁶ lb/in²). The tensile strength ranges from 450 MPa (65,000 lb/in²) for cast monel metal to 860 MPa (125,000 lb/in²) in cold-rolled full-hard strip; yield strength 175 to 725 MPa (25,000 to 115,000 lb/in²). It is highly resistant to corrosion and the action of seawater or mine waters. The industrial uses for it include many applications where its combination of physical properties and corrosion resistance gives it special advantages. (See technological data published by the International Nickel Co.)

17-78. Magnesium Alloys. The outstanding feature of magnesium alloys is their light weight (specific gravity of about 1.8). Alloys containing thorium and rare-earth additions have been developed that retain good strength at temperatures between 260 and 371°C. The correspondingly high strength/weight ratio makes them particularly useful to the aircraft industry. Less exotic alloys, based mainly on alloying with aluminum (up to 10%) and zinc (up to 6%) still have excellent strengths, and are heat-treatable. These alloys have many uses where low density is desired: portable tools, ladders, structural members for trucks and buses, housings, etc. Magnesium alloys are available as castings, forgings, extrusions, and rolled mill products in a variety of shapes. Their thermal coefficient of expansion is about 0.000029/°C, and melting point about 620°C. Tensile strengths of castings range from 145 to 235 MPa (21,000 to 34,000 lb/in²), yield strengths from 62 to 150 MPa (9,000 to 22,000 lb/in^2), and elongation from 1 to 10% in 2 in. Forged or extruded alloys have tensile strengths of 225 to 300 MPa (33,000 to 43,000 lb/in^2), yield strengths 125 to 205 MPa (18,000 to 30,000 lb/in^2), and elongations of 5 to 17% in 2 in. The Brinell hardness ranges from 35 to 78, and the endurance limit from 40 to 115 Mpa (6,000 to 17,000 lb/in^2) depending on the alloy and heat-treatment. Since magnesium is highly anodic to other common metals, care must be taken in designing with this metal. Protective coatings are used, and care must be taken to avoid forming galvanic couples. Finely divided magnesium will burn but massive sections are safely melted and welded. See *Magnesium Alloys and Products,* Dow Chemical Company, Midland, Mich.

17-79. *Lead* is a heavy, soft, malleable metal with a blue-gray color; it shows a metallic luster when freshly cut, but the surface is rapidly oxidized in moist air. It can be easily rolled into thin sheets and foil or extruded into pipes and cable sheaths but cannot be drawn into fine wire. Although in an ordinary tensile test lead may develop a tensile strength of 17 MPa (2400 lb/in^2), it may creep at ordinary room temperatures at stresses as low as 0.34 MPa (50 lb/in^2). Owing to this tendency to creep, it may fracture under long-continued load at stresses as low as 5.5 MPa (800 lb/in^2), and the ordinary static tensile properties do not have much significance. The resistance of pure lead to corrosion makes it useful in the form of sheets, pipes, and cable coverings, and large quantities of lead are used in the manufacture of various alloys, particularly in alloys for bearings. Common alloys of lead for cable sheathings contain (approximately) 0.04% Cu, 0.75% Sb, or 0.03% Ca. The greatest use of metallic lead is in the manufacture of storage batteries. See ASM *Metals Handbook,* 9th ed., vol. 2, 1979.

17-80. *Tin* is a silvery-white, lustrous metal, very soft and malleable and of very low tensile strength. It has a density of about 7.3 and melts at 232°C. In ductility it equals soft steel. The tensile strength varies with the speed of testing. As a metal it has few uses except in sheets, but large quantities of it are used in various industrial alloys. Its chief uses are in tin- and terneplate, solder, babbitt and other bearing metals, brass, and bronze. Tin is very resistant to atmospheric corrosion, and water hardly affects it at all; however, it is electronegative to iron and therefore is not an efficient protective coating under atmospheric exposures. See Symposium on Tin, *ASTM Spec. Tech. Publ.* 141, 1953; ASM *Metals Handbook,* 9th ed., vol. 2, 1979.

17-81. Bearing Metals. Bearing metals are designed to serve as sleeve bearings—to carry loads between surfaces undergoing relative motion. Traditionally, the metallic structure best suited for such service is one in which hard particles are embedded in a soft matrix. The hard particles provide wear resistance and carry the load, while the soft matrix gives conformability to variations in dimensions due to manufacturing tolerances and deflections due to the load, and also embed abrasive particles that otherwise might score the shaft. *White metal or Babbitt bearings* have served this function successfully for years. These are tin- or lead-base alloys containing additions of antimony and copper to form hard precipitate particles in the soft matrix.

Tin babbitts are more expensive than lead, but have better corrosion resistance. In more recent years, other materials have been developed that improve on one or another of the deficiencies of the white metals: low elevated-temperature strength, low thermal conductivity, corrosion resistance, and fatigue strength. Copper-, cadmium-, and aluminum-base alloys in particular are now being used for many applications.

Lead is a common addition to these alloys, and provides softness, conformability, embeddability, and self-lubricating properties. Sintered copper-tin powder-metallurgy bearings are also extensively used as "oilless" bearings for sealed motors. They are made with a high proportion of porosity, then impregnated with oil. In use, they heat up and the oil flows out to the shaft. When they cool, the oil is drawn back into the pores ready for the next period of service. See ASTM Specifications B23 for properties and compositions of babbitt bearings. For general study of bearing materials, see *Bearings;* Cleveland, ASM, 1969.

17-82. *Zinc* is a bluish-white metal, which has a metallic luster on a new fracture. The density of cast zinc ranges from 7.04 to 7.16. At ordinary temperature it is brittle, but in the range of about 100 to 150°C it becomes malleable and can be rolled into sheets and drawn into wire. At 200°C it becomes so brittle that it can be pulverized. The tensile strength of cast zinc ranges from about 55 to 95 MPa (8000 to 14,000 lb/in^2) in an ordinary testing-machine test, and that of drawn zinc from about 150 to 200 MPa (22,000 to 30,000 lb/in^2); it has a poorly defined proportional limit of about 35 MPa (5000 lb/in^2) and exhibits a

STRUCTURAL MATERIALS

17-25

certain amount of creep at room temperatures; hence it may fracture in service under constant stresses below its testing-machine strength. It strongly resists atmospheric corrosion but is readily attacked by acids. The principal industrial uses for it are for galvanizing iron and steel, for plates and sheets for roofing and other applications, and for alloying with copper, tin, and other metals; very large quantities are used in the various types of brass. Next to galvanizing, the greatest use of zinc is in the production of die castings. Because of its moderate melting point, good mechanical properties, and especially because it does not attack steel melting pots and dies, it is the most popular die-casting material (although closely rivaled by aluminum). Zinc alloys for die casting contain some aluminum, copper, and magnesium; all ingredients must be very pure or the casting will have poor corrosion resistance and dimensional stability. See Morgan, S.: *Zinc and Its Alloys;* Plymouth, Macdonald & Evans, 1977; and publications of the New Jersey Zinc Company, New York.

17-83. Titanium and Titanium Alloys. Titanium alloys are important industrially because of their high strength-weight ratio, particularly at temperatures up to 427°C. The density of the commercial titanium alloys ranges from 4.50 to 4.85 g/cm^3, or approximately 70% greater than aluminum alloy and 40% less than steel. The purest titanium currently produced (99.9% Ti) is a soft, white metal. The mechanical strength increases rapidly, however, with an increase of the impurities present, particularly carbon, nitrogen, and oxygen. The commercially important titanium alloys, in addition to these impurities, contain small percentages (1 to 7%) of (1) chromium and iron, (2) manganese, and (3) combinations of aluminum, chromium, iron, manganese, molybdenum, tin, or vanadium. The thermal conductivity of the titanium alloys is low, about 15 W/m·K at 25°C, and the electrical resistivity is high, ranging from 54 $\mu\Omega$ · cm for the purest titanium to approximately 150 $\mu\Omega$ · cm for some of the alloys. The coefficient of thermal expansion of the titanium alloys varies from 2.8 to 3.6 $\times$ 10^{-6} per degree Celsius, and the melting-point range is from 1371 to 1704°C for the purest titanium. The tensile modulus of elasticity varies between 100 to 120 GPa (15 to 17 $\times$ 10^6 lb/in^2). The mechanical properties, at room temperature, for annealed commercial alloys range approximately as follows: yield strength 760 to 965 MPa (110,000 to 140,000 lb/in^2); ultimate strength 800 to 1100 MPa (116,000 to 160,000 lb/in^2); elongation 5 to 18%; hardness 300 to 370 Brinell. On the basis of the strength-weight ratio many of the titanium alloys exhibit superior short-time tensile properties as compared with many of the stainless and heat-resistant alloys up to approximately 427°C. However, at the same stress and elevated temperature, the creep rate of the titanium alloys is generally higher than that of the heat-resistant alloys. Above about 482°C the strength properties of titanium alloys decrease rapidly. The corrosion resistance of the titanium alloys in many media is excellent; for most purposes, it is the equivalent or superior to stainless steel. See *Titanium and Titanium Alloys;* ASM Source Book, American Society for Metals, 1982.

17-84. Mechanical Properties of Nonferrous Metals and Alloys. See Table 17-9.

17-85. Creep Stress for Nonferrous Metals. Figures 17-5 and 17-6 show for several metals the approximate creep strengths to produce a given creep rate per 1000 h. Creep characteristics are influenced by many factors, such as melting practice and grain size as well as chemical composition. Materials of the same composition will not necessarily have the same creep rate. Test data should therefore be regarded only as qualitative.

17-86. Aluminum is an important commercial metal possessing some very unique properties. It is very light (density about 2.7) and some of its alloys are very strong, so its strength-weight ratio makes it very attractive for aeronautical uses and other applications in which weight saving is important. Aluminum, especially in the pure form, has very high electrical and thermal conductivities, and is used as an electrical conductor in heat exchangers, etc. Aluminum has good corrosion resistance, is nontoxic, and has a pleasing silvery white color; these properties make it attractive for applications in the food and container industry, architectural, and general structural fields.

Aluminum is very ductile and easily formed by casting and mechanical forming methods. Aluminum owes its good resistance to atmospheric corrosion to the formation of a tough, tenacious, highly insulating, thin oxide film, in spite of the fact that the metal itself is very anodic to other metals. In moist atmospheres, this protective oxide may not form, and some caution must be taken to maintain this film protection. Although aluminum can be joined by all welding processes, this same oxide film can interfere with the formation of good bonds during both fusion and resistance welding, and special fluxing and cleaning must accompany welding operations.

TABLE 17-9 Approximate Mechanical Properties of Miscellaneous Nonferrous Metals and Alloys
(Compiled from Various Authorities)

Metal	Condition	Approximate composition, %	Strength in tension,[a] 1000 lb/in²*		Endurance limit,[c] 1000 lb/in²*	Brinell hardness no.	Elongation, 2 in, %	Weight, lb/in³†
			Yield[b]	Ultimate				
Aluminum:								
EC-O	Wrought, annealed	99.60% min. Al	4	10	d	d	23[k]	0.098
EC-H19	Wrought, extra hard	99.60% min. Al	24	27	7	d	1.5[k]	0.098
1199-O	Wrought, annealed	99.99 Al	1.5	6.5	d	d	50	0.098
1100-H14	Wrought, half hard	99.0% min. Al	17	18	7	32	20	0.098
2024-T4	Quenched	Al 93, Cu 4.5, Mn 0.6, Mg 1.5	47[i]	68[i]	20	120	19	0.100
5052-H34	Wrought, half hard	Al 97, Mg 2.5, Cr 0.25	31	38	18	68	16	0.097
6061-T6	Quenched and aged	Al 98, Mg 1, Si 0.6, Cu 0.25, Cr 0.25	40	45	14	95	17	0.098
6063-T5	Quenched and aged	Al 99, Mg 0.7, Si 0.4	21	27	17	60	22	0.097
6101-T6	Quenched and aged	Al 99, Mg 0.6, Si 0.5	28	32	9	71	20	0.097
7075-T6[j]	Quenched and aged	Al 90, Zn 5.6, Mg 2.5, Cu 1.6, Cr 0.3	73	83	22	150	11	0.101
A356-T61	Permanent mold, quenched and aged	Al 9, Si 7, Mg 0.3	30	41	13	90	10	0.097
Brass	Annealed, cold-drawn	Cu 60; Zn 40	18	54	22	72	56	0.30
		Cu 60; Zn 40	49	97	26	179	13	0.30

Bronze	Annealed, cold-drawn	Cu 95; Sn 5	13	46	23	74	67	0.32
Bronze, phosphor	Rolled sheet	Cu95; Sn 5 Sn 8; Zn 0.2; P 0.1, Cu 91	59 _d_	85 50–100	27 _d_	166 35–90_g_	12 5–50	0.32 _d_
Bronze, aluminum	As rolled	Cu 88; Al 9; Fe 3	28	70	_d_	125	30	0.29
Bronze, manganese	Cast	Cu 57, Zn 41, +Mn, Al, Fe & Sn	30	70	17	115	33	0.30
Copper	Annealed	Commercially pure	5	32	10	47	56	0.32
Copper alloy	Wire	Cu 95; +Si, Sn, Al, Fe	_d_	68–120	_d_	_d_	0.8–3.8_e_	0.32
Copper-silicon alloy	Half hard	Cu 94, Si 3, +Mn, Zn, & Fe	40–50	71–81	_d_	80–90_g_	10–20	0.32
Copper, beryllium	Heat-treat, half hard	Be 2, Ni 0.5, Fe 0.2, Cu 97	132	173	35	340	4.8	0.297
Gunmetal	Cast	Cu 88, Sn 10, Zn 2–4	14–20	30–45	_d_	50–80	15–40	0.314
Magnesium alloy:								
AZ31C	Sheet, hard	Mg 96, Al 3, Zn 1, Mn 0.2	32	42	14	73	15	0.064
AZ80A-T5	Extrusions, aged	Mg 91, Al 8.5, Zn 0.5, Mn 0.2	39	54	19	82	7	0.065
ZK60A	Extrusions, aged	Mg 94, Zn 5.5, Zr 0.6	43	52	16	82	12	0.066
AZ63A-T6	Sand cast, quenched, aged	Mg 91, Al 6, Zn 3, Mn 0.2	19	40	17	73	5	0.067
AZ92A-T6	Sand cast, quenched and aged	Mg 89, Al 9, Zn 2, Mn 0.1	23	40	13	84	2	0.066

TABLE 17-9 Approximate Mechanical Properties of Miscellaneous Nonferrous Metals and Alloys (*Continued*)
(Compiled from Various Authorities)

Metal	Condition	Approximate composition, %	Strength in tension,[a] 1000 lb/in²*		Endurance limit,[c] 1000 lb/in²*	Brinell hardness no.	Elongation, 2 in, %	Weight, lb/in³†
			Yield[b]	Ultimate				
Monel metal (B)	Cold-drawn	Ni 67, Cu 30, +Fe & Mn	60-&95	85–125	d	160–220	35–15	0.318
	Annealed		30–40	70–85		120–160	50–35	
Nickel	Cast	Ni 97, Si 1.2, +Mn & Fe	21	55	d	120	22	0.319
	Hot-rolled	Ni 99, +Fe, Cu, Mn, C, & Co	27	73	33	109	46	0.319
Zinc	Die-cast	Zn 95, Al 4, Cu 1	f	45[h]	d	70–85	3	0.242
	Rolled sheet	Zn 99.9	f	16–19[h]	d	d	40-70	0.242

*1 lb/in² = 6.895 kPa.
†g/c³ = 27.68 × lb/in³.
[a]The yield strength in compression may be usually assumed approximately equal to the yield strength in tension; the ultimate in compression may be considered slightly above the yield strength; the strength in shear may be considered about 0.6 the strength in tension.
[b]Yield strength determined as stress corresponding to approximately 0.2% permanent set.
[c]For completely reversed cycles of flexural stress; polished specimens.
[d]Test data lacking.
[e]In 152 cm (60 in).
[f]Not clearly defined owing to creep at low loads.
[g]Rockwell B scale hardness.
[h]Values for ordinary testing machine tests; much lower values probable for long-time tests under steady load.
[i]The strengths of extrusions more than about 20 cm (⅜ in) thick will be 15 to 20% higher.
[j]These values are for other products than extrusions, which will have strengths 8 to 10% higher.
[k]Value for wire in 254-cm (10-in) gage length.

STRUCTURAL MATERIALS

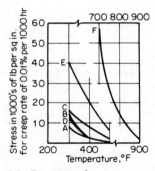
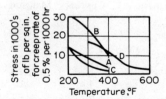

FIG. 17-5 Creep stress for a creep rate of 0.01% per 1000 h for several nonferrous metals. *A.* Copper, deoxidized and annealed, grain size 0.013 mm. *B.* Aluminum brass (76% Cu, 22% Zn, and 2% A.), annealed, grain size 0.015 mm. *C.* 70-30 brass, annealed, grain size 0.016 mm. *E.* Copper-nickel-phosphorus-tellurium alloy, quenched and aged (98.1% Cu, 1.11% Ni, 0.28% P, 0.51% Te). *F.* Monel metal, cold-drawn 40%.

FIG. 17-6 Creep stress for a creep rate of 0.5% per 1000 h for several aluminum and magnesium alloys. *A.* 3003-H14 aluminum alloy. *B.* 6061-T6 aluminum alloy (see Par. **17-88**). *C.* AZ-63 HTS magnesium alloy (5.3% to 6.7% Al, 2.5% to 3.5% Zn, 0.15% Mn). *D.* EM 51 HTA magnesium alloy (1.2% Mn, 3.8 to 6.2% Ce).

Commercially pure aluminum (99+%) is very weak and ductile: tensile strength of 90 MPa (13,000 lb/in^2), yield strength of 34.5 MPa (5000 lb/in^2), and shearing strength of 62 MPa (9500 lb/in^2). Extra-pure grades (electrical conductor grade) are 99.7+% pure, and are even weaker, but have better conductivity. A very comprehensive review of production, properties, processing, design, and applications of aluminum is given in a three-volume series, *Aluminum,* edited by K. R. Van Horn, Cleveland, American Society for Metals, 1966.

17-87. Structural aluminum alloys are alloys of aluminum with relatively small additions of other elements, including copper, manganese, silicon, chromium, and zinc. Very small amounts of boron, titanium, and zirconium are sometimes added for grain-size control. The approximate compositions of alloys frequently used for structural applications are listed in Table 17-10 (see ASTM Specifications B209 and B211). The aluminum industry uses a four-part designation system to identify wrought alloys. The first number indicates the alloy type (pure aluminum, copper-bearing, etc.), the second number identifies alloy modifications, and the last two digits indicate the purity of the specific alloy. Following these four digits is a letter to show the temper of the alloy: O means annealed, F mean as fabricated, H means the alloy has been cold-worked, and T means the alloy has been heat-treated. Numbers following these letters specify the details of the treatment—the extent of cold working, or the type of heat-treatment.

17-88. Heat-Treatment of Aluminum Alloys. Alloys of the 1000, 3000, and 5000 series cannot be hardened by heat-treatment. They can be hardened by cold working and are available in annealed (recrystallized) and cold-worked tempers. The 5000-series alloys are the strongest non-heat-treatable alloys, and are frequently used where welding is to be employed, since welding will generally destroy the effects of hardening heat-treatment. The remaining wrought alloys can be hardened by controlled precipitation of alloy phases. The precipitation is accomplished by first heating the alloy to dissolve the alloying elements, followed by quenching to retain the alloy in supersaturation. The alloys are then "aged" to develop a controlled size and distribution of precipitate that produces the desired level of hardening. Some alloys naturally age at room temperature; others must be artificially aged at elevated temperatures. Table 17-11 specifies recommended heat-treatments for hardening selected representative alloys. Additional hardening can be obtained in these alloys by cold working the quenched alloy before the aging begins in order to realize the combined strengthening of the two hardening mechanisms. The alloy additions that produce hardening have a deleterious effect on corrosion resistance; the phases introduced by alloying are generally cathodic to the metal, and set up galvanic corrosion cells. To protect high-strength alloys, they are commonly clad with a higher-purity aluminum alloy. The clad composition

ELECTRICAL ENGINEERING MATERIALS REFERENCE GUIDE

TABLE 17-10 Nominal Compositions of Structural Aluminum Alloys
(Aluminum Company of America)

Alloy designation	Other elements added to aluminum, %					
	Copper	Manganese	Magnesium	Silicon	Chromium	Zinc
Wrought:						
1100	(Minimum 99.0 aluminum)					
3003	. . . .	1.2				
3004	. . . .	1.2	1.0			
2014	4.4	0.8	0.5	0.8		
2024	4.5	0.6	1.5			
5052	. . . .	. . . .	2.5	. . .	0.25	
6061	0.28	. . . .	1.0	0.6	0.2	
6063	. . . .	. . . .	0.7	0.4		
6070	0.28	0.7	0.8	1.4		
7075	1.6	. . . .	2.5	. . .	0.23	5.6
Cast:						
A443	0.3 max	. . . .	. . . .	5.2		
A356	0.25 max	0.35 max	0.32	7.0		0.35 max
380	3.5	. . . .	. . . .	8.5		

TABLE 17-11 Age-Hardening Heat-Treatment of Aluminum Alloys

Alloy designation	Heat-treating temperature, deg F	Approx duration of heating,* min	Quench†	Aging temperature, deg F	Time of aging
2014	930–950	15–60	Water	340	10 hr
2024	910–930	15–60	Cold water	Room	4 days‡
6061	980–990	15–60	Cold water	315–325	18 hr
7075	870–900	15–60	Cold water	250	24 hr

*This depends on the size and amount of material. In some cases, even longer times may be needed.
†The quench should be made with minimum time loss in transfer from furnace.
‡More than 90% of the maximum properties are obtained during the first day of aging.
NOTE: $t\cdot_C = t\cdot_F - 32)/1.8$.

should be designed to be anodic to the base metal in order to provide sacrificial protection in case the clad is penetrated and the core metal exposed.

Stone, Brick, Concrete, and Glass Brick

17-89. Building Stone. Stone is any natural rock deposit or formation of igneous, sedimentary, and/or metamorphic origin, in either its original or its altered form. Building stone is the quarried product of such deposit or formation which is suitable for structural and ornamental purposes. Igneous or volcanic rock, such as granite or basalt, is rock of plutonic or volcanic origin, formed from a fused condition and crystalline in structure. Sedimentary rock, such as limestone, dolomite, and sandstone, is formed by the deposition of particles from water and laminated in structure. Metamorphic rock, such as gneiss, marble, and slate, is rock formation which, in the natural ledge, has undergone marked change in microstructure or character due to heat, pressure, or moisture and therefore exists in form different from the original. (See ASTM Designation C119 for additional definitions.)

STRUCTURAL MATERIALS

17-90. Weight and Strength of Stone. The properties of stone from different quarries vary over a considerable range. Average values, from Everett: *Materials;* New York, John Wiley & Sons, Inc., 1978, are listed in Table 17-12. These values should be used only with great caution; those of limestones and sandstones particularly depend on the composition and location of the deposit. (See Mantell, C. L., *Engineering Materials Handbook,* New York, McGraw-Hill, 1958.)

17-91. Building brick made from clay or shale is required to have certain physical properties (Table 17-13) under ASTM Specification C62-58: for sand-lime building brick see ASTM Specification C73-51. (Also see Miner, D. F., and Seastone, J. B.: *Handbook of Engineering Materials;* New York, John Wiley & Sons, Inc., 1955.)

17-92. Clay firebrick for stationary and marine boiler service is covered by ASTM Standard Specification C64-61. *Refractory brick* for resisting high temperatures and the effects of very hot gases and molten slag and clinker is made of various special compositions known as "firebrick," "silica brick," "magnesia brick," "bauxite brick," "chromite brick," etc. (See ASTM specifications for brick and clay products; also see Dagostino, F. R.: *Materials of Construction;* Englewood Cliffs, N.J., Prentice-Hall, 1982.)

TABLE 17-12 Summary of Properties of Building Stone

	Density, kg/m^3	Failing stress in compression, N/mm^3	Thermal movement, mm/m per 90°C, % approx.	Moisture movement, mm/m for dry-wet change
Granites	2560–3200	105 335	0·93	None
Sandstones	2130 2750	27·5 195	1·0	Approx. 0·7
Limestones	1950–2400	16·5 42·5	0·25 (porous limestone) 0·34 (dense limestone)	0·8 Negligible
Slates	2800–3040	42·5 216	0·93	Negligible
Marbles	2880		0·34	Negligible
Quartzites	2630		0·90	None

TABLE 17-13 Physical Properties of Building Bricks

Designation	Minimum compressive strength (brick flatwise), lb per sq in., average, gross area		Maximum water absorption by 5-hr boiling, %		Maximum saturation coefficient*	
	Average of 5 bricks	Individual	Average of 5 bricks	Individual	Average of 5 bricks	Individual
Grade SW...................	3000	2500	17	20	0.78	0.80
Grade MW...................	2500	2200	22	25	0.88	0.90
Grade NW...................	1400	1250	No limit	No limit	No limit	No limit

*The saturation coefficient is the ratio of absorption by 24-hr submersion in cold water to that after 5-hr submersion in boiling water.

17-93. Structural gypsum products include gypsum tile, plaster board, wallboard, and plain or reinforced members cast in place. (For data on physical properties and specifications see ASTM specifications for gypsum and various gypsum products, and *Testing and Inspection of Engineering Materials*, McGraw-Hill, 1964.)

17-94. Portland cement is produced by sintering a proportional mixture of lime and clay, which is subsequently ground with the addition of gypsum (to retard the rate of setting). The properties of the clay and limestone determine the principal characteristics: fineness, soundness, time of set, and strength. Strength is measured from briquettes made of a mortar of 1 part cement to 3 parts sand, and measured under specified conditions. Minimum tensile strengths are 2 MPa (275 lb/in^2) at 7 days, and 2.6 MPa (350 lb/in^2) at 28 days. See ASTM Specs. C109, C115, C151, C170, C191, and C126.

17-95. Mortar is a mixture of sand, screenings, or similar inert particles, with cement and water, which has the capacity of hardening into a rocklike mass. The inert particles are usually less than ¼ in in size. The proportions of cement to sand range all the way from 1:0 to 1:4 for various purposes.

17-96. Concrete is a mixture of crushed stone, gravel, or similar inert material with a mortar. The maximum size of inert particles is variable but usually less than 2 in. These inert constituents of mortar and concrete are known as the "aggregate." In making good concrete, the properties of the aggregate are as important as those of the cement. The fine aggregates consist of sand, screenings, mine tailings, pulverized slag, etc., with particle sizes less than ¼ in; the coarse aggregates consist of crushed stone, gravel, cinders, slag, etc. Rubble concrete is made by embedding a considerable proportion of boulders or stone blocks in concrete. The proportions by volume of cement, sand, and coarse aggregate range all the way from 1:1:2 for high compressive strength to 1:4:8 for structures requiring mass more than strength. In general, the strength of concrete increases with the density and richness of mix (proportion of cement) but is decreased in proportion to the amount of mixing water that is added beyond that required to produce a plastic workable mixture. In controlling the quality of a concrete of given mix, the ratio of the volume of mixing water to the volume of cement (water-cement ratio) is often used as a criterion of the strength. For proper curing the concrete should be kept moist for at least a week after placing, and care should be taken to prevent its freezing in cold weather during the early stages of curing. Freshly poured concrete gains strength very slowly in cold weather. Various admixtures are often added in small amounts to modify pouring characteristics and setting time as well as physical characteristics such as resistance to freezing and thawing cycles, wear, abrasion, and permeability. (For systematic information see publications of the Portland Cement Association and ASTM Standards.)

17-97. Compressive Strength of Concrete. For concrete of common proportions cured under good conditions, 25 to 40% of the 2-year strength is developed in 7 days, 50 to 65% in 1 month, and 70 to 90% in 6 months. The tensile strength of concrete is very low (about one-tenth its compressive strength), and hence in structural members the concrete is usually designed to resist the compressive stresses only, the tensile strength of the concrete being considered negligible. For flexural members, steel reinforcing bars are usually inserted on the tensile side of the beam to resist the tensile stresses. The curves in Fig. 17-7 show typical variations of the compressive strength of concrete with the mix, density, and amount of mixing water used. The mix in each case is given as the proportion of cement to the total volume of aggregate used, and the water-cement ratio is plotted as the number of gallons of water per sack of cement. (One sack is assumed to be 1 ft^3 of cement.)

17-98. Flexural Strength of Concrete. Slabs and pavements are often designed on the basis of the flexural strength of the concrete. Typical variation of the modulus of rupture with the compressive strength of concrete for a number of different mixes and water-cement ratios is shown in Fig. 17-8. This curve is adapted from data compiled by the Portland Cement Association on beams 7 in deep and 10 in wide, loaded with equal loads at the ⅓ points of a 36-in span.

17-99. Glass bricks are not primarily intended as load-bearing structural units but are used mainly for partition walls and outside walls of buildings in which the main loads are carried by other structural units. Safe working stresses in compression (obtained from data published by the manufacturers) would probably be less than about 3 MPa (400 lb/in^2). The blocks are ribbed in various ways to transmit a large proportion of the incident light with almost complete diffusion to eliminate glare. The interior of these blocks contain a thoroughly dry air under partial vacuum, and they are therefore excellent thermal insulating materials for wall construction.

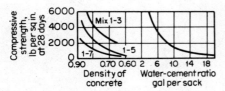

FIG. 17-7 Typical variations of strength of concrete with density and water-cement ratio.

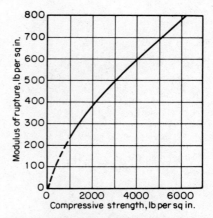

FIG. 17-8 Average relation of modulus to rupture to compressive strength of concrete. Mix 1:4 by volume. Age of test, grading, and source of aggregate variable.

Chapter 18

WOOD PRODUCTS

By DUANE E. LYON

18-1	Wood 18-1	18-22	Strength of Treated Lumber	18-8
18-2	Types of Wood 18-1	18-24	American Lumber Standards	18-8
18-3	Wood Structure 18-1	18-28	Panel Materials 18-9	
18-11	Properties of Wood 18-3	18-29	Wood Poles and Crossarms	18-10
18-14	Strength of Wood 18-4			
18-17	Decay of Wood and Its Prevention 18-4			

18-1. Wood is the vascular tissue of trees. Woods presently considered to have major commercial value in the United States are obtained from approximately 60 native tree species and 30 imported species. The varied structure and physical characteristics of these species make wood suitable for a wide variety of products ranging from decorative veneers to utility poles.

18-2. Hardwoods and Softwoods. Wood is obtained from two classes of trees, commonly referred to as hardwoods and softwoods. The hardwoods are angiosperms and are characterized by broad leaves and seeds enclosed in a fruit. Most hardwoods lose their leaves in the fall or winter. Oaks, maple, ash, and walnut are examples of common hardwood trees. The softwoods are members of the gymnosperms classified as conifers. The conifers are characterized by exposed seeds, usually in cones. Typical softwoods include the pines, spruces, and cedars. The terms hardwood and softwood do not indicate the relative hardness of the two types of wood, since many conifers are actually harder than trees of the angiosperm type.

18-3. Gross wood structure may be used to differentiate between the hardwood and softwood classes, and to identify species within each group. The cross section of a log shows several well-defined features from the outer bark, through the wood, to the central pith. The purpose of the bark is to protect the inner living tissues from injury. The inner portion of the bark is a conductive tissue that transports foodstuff from the leaves to the living cells. The wood portion of the tree has outer sapwood and inner heartwood regions, which often are distinguishable by a difference in color. The lighter-colored sapwood is a living tissue. The darker-colored heartwood consists of entirely dead cells, and serves primarily for mechanical support in the tree. The heartwood contains deposits of gums, oils, and other organic infiltrations in the cells. These materials are called extractives, and import the darker color (ranging from light brown to black) to heartwood. Other differences include durability and permeability. Sapwood of all species is readily destroyed by insects and the decay fungi. The heartwoods of many species are also nondurable, but others are very durable (for example cypress, the cedars, and redwood). Sapwood is usually more permeable to liquids, owing largely to the absence of extractives. For this reason, sapwood is more easily treated with preservatives. Wood near the center of the tree is often characterized by fast

18-2 ELECTRICAL ENGINEERING MATERIALS REFERENCE GUIDE

growth, and is termed juvenile wood. The properties of juvenile wood are inferior to those of normal heartwood. The strength properties of heartwood and sapwood are the same.

18-4. Annual Rings. In temperate zones, the tree increases in diameter and height by division of cells in a unicell layer called the cambium. The growth increment for each year is called an annual ring. Many North American woods have well-defined annual rings consisting of an inner light-colored zone called springwood and an outer darker-colored zone called summerwood. The summerwood, formed in the later part of the growing season, is denser than the springwood. The relative portion of summerwood present in an annual ring is an important factor in determining the mechanical behavior of wood. Many commercial species of wood imported from tropical zones do not have a well-defined annual ring structure.

18-5. Minute wood structure differs for hardwood and softwood species. The conifers are composed primarily of long hollow fibers oriented with their long axis parallel to the length of the tree. The ends of these fibers are tapered and the cells overlap at the ends. The size, shape, and structure of these fibers vary considerably from species to species, which accounts for much of the variation in properties of wood. They also give rise to wood's anisotropic nature, having different properties along its three principal directions, longitudinal, tangential, and radial. The minute structure of the hardwood is more varied and consists of many types of fiberlike cells. Individual fibers are composed of a multilayer cell wall enclosing a central cavity. For more information on wood, refer to Ref. 13.

18-6. Chemically wood consists of approximately 70% cellulose, 25% lignin, and about 5% extractives. The strength of wood may be attributed almost entirely to the cellulose and lignin present in the cell wall. The extractives do not contribute to the cell-wall structure, but do contribute to such properties as color, odor, taste, and resistance to decay. Additional information of the chemistry of wood may be found in Ref. 12.

18-7. Specific gravity is a measure of the amount of material contained in a piece of wood. It is calculated by dividing the weight of a given volume of wood by the weight of an equal volume of water. Specific gravity varies according to the amount of water present in wood. For this reason the weight of bone-dry wood is usually used for reported values of specific gravity. The specific gravity of some commercially imported woods is listed in Table 18-1, which was abstracted from Ref. 12. For clear, straight-grained wood at a known moisture content, specific gravity is positively correlated with several important properties, including strength and stiffness in bending, tension, and compression. Approximate functions for predicting the mechanical properties of wood for a known specific gravity are given in the Wood Handbook (Ref. 12). When wood contains defects or natural growth imperfections, the relationship between properties and specific gravity may be less pronounced.

18-8. Moisture in Wood. Wood is a hygroscopic material. Moisture in wood occurs in three forms: water vapor in air spaces in the cell cavities, capillary water in the cell cavities, and water molecules bound to the hydroxyl groups of the cellulose in the cell wall. In most

TABLE 18-1 Specific Gravity and Shrinkage of Common Woods

Species	Average specific gravity*	Shrinkage, %	
		Radial	Tangential
Douglas fir, various regions	0.43–0.48	3.6–5.0	6.2–7.8
White ash	0.60	4.8	7.8
Aspen	0.38	3.3	7.9
Yellow birch	0.55	7.2	9.2
White fir	0.37	3.2	7.1
Western red cedar	0.33	2.4	5.0
Northern white cedar	0.31	2.2	4.9
Western hemlock	0.42	4.3	7.9
Southern yellow pines	0.51–0.61	4.4–5.5	7.4–7.8
Tamarack	0.53	3.7	7.4
White oak species	0.63–0.68	4.1–5.5	7.2–10.8
Red oak species	0.59–0.69	4.0–5.5	8.2–10.6
Hickory species	0.69–0.75	7.0–7.8	10.0–12.6

*Weight ovendry and volume at 12% moisture content.
†From green to ovendry condition, based on green dimension.

WOOD PRODUCTS

18-3

end-use conditions, when wood is not in contact with water, nearly all the moisture present is bound water and is usually between 3 and 30% of the dry weight of the wood. Since this bound water tends to be at equilibrium with the vapor pressure of the surrounding atmosphere, the maximum amount of bound water in wood occurs in a saturated atmosphere. Any increase in moisture content above this maximum is due to capillary water, acquired from contact with liquid water.

The moisture content of wood is expressed as a percent of the ovendry weight of wood. It can be measured by weighing a wood sample before and after drying to constant weight at 210°F, using the relationship:

$$\text{Moisture content, \%} = \frac{\text{moist weight} - \text{dry weight}}{\text{dry weight}} \times 100$$

When moist wood dries, the liquid water present in the cell capillaries evaporates before the bound water leaves the cell wall. The fiber-saturation point is defined as the moisture content of wood that is in equilibrium with a saturated atmosphere, with no free water in the cell cavities. This moisture content is around 30% but varies with species.

18-9. Volumetric Changes. Below the fiber-saturation point, water that evaporates from wood results in a reduction in wood volume. The amount of volumetric change is positively related to the changes in moisture content and density. The anisotropic nature of wood results in unequal shrinkage in the three principal grain directions. Shrinkage is greatest in the transverse grain direction parallel to the growth rings (tangential), and the total shrinkage from green to the ovendry condition ranges from 4 to 13%, depending upon the species and density. In general, shrinkage increases with density. The transverse shrinkage perpendicular to the growth rings (radial) is usually about one-half that parallel to the growth rings and ranges from 2 to 8%. Total shrinkage along the grain ranges from 0.1 to 0.3%. Some representative shrinkage values are given in Table 18-1. In poles, the tangential shrinkage results in seasoning checks, while radial shrinkage is the dominant factor in reducing pole circumference. Circumferential shrinkage is about 1% for poles dried to approximately 20% moisture content.

18-10. Wood Seasoning. Most wood products are dried prior to use to remove the large amount of moisture present in freshly cut wood. Wood that has been dried offers a number of advantages, including reduced weight and shrinkage, and increased strength and durability. Drying may be accomplished by one of several procedures. Two of the most common are air drying and kiln drying. A number of defects may develop during drying if the process is not carefully controlled. These defects are the result of drying stresses due to unequal shrinkage. Most kiln-drying procedures include moisture-equalizing and conditioning treatments to improve moisture uniformity throughout the thickness of the wood product, and to relieve residual stresses. Improper drying may result in warping, checking, or more severe defects. For a more complete discussion of wood drying, see Ref. 10.

18-11. Thermal Properties of Wood. Temperature affects several properties of wood. As wood is heated, it expands. The coefficient of thermoexpansion for wood averages near 1.1×10^{-6} per degree Celsius for most native species. Wood is a good insulator, and does not respond very fast to a change in environmental temperature. The coefficient of thermoconductivity for wood ranges from 0.4 to 0.7 Btu/(h)(°C) for a 1 ft² area 1 in in thickness at a moisture content of 12%. The thermoconductivity of wood increases with increasing specific gravity and moisture content.

18-12. Electrical Properties of Wood. Three important electrical properties of wood are resistivity, dielectric constant, and power factor. Wood is an excellent insulator. The resistivity of dry wood to the flow of direct current is high, approximately $3 \times 10^{17} \, \Omega \cdot \text{cm}/ \text{cm}^3$ parallel to the grain. The presence of moisture lowers resistivity. The dielectric constant for wood determines the amount of stored electric potential energy when it is placed in a high-frequency alternating current. The dielectric constant for wood varies over a range from 2.0 for dry wood to 8 for wood above the fiber-saturation point. The dielectric constant is affected by density and grain direction. The power factor determines the amount of energy that is dissipated as heat when wood absorbs power in a high-frequency dielectric field. The power factor for wood is about 2 to 6% at low moisture contents for frequencies between 2 and 15 Hz. Additional information on the electrical properties may be found in Refs. 12 and 13.

18-13. Specific Gravity and Fiber-Saturation Point. The specific gravity of wood is the weight of wood divided by the weight of an equal volume of water. As both the weight and volume of wood vary with the moisture content, specific gravity of wood is an indefinite quantity unless the conditions under which it is obtained are clearly specified. Most commonly, specific gravity of wood is based on weight ovendry and volume green, ovendry, or some intermediate moisture content. As wood dries, most of the liquid water held in the capillaries evaporates before the bound-water molecules begin to leave the cell wall. The fiber-saturation point is defined as the moisture content of the wood at this transition point and is the moisture content at which shrinkage begins. The radial and tangential shrinkage and specific gravity of a number of woods are listed in Table 18-1, which was abstracted from *U.S. Dept. Agr., 72,* Wood Handbook, Forest Products Laboratory, Forest Service.

18-14. Effect of Moisture on Strength. Clear wood increases in strength as it dries below the fiber-saturation point. The change in strength resulting from a 1% change in moisture content, expressed in percent, is approximately 4 for modulus of rupture, 2 for modulus of elasticity, and 5 for compression parallel to the grain. More exact relationships may be found in Ref. 12.

For structural lumber, the increase in strength of the clear wood is partly offset by the development of seasoning defects such as checks and splits. For this reason, the properties in the green condition are generally used as the base for the development of design stresses for wood. For lumber that is nominally 2 in in thickness, the design stress is increased by up to 25% in bending, 20% in modulus of elasticity, and 37.5% in compression parallel to grain when the moisture content is at or less than 15%.

18-15. Effect of Temperature on Strength. In general, heating reduces and cooling increases the mechanical properties of wood. The change is immediate, and irreversible for temperatures remaining above 93°C for any appreciable period of time. The adverse effect of high temperature is more pronounced at high moisture contents. For elevated temperatures below 93°C, the immediate loss of strength is recovered when the wood is cooled to ambient conditions. When wood is repeatedly exposed to high temperature, the adverse effect on properties is cumulative.

18-16. The mechanical properties of the commercially important woods of the United States have been evaluated in accordance with ASTM Standard D143, which specifies small, clear specimens to eliminate the influence of naturally occurring physical defects in the wood.

Tables 18-2 and 18-3 show some mechanical properties for wood in the green condition and at 12% moisture content, respectively. The green properties are obtained from specimens at essentially the same moisture content as in the living tree, well above the fiber-saturation point.

Tables 18-2 and 18-3 have been abstracted to include several of the more important hardwood and softwood species and the mechanical properties of each which are likely to be uniquely important for specific uses encountered in electrical engineering applications. For additional data on other strength properties and other species, see the references.

18-17. Decay and Its Prevention. At ordinary temperatures, wood is very stable and unless attacked by living organisms remains the same for centuries, either in air or under water. Fungi are the chief enemies of wood, and they thrive best with warmth and abundance of moisture and air, for example, in contact with the ground. Higher temperatures near the surface of the ground, together with adequate air and a greater prevalence of fungi, cause decay to progress faster near the ground line than at several feet below. Proper seasoning, together with protection against the entrance of moisture and impregnating with fungus-inhibiting compounds (see Par. **18-18**), which prevent fungi from feeding on the wood, is the best means of preservation. Only the heartwood is resistant, however; consequently, the sapwood should be preservative-treated, irrespective of the species of wood, if decay resistance is needed. Various species differ materially in their natural resistance to decay. For systematic presentation of the subject of wood preservation, see the references.

18-18. Wood Preservatives. Wood preservatives fall into two main classes: (1) oilborne preservatives and (2) water-borne metallic salts. The former may be further subdivided into (a) coal-tar creosote with and without the mixture of cheaper materials such as petroleum or coal tar and (b) solutions of toxic organic chemicals such as pentachlorophenol dissolved in petroleum oils. Oil-type preservatives are used extensively for products that are exposed to ground contact whereby resistance to leaching is an important requirement of the preservative. These products include poles, crossties, piling, bridge timbers, and fence

TABLE 18-2 Mechanical Properties of Various Woods in the Green Condition Grown in the United States

Species	Moisture content, %	Specific gravity*	Static bending		Compression parallel to grain maximum crushing strength, lb/in²	Compression perpendicular to grain stress at proportional limit, lb/in²	Tension perpendicular to grain maximum tensile strength, lb/in²	Hardness†		Maximum shearing strength parallel to grain, lb/in²
			Modulus of rupture, lb/in²	Modulus of elasticity, 1,000 lb/in²				End, lb	Side, lb	
Ash, black	85	0.45	6,000	1,040	2,300	350	490	590	520	860
Ash, white	42	0.55	9,600	1,460	3,990	670	590	1,010	960	670
Aspen	94	0.35	5,100	860	2,140	180	230	280	300	660
Basswood	105	0.32	5,000	1,040	2,220	170	280	290	250	600
Beech	54	0.56	8,600	1,380	3,550	540	720	970	850	1,290
Birch, yellow	67	0.55	8,300	1,500	3,380	430	430	810	780	1,110
Cottonwood, eastern	111	0.37	5,300	1,010	2,280	200	410	380	340	680
Elm, American	89	0.46	7,200	1,110	2,910	360	590	680	620	1,000
Elm, slippery	85	0.48	8,000	1,230	3,320	420	640	750	660	1,110
Hickory, shagbark	60	0.64	11,000	1,570	4,580	840		1,640	1,570	1,520
Locust, black	40	0.66	13,800	1,850	6,800	1,160	770			1,760
Maple, silver	66	0.44	5,800	940	2,490	370	560	670	590	1,050
Maple, sugar	58	0.56	9,400	1,550	4,020	640		1,070	970	1,460
Oak, red	80	0.56	8,300	1,350	3,440	610	750	1,060	1,000	1,210
Oak, white	68	0.60	8,300	1,250	3,560	670	770	1,120	1,060	1,250
Sweetgum	115	0.46	7,100	1,200	3,040	370	540	670	600	990
Sycamore	83	0.46	6,500	1,060	2,920	360	630	700	610	1,000
Yellow poplar	83	0.40	6,000	1,220	2,660	270	510	480	440	790
Baldcypress	91	0.42	6,600	1,180	3,580	400	300	440	390	810
Cedar, northern white	55	0.29	4,200	640	1,990	230	240	320	230	620
Cedar, Port Orford	43	0.40	6,200	1,420	3,130	280	180	460	400	830
Cedar, western red	37	0.31	5,100	920	2,750	270	230	430	270	710
Douglas fir, coast†	38	0.45	7,700	1,560	3,780	380	300	570	500	900
Fir, white	110	0.37	5,900	1,160	2,900	280	300	410	340	760
Hemlock, western	77	0.42	6,600	1,310	3,360	280	290	500	410	860
Larch, western	58	0.48	7,700	1,460	3,760	400	330	580	510	870
Pine, lodgepole	65	0.38	5,500	1,080	2,610	250	220	320	330	680
Pine, ponderosa	91	0.38	5,100	1,000	1,940	280	310	310	320	700
Pine, loblolly	81	0.47	7,300	1,410	3,490	390	260	420	450	850
Pine, longleaf	62	0.54	8,700	1,600	4,300	480	330	550	590	1,040
Pine, shortleaf	81	0.46	7,300	1,390	3,430	350	320	410	440	850
Pine, western white	54	0.36	5,200	1,170	2,650	240	260	310	310	640
Spruce, Engelmann	80	0.32	4,500	960	2,190	220	240	310	260	590
Spruce, Sitka	42	0.37	5,700	1,230	2,670	280	250	430	350	760

*Specific gravity based on green volume and ovendry weight.
†Load required to embed a 0.444-in ball to half its diameter.
‡Coast Douglas fir is defined as that coming from counties in Oregon and Washington west of the summit of the Cascade Mountains. For Douglas fir from other sources, see Western Wood Density Survey, *U.S. Forest Service Res. Paper FPL 27.*

NOTE: 1 lb/in² = 6.895 kPa; 1 lb = 0.4536 kg.

18-5

TABLE 18-3 Mechanical Properties of Various Woods in the Air-Dry Condition Grown in the United States

Species	Moisture content, %	Specific gravity*	Static bending		Compression parallel to grain maximum crushing strength, lb/in²	Compression perpendicular to grain stress at proportional limit, lb/in²	Tension perpendicular to grain maximum tensile strength, lb/in²	Hardness†		Maximum shearing strength parallel to grain, lb/in²
			Modulus of rupture, lb/in²	Modulus of elasticity, 1,000 lb/in²				End, lb	Side, lb	
Ash, black	12	0.49	12,600	1,600	5,970	760	700	1,150	850	1,570
Ash, white	12	0.60	15,400	1,770	7,410	1,160	940	1.720	1,320	1,160
Aspen	12	0.38	8,400	1,180	4,250	370	260	510	350	850
Basswood	12	0.37	8,700	1,460	4,730	370	350	520	410	990
Beech	12	0.64	14,900	1,720	7,300	1,010	1,010	1,590	1,300	2,010
Birch, yellow	12	0.62	16,600	2,010	8,170	970	920	1,480	1,260	1,880
Cottonwood, eastern ...	12	0.40	8,500	1,370	4,910	380	580	580	430	930
Elm, American	12	0.50	11,800	1,340	5,520	690	660	1,110	830	1,510
Elm, slippery	12	0.53	13,000	1,490	6,360	820	530	1,120	860	1,630
Hickory, shagbark	12	0.72	20,200	2,160	9,210	1,760	.....	.....	.....	2,430
Locust, black	12	0.69	19,400	2,050	10,180	1,830	640	1,580	1,700	2,480
Maple, silver	12	0.47	8,900	1,140	5,220	740	500	1,140	700	1,480
Maple, sugar	12	0.63	15,800	1,830	7,830	1,470	.....	1,840	1,450	2,330
Oak, red	12	0.63	14,300	1,820	6,760	1,010	800	1,580	1,290	1,780
Oak, white	12	0.68	15,200	1,780	7,440	1,070	800	1,520	1,360	2,000
Sweetgum	12	0.52	12,500	1,640	6,320	620	760	1,080	850	1,600
Sycamore	12	0.49	10,000	1,420	5,380	700	720	920	770	1,470
Yellow poplar	12	0.42	10,100	1,580	5,540	500	540	670	540	1,190
Baldcypress	12	0.46	10,600	1,440	6,360	730	270	660	510	1,000
Cedar, northern white ..	12	0.31	6,500	800	3,960	310	240	450	320	850
Cedar, Port Orford ...	12	0.42	11,300	1,730	6,470	620	400	730	560	1,080
Cedar, western red	12	0.33	7,700	1,120	5,020	490	220	660	350	860
Douglas fir, coast‡	12	0.48	12,400	1,950	7,240	800	340	900	710	1,130
Fir, white	12	0.39	9,800	1,490	5,810	530	300	780	480	1,100
Hemlock, western	12	0.45	11,300	1,640	7,110	550	340	900	540	1,250
Larch, western	12	0.52	13,100	1,870	7,640	930	430	1,120	830	1,360
Pine, lodgepole	12	0.41	9,400	1,340	5,370	610	290	530	480	880
Pine, ponderosa	12	0.40	9,400	1,290	5,320	580	420	570	460	1,130
Pine, loblolly	12	0.51	12,800	1,800	7,080	800	470	750	690	1,370
Pine, longleaf	12	0.58	14,700	1,990	8,440	960	470	920	870	1,500
Pine, shortleaf	12	0.51	12,800	1,760	7,070	810	470	750	690	1,310
Pine, western white	12	0.38	9,500	1,510	5,620	440	.....	440	370	850
Spruce, Engelmann	12	0.34	8,700	1,280	4,770	470	350	560	350	1,030
Spruce, Sitka	12	0.40	10,200	1,570	5,610	580	370	760	510	1,150

*Specific gravity based on green volume and ovendry weight.

†Load required to embed a 0.444-in ball to half its diameter.

‡Coast Douglas fir is defined as that coming from counties in Oregon and Washington west of the summit of the Cascade Mountains. For Douglas fir from other sources, see Western Wood Density Survey, *U.S. Forest Service Res. Paper FPL 27.*

1 lb/in² = 6.895 kPa; 1 lb = 0.4536 kg.

WOOD PRODUCTS 18-7

posts. Water-borne preservatives are used mainly for the treatment of lumber. Wood treated with a water-borne preservative is clean, paintable, and odorless.

Creosote is a distillate of coal tar formed during the coking of coal. On the basis of the quantity of wood treated, it is the most important preservative. Much of the treated wood is used in ground contact; appreciable amounts are also used in coastal waters infested with marine organisms that bore into and destroy untreated wood.

Pentachlorophenol dissolved in petroleum oils of varied nature has come into wide use. As a general rule, the effectiveness is highest when high-boiling oils are used as solvents, but relatively low-boiling oils are sometimes used in the treatment of products, such as millwork, having high cleanliness requirements. Water-repellent materials are generally added to the preservative in millwork treatments to minimize the dimensional changes that accompany fluctuations in the moisture content of the wood. A relatively new type of treatment comprises the solution of pentachlorophenol in a liquefied petroleum gas which is subject to practically complete removal by evaporation, leaving the treated wood very clean and readily paintable. Other standard oil-borne preservatives are copper-8-quinolinolate, and tributyltin oxide.

Water-borne preservatives are generally mixtures of several inorganic salts, the most important of which are salts of copper, chromium, arsenic, and zinc. Sodium fluroide is an ingredient of two widely used commercial preservatives. Traditionally, the use of water-borne preservatives has been restricted to situations where resistance to leaching is not required; however, several formulations now available comprise mixtures of salts that undergo chemical reaction within the wood with the formation of relatively insoluble toxic compounds. Such preservatives give good protection to wood exposed to wet conditions. Wood treated with water-borne preservatives is clean and paintable after drying to below 25% moisture content.

Paints, varnishes, and stains are used for decorative effects, but they also afford surface protection by retarding moisture changes and thus decreasing checking, warping, and weathering. Such protection is only superficial, however, and internal decay may be expected unless the wood is kept dry.

Fire-retardant chemicals such as ammonium phosphate and sulfate and salts of zinc and boron are used to decrease the flammability of wood. Some fire-retardant formulations also give protection against decay.

18-19. Methods of Treating Wood. The methods of preservative treatment may be divided into two classes, pressure and nonpressure. Pressure methods are by far the most effective for protecting wood. In pressure methods the wood is enclosed in a vessel, and the liquid preservative is forced into the wood under considerable hydrostatic pressure. Nonpressure methods do not utilize artificial pressure, the preservative being applied by dipping, soaking, brushing, or spraying. A third method, somewhat distinct from the others and called the thermal method, may be mentioned. It consists in heating the wood to expel air and then allowing the wood to cool in the liquid, whereby a partial vacuum forms in the internal spaces. Although movement of the liquid into the wood is due to atmospheric pressure, the process is not classed among pressure processes.

There are several modifications of the pressure process. In the full-cell process, the wood is subjected to a vacuum in order to evacuate the internal cavities used for the treatment of marine piling, which requires high retention of creosote for protection against wood-boring animals. The process is also used commonly in treatments of lumber with water-borne preservatives. Much wood for land use is treated with oil-type preservatives by one of the so-called empty-cell methods, whereby it is possible to increase the depth of penetration obtained with a limited retention of preservative. In the Rueping process, air is first injected to create within the wood a pressure greater than atmospheric. The cylinder is filled with preservative in such a way that the injected air is trapped in the wood. The pressure is then increased to force preservative into the wood. After the pressure is released and the cylinder drained, the compressed air in the wood expands to expel some of the preservative. The recovered preservative is called kickback. The Lowry process differs from the Rueping process in that no initial air pressure is applied. The air normally present is compressed during the pressure cycle and produces some kickback when pressure is released.

The conditioning of the wood prior to treatment is an important step. Air seasoning, kiln drying, and various processes of cylinder conditioning are employed. The latter include steaming plus vacuum, boiling in oil under vacuum, and vapor drying, in which green wood is surrounded by hot vapors of distillates of coal tar or petroleum.

18-8   ELECTRICAL ENGINEERING MATERIALS REFERENCE GUIDE

When oil preservatives are applied by simple soaking methods, the wood should be well seasoned in order to provide air spaces into which the oil may move. Oil preservatives of low viscosity are preferable. The results attainable vary greatly with the species of wood.

Diffusion methods depend upon the diffusion of water-soluble chemicals into the moisture present in green wood. Here again, the species of wood is an important factor, but the results are affected by other factors such as the nature of the chemical, the concentration of the solution, and the duration of the soaking period.

18-20. Applications of Preservative Treatment. Preservative treatments are applied to many wood products, the most important being poles, crossties, lumber and structural timbers, fence posts, piling, and crossarms. Approximately 85% of all crossties treated in 1970 were of hardwood species, with oak accounting for 53%. The coniferous species dominated the treatment of other wood items, with southern pine being the most important, followed by Douglas fir.

18-21. Advantages of Preservative Treatment. In addition to the conservation of a natural resource, preservative treatment results in economic savings due to increased service life and reduced maintenance costs. This has been recognized for many years by railroad companies, utility companies, and other large users of wood products. Because of demonstrated savings, practically all crossties and poles are now given a preservative treatment before installation. There has been a gradual increase in the volume of lumber treated annually, due to more widespread knowledge of the need for such treatment when the wood is to be used under conditions favorable to attack by decay or insects. For best performance it is desirable that all machining operations be completed before treatment.

18-22. Strength of Treated Lumber. The effect of a preservative such as creosote or pentachlorophenol, in and of itself, on the strength of treated lumber appears to be negligible. It may be necessary, however, in establishing design stress values, to take into account possible reductions in strength that may result from temperatures or pressures used in the conditioning or treating processes. Results of tests of treated wood show reductions of stress in extreme fiber in bending and in compression perpendicular to grain, ranging from a few percent up to 25%, depending on the processes used. Compression parallel to grain is affected less and modulus of elasticity very little. The effect on resistance to horizontal shear can be estimated by inspection for shakes and checks after treatment. Strength reductions for wood poles agreed upon in formulating fiber-stress recommendations in American Standard Specifications and Dimensions for Wood Poles, ANSI 05.1-1979, range from 0 to 15% in various species, depending upon the conditioning and treating processes. Treating conditions specified by the American Wood Preservers' Association should never be exceeded. Reductions of strength can be minimized by restricting temperatures, heating periods, and pressures as much as is consistent with obtaining the absorption and penetration required for proper treatment.

18-23. Effect of Preservative Treatment on Electrical Resistivity. The electrical resistivity of wood depends on its moisture content to a much greater degree than any other single variable. Ovendry wood is an excellent insulator, but as the wood absorbs moisture, its resistivity decreases rapidly. Wood in normal use, however, where its moisture content may range from about 6 to 14%, is still a good enough insulator for many electrical applications.

When wood has been treated with salts for preservative or fire-retardant purposes, its electrical resistivity may be markedly reduced. The effect of such salt treatment is small when the wood moisture is below about 8% but increases rapidly as the moisture content exceeds about 10%. Treatment with creosote or pentachlorophenol has practically no effect on the resistivity of wood.

The resistivity of wood decreases by about a factor of 2 for each increase of 10°C in the temperature and is about half as great for current flow along the grain as across the grain.

18-24. American Lumber Standards. Simplified Practice Recommendation 16, American Lumber Standards for Softwood Lumber, is a voluntary standard of manufacturers, distributors, and users, promulgated in cooperation with the U.S. Department of Commerce. It provides for use classifications of (1) yard lumber, (2) structural lumber, and (3) factory and shop lumber. Different grading rules apply to each class. Size standards and generalized grade descriptions are part of SPR 16, but details of grading rules are left to the organized agencies of the lumber manufacturing industry. The grades and working stresses for structural lumber are referred by SPR 16 to the authority of ASTM D245, Methods for Establishing Structural Grades of Lumber, or D2018, Recommended for Determining Design Stresses for Load-Sharing Lumber Members.

WOOD PRODUCTS

18-9

18-25. Standard Commercial Names. Standard commercial names of the most commonly used structural softwood from ASTM D1165, Standard Nomenclature of Domestic Hardwood and Softwoods, are as follows:

Cedar:
  Alaska cedar
  Port Orford cedar
  Western red cedar
Fir:
  Douglas fir
  White fir
Hemlock:
  Eastern hemlock
  West Coast hemlock
Larch, Western

Pine:
  Jack pine
  Lodgepole pine
  Norway pine
  Ponderosa pine
  Southern yellow pine
Redwood
Spruce:
  Eastern spruce
  Engelmann spruce
  Sitka spruce

18-26. Standard Structural Grades. Detailed descriptions of the standard structural grades are published in the grading rule books of the organized regional agencies of the lumber manufacturing industry. These are subject to review for compliance with the general requirements of SPR 16, American Lumber Standards for Softwood Lumber. The principal-use classes of structural lumber are: (1) *joists and planks,* pieces of rectangular cross section 2 to 4 in thick and 4 or more in wide (nominal dimensions), graded primarily for bending strength edgewise or flatwise; (2) *beams and stringers,* pieces of rectangular cross section 5 by 8 in (nominal dimensions) and up, graded for strength in bending when loaded on the narrow face; and (3) *posts and timbers,* pieces of square or nearly square cross section, 5 by 5 in (nominal dimensions) and larger, graded primarily for use as posts and columns.

18-27. Working Stresses. Working stresses recommended by the lumber industry for their structural grades are found with the detailed grade descriptions in the grading rule books of the organized regional agencies of the industry. A complete listing of all structural grades and their working stresses is found in the "National Design Specification for Stress-Grade Lumber and Its Fastenings," published by the National Forest Products Association. Values for a few typical grades are shown in Table 18-4. Working stresses vary according to the grades and sizes of lumber and their condition with respect to moisture content. Stresses are adjustable also for duration of load and for special conditions such as extreme temperature. Stress increases are provided for "load-sharing members" in which the safety of the structure depends upon the strength of the assemblage of members rather than upon the lowest strength value for any single member. These stress modifications are described in ASTM standards.

Allowable working stresses for the structural grades of lumber are also a part of certain use specifications, such as the Minimum Property Standards of the Federal Housing Administration, the American Railway Engineering Association Manual, and various local or regional building codes. These allowable values may or may not coincide with the lumber industry stress recommendations for the same species and grade.

18-28. Wood-Base Panel Materials. Included in this category are plywood, insulating board, hardboard, particle board, waferboard, and the medium-density building fiberboards. Plywood, normally fabricated by bonding an odd number of layers of veneers together with the grain direction in adjacent plies at right angles to each other, is more dimensionally stable and more uniform in strength in the plane of the sheet than wood. Qualities of glue line and veneer permitted are set by the various commercial standards for plywood and determine the grades under which plywood is sold. In general, glue-line quality determines whether plywood is classed as being suitable for interior or exterior use.

U.S. Product Standard PS1-83 covers the basic specifications for the manufacture of construction plywood. Decorative hardwood plywood is described by U.S. Product Standard PS51(5.2). Plywood manufactured according to this standard will carry a grade trademark of a qualified testing agency.

Insulation board, hardboard, and medium-density building fiberboard are panel products made by reducing wood substance to particles or fiber and reconstituting the fiber into stiff panels 4 by 8 ft in area or larger. Insulation board is of either interior or water-resistant quality and is usually manufactured for use where combinations of thermal and sound-insulating properties and stiffness and strength are desired. Hardboard with a density of 50 lb/

18-10   ELECTRICAL ENGINEERING MATERIALS REFERENCE GUIDE

TABLE 18-4   Typical Stress Grades and Working Stresses for Structural Lumber*
(Normal duration of load and dry conditions of use)

Species	Grade	Allowable working stress				
		Bending or tension parallel, lb/in²	Horizon-tal shear, lb/in²	Com-pression perpen-dicular to grain, lb/in²	Com-pression parallel to grain, lb/in²	Modulus of elasticity, lb/in²
Douglas fir.............	Select structural beams and stringers	1,900	120	415	1,400	1,760,000
	Construction joists and planks	1,500	120	390	1,200	1,760,000
	Standard joists and planks	1,200	95	390	1,000	1,760,000
	Construction posts and timbers	1,200	120	390	1,200	1,760,000
West Coast hemlock......	Construction joists and planks	1,500	100	365	1,100	1,540,000
	Construction, MC joists and planks	1,650	105	365	1,250	1,540,000
	Standard joists and planks	1,200	80	365	1,000	1,540,000
Western larch...........	Standard joists and planks	1,200	95	390	1,000	1,760,000
Southern pine..........	Dense structural 72 beams and string-ers	2,000	135	455	1,550	1,760,000
	Dense structural 58 beams and string-ers	1,600	105	455	1,300	1,760,000
	No. 1 dimension, 2-in	1,500	120	390	1,350	1,760,000
	No. 2 dimension, 2-in	1,200	105	390	900	1,760,000
Norway pine...........	Common structural joists and planks	1,100	75	360	775	1,320,000
Redwood..............	Heart structural joists and planks	1,300	95	320	1,100	1,320,000
Eastern spruce..........	1300f structural joists and planks	1,300	95	300	975	1,320,000

*Compiled from "National Design Specification for Stress-Grade Lumber and Its Fastenings"; Washington, D.C., National Forest Products Association, 1962.
NOTE: 1 lb /in² = 6.895 kPa.

ft³ or more is used in many applications where a relatively thin, hard, uniform panel material is required. Of great importance in the electrical field are special high-density hardboard products expressly manufactured with high dielectric properties. Medium-density fiberboards with a density between that of insulation board and hardboard are new products.

Particle boards are panel products made by gluing small pieces of wood in a form such as flakes (this product is called waferboard) and shavings into relatively thick, rigid panels. Thermosetting resins, usually urea or phenolformaldehyde, are used to provide bonds of either interior or water-resistant quality. Standards ASTM C208-72 and ANSF A208.1 and PS58-73 govern minimum qualities of regular insulation board, particle board, and hardboard. The important physical and strength properties of various board products are indicated by Table 18-5.

18-29. Wood Poles and Crossarms.   Western red cedar and southern yellow pine are two species most commonly used in the United States for poles to support electric supply and communication equipment. In the northeast part of the country there are still a number of chestnut and northern white cedar poles in service, but these species are no longer available for purchase. Douglas fir, lodgepole pine, and western larch are used in considerable numbers, particularly in the western states. Other species that can be used for poles but not considered so desirable as western cedar or southern yellow pine are eastern hemlock, eastern larch, jack pine (large usage in Canada), northern white pine, ponderosa pine, red (Norway) pine, southern white cedar, spruce, sugar pine, western helmock, western white pine, and white fir.

18-30. Standards for Wood Poles.   The ANSI specifications for wood poles serve as a basis for purchasing and use. The ANSI specifications cover fiber stresses, dimensions, defect limitations, and manufacturing requirements. These specifications are also the basis for standards and specifications of organizations such as Edison Electric Institute and American Telephone and Telegraph Co. EEI Specification TD-100 for non-pressure-treated cedar poles and AT&T Specification AT-7312 are good examples of users' pole-purchasing specifications.

TABLE 18-5 Strength and Mechanical Properties of Wood-Base Fiber and Particle Panel Materials*

Material	Density, lb/ft	Specific gravity	Modulus of rupture, lb/in²	Modulus of elasticity (bending), 1,000 lb/in²	Tensile strength parallel to surface, lb/in²	Tensile strength perpendicular to surface, lb/in²	Compression strength parallel to surface, lb/in²	24-hr water absorption % by volume	24-hr water absorption % by weight	Thickness swelling, 24-h soak, %	Maximum linear expansion,† %	Thermal conductivity, Btu/(ft²)(h)(°F)(in thickness)
Fibrous-felted boards:												
1. Structural insulating board....	10–26	0.16–0.42	200–800	25–125	200–500	10–25	....	1–10	....	.....	0.5	0.27–0.45
2. Medium-density building fiberboard	26–50	0.42–0.80	400–4,000	90–700	800–2,000	....	500–3,400	....	6–150	.....	0.2–1.30‡	0.50–0.60
3. Hardboard:												
a. Untempered...........	50–80	0.80–1.28	3,000–7,000	400–800	3,000–6,000	....	1,800–6,000	....	3–30	10–25	0.6	0.80–1.40
b. Tempered...........	60–80	0.96–1.28	6,500–10,000	800–1,000	4,000–7,800	....	4,200–6,000	....	3–20	8–15	0.4	1.10–1.50
4. Super hardboard............	85–90	1.36–1.44	10,000–12,500	1,250	7,800	500	26,500	....	0.3–1.2	.....	.....	1.85
Particle boards:												
1. Insulating type.............	10–26	0.16–0.42	700	....	....	....	....	....	.....	.....	....	0.36
2. Medium-density type.........	26–50	0.42–0.80										
a. Extrusion....................	....	....	Values not presented because extruded boards are always used and tested with facings applied									
b. Flat-platen pressed.........	....	....	1,500–8,000	150–700	500–4,000	40–400	1,400–2,800	....	20–75	20–75	0.6	0.40–1.00
3. Hard-pressed type...........	50–80	0.80–1.28	3,000–7,500	400–1,000	1,000–5,000	275–400	3,500–4,000	....	15–40	15–40	0.85	1.10–1.50
4. Waferboard	26–50	0.42–0.80	2,500–3,200	450–650	500–5,000	50	2,000–5,000	....	.....	20–25	0.20	....

*The data presented are general round-figure values, accumulated from numerous sources; for more exact figures on a specific product, individual manufacturers should be consulted or actual tests made. Values are for general laboratory conditions of temperature and relative humidity.

†Expansion resulting from a change in moisture content from equilibrium at 50% relative humidity to equilibrium at 90% relative humidity.

‡For homogeneous and laminated boards, respectively.

NOTE: 1 lb/ft = 1.488 kg/m; 1 lb/in² = 6.895 kPa.

18-12   ELECTRICAL ENGINEERING MATERIALS REFERENCE GUIDE

18-31. Ultimate Fiber Stresses. The ultimate fiber stresses approved by the ANSI and contained in its Standard 05.1 are as shown in Tables 18-6, 18-7, and 18-8.

These tables cover all species of poles normally used in communication and electrical power construction.

18-32. Pole Dimensions. The circumference at "6 ft from butt" in Standard 05.1 is based on the following principles:

a. The classes from the lowest to the highest were arranged in approximate geometric progression, the increments in breaking load between classes being about 25%.

b. The dimensions were specified in terms of circumference in inches at the top and circumference in inches at 6 ft from the butt for poles of the respective classes and lengths, except for three classes having no requirement for butt circumference.

c. All poles of the same class and length were to have, when new, approximately equal strength or, in more precise terms, equal moments of resistance at the ground line.

d. All poles of different lengths within the same class were of sizes suitable to withstand approximately the same breaking load, on the assumption that the load is applied 2 ft from the top and that the break (failure) would occur at the ground line.

The breaking loads referred to in (d) above for the classes for which "6 ft from butt" circumferences are given are as follows: Class 1, 4,500 lb; Class 2, 3,700 lb; Class 3, 3,000 lb; Class 4, 2,400 lb; Class 5, 1,900 lb; Class 6, 1,500 lb; Class 7, 1,200 lb; Class 9, 740 lb; Class 10, 370 lb.

TABLE 18-6   Dimensions of Northern White Cedar and Engelmann Spruce Poles

Class		1	2	3	4	5	6	7	9	10
Minimum circumference at top, in		27	25	23	21	19	17	15	15	12
Length of pole, ft	Ground-line distance from butt,* ft	Minimum circumference at 6 ft from butt, in								
Northern white cedar poles (based on a fiber stress of 4,000 lb/in²)										
20	4	38.0	35.5	33.0	30.5	28.0	**26.0**	**24.0**	**22.0**	**17.5**
25	5	.42.0	39.5	36.5	34.0	31.5	**29.0**	**27.0**	**24.0**	**19.5**
30	5½	45.5	**43.0**	**40.0**	37.0	**34.5**	**32.0**	**29.5**	**26.0**	
35	6	49.0	**46.0**	**42.5**	**39.5**	**37.0**	**34.0**	**31.5**		
40	6	**51.5**	**48.5**	**45.0**	42.0	**39.0**	**36.0**			
45	6½	**54.5**	**51.0**	**47.5**	**44.0**	**41.0**				
50	7	**57.0**	**53.5**	**49.5**	**46.0**	**43.0**				
55	7½	**59.0**	**55.5**	**51.5**	48.0	**44.5**				
60	8	**61.0**	**57.5**	**53.5**	50.0					
Engelmann spruce poles (based on a fiber stress of 5,600 lb/in²)										
20	4	34.5	32.0	30.0	28.0	25.5	**23.5**	**22.0**	**19.0**	**15.0**
25	5	38.0	35.5	33.0	30.5	28.5	**26.0**	**24.5**	**21.0**	**16.5**
30	5½	41.0	**38.5**	**35.0**	33.0	**30.5**	**28.5**	**26.5**	**22.5**	
35	6	43.5	**41.0**	**38.0**	35.5	**32.5**	**30.5**	**28.0**		
40	6	**46.0**	**43.5**	**40.5**	**37.5**	**34.5**	**32.0**			
45	6½	**48.5**	**45.5**	**42.5**	**39.5**	**36.5**				
50	7	**50.5**	**47.5**	**44.5**	41.0	**38.0**				
55	7½	**52.5**	**49.5**	**46.0**	42.5	**39.5**				
60	8	**54.5**	**51.0**	**47.5**	44.0					
65	8½	**56.0**	**52.5**	**49.0**	45.5					
70	9	**57.5**	**54.0**	**50.5**	47.0					
75	9½	**59.5**	**55.5**	**52.0**	48.5					
80	10	**61.0**	**57.0**	**53.5**	49.5					
85	10½	**62.5**	**58.5**	54.5						
90	11	**63.5**	**60.0**	56.0						
95	11	**65.0**	**61.0**	57.0						
100	11	**66.0**	**62.0**	58.0						

*The figures in this column are intended for use only when a definition of ground line is necessary in order to apply requirements relating to scars, straightness, etc.

NOTES: Classes and lengths for which circumferences at 6 ft from the butt are listed in boldface type are the preferred standard sizes. Those shown in light type are included for engineering purposes only.

1 in = 2.54 cm; 1 ft = 0.3048 m; 1 lb/in² = 6.895 kPa.

WOOD PRODUCTS 18-13

TABLE 18-7 Western Red Cedar, Ponderosa Pine, Douglas Fir, and Southern Pine

Class		1	2	3	4	5	6	7	9	10
Minimum circumference at top, in		27	25	23	21	19	17	15	15	12
Length of pole, ft	Ground-line distance from butt,* ft	Minimum circumference at 6 ft from butt, in								
Western red cedar and ponderosa pine poles (based on a fiber stress of 6,000 lb/in²)										
20	4	33.5	31.5	29.5	27.0	25.0	**23.0**	**21.5**	**18.5**	**15.0**
25	5	37.0	34.5	32.5	30.0	28.0	**25.5**	**24.0**	**20.5**	**16.5**
30	5½	40.0	**37.5**	**35.0**	**32.5**	**30.0**	**28.0**	**26.0**	**22.0**	
35	6	42.5	40.0	**37.5**	**34.5**	**32.0**	**30.0**	**27.5**		
40	6	**45.0**	**42.5**	**39.5**	**36.5**	**34.0**	**31.5**	29.5		
45	6½	**47.5**	**44.5**	**41.5**	**38.5**	**36.0**	33.0	31.0		
50	7	**49.5**	**46.5**	**43.5**	40.0	37.5	34.5	32.0		
55	7½	**51.5**	**48.5**	**45.0**	42.0	39.0	36.0			
60	8	**53.5**	**50.0**	**46.5**	43.5	40.0	37.0			
65	8½	**55.0**	**51.5**	**48.0**	45.0	41.5				
70	9	**56.5**	**53.0**	**49.5**	46.0	42.5				
75	9½	**58.0**	**54.5**	**51.0**	47.5					
80	10	**59.5**	**56.0**	52.0	48.5					
85	10½	**61.0**	**57.0**	53.5						
90	11	**62.5**	**58.5**	54.5						
95	11	**63.5**	**59.5**	56.0						
100	11	**65.0**	**61.0**	57.0						
105	12	**66.0**	**62.0**	58.0						
110	12	**67.5**	**63.0**	59.0						
115	12	**68.5**	**64.0**							
120	12	**69.5**	**65.0**							
125	12	**70.5**	**66.0**							
Douglas fir and southern pine poles (based on a fiber stress of 8,000 lb/in)²										
20	4	31.0	29.0	27.0	25.0	23.0	**21.0**	**19.5**	**17.5**	**14.0**
25	5	33.5	31.5	29.5	27.5	25.5	**23.0**	**21.5**	**19.5**	**15.0**
30	5½	36.5	34.0	32.0	**29.5**	**27.5**	**25.0**	**23.5**	**20.5**	
35	6	39.0	36.5	34.0	**31.5**	**29.0**	**27.0**	**25.0**		
40	6	**41.0**	**38.5**	**36.0**	**33.5**	**31.0**	**28.5**	26.5		
45	6½	**43.0**	**40.5**	**37.5**	**35.0**	**32.5**	30.0	28.0		
50	7	**45.0**	**42.0**	**39.0**	36.5	34.0	31.5	29.0		
55	7½	**46.5**	**43.5**	**40.5**	38.0	35.0	32.5			
60	8	**48.0**	**45.0**	**42.0**	39.0	36.0	33.5			
65	8½	**49.5**	**46.5**	**43.5**	40.5	37.5				
70	9	**51.0**	**48.0**	**45.0**	41.5	38.5				
75	9½	**52.5**	**49.0**	**46.0**	43.0					
80	10	**54.0**	**50.5**	47.0	44.0					
85	10½	**55.0**	**51.5**	48.0						
90	11	**56.0**	**53.0**	49.0						
95	11	**57.0**	**54.0**	50.0						
100	11	**58.5**	**55.0**	51.0						
105	12	**59.5**	**56.0**	52.0						
110	12	**60.5**	**57.0**	53.0						
115	12	**61.5**	**58.0**							
120	12	**62.5**	**59.0**							
125	12	**63.5**	**59.5**							

*The figures in this column are intended for use only when a definition of ground line is necessary in order to apply requirements relating to scars, straightness, etc.

NOTES: Classes and lengths for which circumferences at 6 ft from the butt are listed in boldface type are the preferred standard sizes. Those shown in light type are included for engineering purposes only.

1 in = 2.54 cm; 1 ft = 0.3048 m; 1 lb/in² = 6.895 kPa.

Minimum top circumferences and minimum circumferences at 6 ft from butt are given in Tables 18-6, 18-7, and 18-8.

Length. Poles under 50 ft in length should not be more than 3 in shorter or 6 in longer than nominal length. Poles 50 ft or over in length should not be more than 6 in shorter or 12 in longer than nominal length.

Length should be measured between the extreme ends of the pole.

Circumference. The minimum circumference at 6 ft from the butt and at the top, for each length and class of pole, is listed in the tables of dimensions. The circumference at 6 ft from the butt of poles should be not more than 7 in or 20% larger than the specified minimum, whichever is greater.

The top dimensional requirement should apply at a point corresponding to the minimum length permitted for the pole.

TABLE 18-8 Additional Wood Species

	Class	1	2	3	4	5	6	7	9	10
	Minimum circumference at top, in	27	25	23	21	19	17	15	15	12

Length of pole, ft	Ground-line distance from butt,* ft	Minimum circumference at 6 ft from butt, in								
Jack pine, lodgepole pine, red pine, redwood, Sitka spruce, western fir, and white spruce poles (based on a fiber stress of 6,600 lb/in²)										
20	4	32.5	30.5	28.5	26.5	24.5	**22.5**	**21.0**	**18.0**	**14.5**
25	5	36.0	33.5	31.0	29.0	27.0	**25.0**	**23.0**	**20.0**	**15.5**
30	5½	39.0	**36.5**	**34.0**	31.5	29.0	27.0	**25.0**	**21.0**	
35	6	41.5	**38.5**	**36.0**	**33.5**	31.0	**28.5**	**26.5**		
40	6	**44.0**	**41.0**	**38.0**	**35.5**	**33.0**	**30.5**	28.0		
45	6½	**46.0**	**43.0**	**40.0**	37.0	**34.5**	32.0	29.5		
50	7	**48.0**	**45.0**	**42.0**	39.0	36.0	33.5	31.0		
55	7½	**49.5**	**46.5**	**43.5**	40.5	37.5	34.5			
60	8	**51.5**	**48.0**	**45.0**	42.0	38.5	36.0			
65	8½	**53.0**	**49.5**	**46.0**	43.0	40.0				
70	9	**54.5**	**51.0**	**47.5**	44.5	41.0				
75	9½	**56.0**	**52.5**	**49.0**	45.5					
80	10	**57.5**	**54.0**	50.5	47.0					
85	10½	**58.5**	**55.0**	51.5						
90	11	**60.0**	**56.5**	52.5						
95	11	**61.5**	**57.5**	54.0						
100	11	**62.5**	**58.5**	55.0						
105	12	**63.5**	**60.0**	56.0						
110	12	**65.0**	**61.0**	57.0						
115	12	**66.0**	**62.0**							
120	12	**67.0**	**63.0**							
125	12	**68.0**	**64.0**							
Alaska yellow cedar and western hemlock poles (based on a fiber stress of 7,400 lb/in²)										
20	4	31.5	29.5	27.5	25.5	23.5	**22.0**	**20.0**	**17.5**	**14.0**
25	5	34.5	32.5	30.0	28.0	26.0	**24.0**	**22.0**	**19.5**	**15.0**
30	5½	37.5	**35.0**	**32.5**	30.0	28.0	26.0	**24.0**	**20.5**	
35	6	40.0	**37.5**	**35.0**	32.0	30.0	27.5	**25.5**		
40	6	**42.0**	**39.5**	**37.0**	**34.0**	31.5	29.0	27.0		
45	6½	**44.0**	**41.5**	**38.5**	**36.0**	**33.0**	30.5	28.5		
50	7	**46.0**	**43.0**	**40.0**	37.5	34.5	32.0	29.5		
55	7½	**47.5**	**44.5**	**41.5**	39.0	36.0	33.5			
60	8	**49.5**	**46.0**	**43.0**	40.0	37.0	34.5			
65	8½	**51.0**	**47.5**	**44.5**	41.5	38.5				
70	9	**52.5**	**49.0**	**46.0**	42.5	39.5				
75	9½	**54.0**	**50.5**	**47.0**	44.0					
80	10	**55.0**	**51.5**	48.5	45.0					
85	10½	**56.5**	**53.0**	49.5						
90	11	**57.5**	**54.0**	50.5						
95	11	**58.5**	**55.0**	51.5						
100	11	**60.0**	**56.0**	52.5						
105	12	**61.0**	**57.0**	53.5						
110	12	**62.0**	**58.0**	54.5						
115	12	**63.0**	**59.0**							
120	12	**64.0**	**60.0**							
125	12	**65.0**	**61.0**							
Western larch poles (based on a fiber stress of 8,400 lb/in²)										
20	4	30.0	28.5	26.5	24.5	22.5	**21.0**	**19.0**	**17.0**	**13.5**
25	5	33.0	31.0	29.0	26.5	24.5	**23.0**	**21.0**	**18.5**	**14.5**
30	5½	35.5	**33.5**	**31.0**	29.0	**26.5**	**24.5**	**23.0**	**19.5**	
35	6	38.0	**35.5**	**33.0**	31.0	**28.5**	**26.5**	**24.5**		
40	6	**40.0**	**37.5**	**35.0**	**32.5**	**30.0**	**28.0**	26.0		
45	6½	**42.0**	**39.5**	**37.0**	**34.0**	**31.5**	29.0	27.0		
50	7	**44.0**	**41.0**	**38.5**	35.5	33.0	30.5	28.5		
55	7½	**45.5**	**42.5**	**40.0**	37.0	34.5	31.5			
60	8	**47.0**	**44.0**	**41.0**	38.5	35.5	33.0			
65	8½	**48.5**	**46.0**	**42.5**	39.5	36.5				
70	9	**50.0**	**47.0**	**44.0**	41.0	38.0				
75	9½	**51.5**	**48.0**	**45.0**	42.0					
80	10	**52.5**	**49.5**	46.0	43.0					
85	10½	**54.0**	**50.5**	47.0						
90	11	**55.0**	**51.5**	48.5						
95	11	**56.5**	**53.0**	49.5						
100	11	**57.5**	**54.0**	50.5						
105	12	**58.5**	**55.0**	51.5						
110	12	**59.5**	**56.0**	52.5						
115	12	**60.5**	**57.0**							
120	12	**61.5**	**58.0**							
125	12	**62.5**	**58.5**							

*The figures in this column are intended for use only when a definition of ground line is necessary in order to apply requirements relating to scars, straightness, etc.

NOTES: Classes and lengths for which circumferences at 6 ft from the butt are listed in boldface type are the preferred standard sizes. Those shown in light type are included for engineering purposes only.

1 in = 2.54 cm; 1 ft = 0.3048 m; 1 lb/in² = 6.895 kPa.

WOOD PRODUCTS

Classification. The true circumference class should be determined as follows: Measure the circumference at 6 ft from the butt. This dimension will determine the true class of the pole, provided that its top (measured at the minimum length point) is large enough. Otherwise the circumference at the top will determine the true class, provided that the circumference at 6 ft from the butt does not exceed the specified minimum by more than 7 in or 20%, whichever is greater.

The above information relating to the pole standards approved by the ANSI does not constitute the complete standards. For further information, consult the standards, which may be obtained at a nominal charge.

18-33. Machine shaving of poles has increased as a practice of producers. Approximately 85% of present production is so shaved. Some producers also turn the pole down in the process, thereby obtaining a straighter pole with a specific taper. The machine-processed poles season more rapidly, which is particularly important with species like southern yellow pine which are susceptible to fungus attack before treatment. Machine shaving makes for easier detection of defects and provides a pole of improved appearance. If poles having normally thin sapwood (such as western red cedar and larch) are to be full-length when treated with preservative, it is undesirable to reduce the thickness of the sapwood more than necessary to obtain a dressed pole.

18-34. Preservative Treatment. For a nominal cost the service life of wood can be greatly increased by the use of preservative treatment. Creosote and pentachlorophenol are extensively used for the protection of poles and crossarms. Southern yellow pine because of its thick sapwood requires a pressure treatment. Species with intermediate sapwood thickness such as Douglas fir, lodgepole pine, and jack pine are treated by either pressure or nonpressure processes. Thin sapwood species such as western red cedar and larch are generally treated by nonpressure processes.

The American Wood-Preservers' Association Standards are used to specify preservative chemicals and treatment methods for wood products.

18-35. Inspection. Poles are inspected prior to treatment for physical defects and decay and after treatment for penetration and retention of preservative and for cleanliness. Inspection is most effective when made at vendors' plants, because defects that may be hidden by preservative are detected and freight is saved on rejects. Commercial inspection agencies are available at most producing locations, and it is normally economical to utilize their services. Quantity users may have their own trained inspectors. *Crossarm inspection* is important because safety of linemen is a consideration in addition to quality of timber. As with poles, inspection should be made before treatment for defects and after treatment for penetration, retention, and cleanliness. Inspection should be done by qualified timber specialists.

18-36. Conductivity is of concern to many electric-utility companies. Pole resistance varies greatly with moisture content. Dry wood of all species exhibits high resistance. Surface absorption of rainwater by untreated wood may vary the resistance over a wide range. Full-length-treated poles thoroughly dried before treatment generally show only moderate reduction in resistance following a rain. A rough correlation between resistance and moisture will show that 500,000 Ω over a 20-ft length of pole between contacts driven 3½ in deep corresponds to a moisture content of about 25%. Other average points on the curve band are 50,000,000 Ω 15% moisture and 20,000 Ω 40% moisture.

18-37. Depth of Pole Setting. The values in the column headed "Ground-line distance from butt" in Tables 18-6 to 18-8 may be accepted as a guide for a satisfactory depth of pole settings in ordinary firm soil. In marshy soil and at unguyed angles in lines, setting depths should be increased 1 to 2 ft. In rock, the indicated settings may be reduced one-half for that part of the pole set in rock. Rock backfill in ordinary earth locations is not considered as set in rock.

18-38. Pole stubbing can frequently be employed to effect substantial money savings. An otherwise good pole that is decayed at or below the ground line is fastened securely to a new preservative-treated stub set in the ground alongside it. The major part of the savings resulted from avoidance of transferring wires and equipment.

18-39. Salvaging. Poles removed for any reason can frequently be salvaged for future use. Users of large quantities can economically do this. One or more of the following operations may be employed: cut off top, cut off butt, remove old hardware, shave, reframe, re-treat.

18-16 ELECTRICAL ENGINEERING MATERIALS REFERENCE GUIDE

TABLE 18-9 Bending Load and Crushing Strength of Crossarms

Species	Rings per in.	Per cent			Density (dry)	Max. bending load, lb	Maximum crushing strength lb per sq in.
		Summer wood	Sap wood	Moisture			
Douglas fir	20	40	0	11.5	0.48	7,590	7,080
Longleaf pine (50 % heart)	18	44	55	13.4	0.54	8,984	5,425
Longleaf pine (75 % heart)	19	53	32	13.5	0.63	10,180	8,950
Longleaf pine (100 % heart)	16	44	1	12.8	0.63	9,782	8,940
Shortleaf pine	11	46	79	13.3	0.52	9,260	7,300
Shortleaf pine, creosoted	11	49	..	....	....	7,649	5,770
White cedar	12	45	2	14.3	0.36	5,200	4,700

NOTE: 1 in = 2.54 cm; 1 lb = 0.4536 kg; 1 lb/in^2 = 6.895 kPa.

18-40. Kinds of Timber for Crossarms. Two kinds of timber are in general use for crossarms, Douglas fir and southern yellow pine. All pine crossarms are treated with creosote or pentachlorophenol. The practice of treating fir crossarms is increasing rapidly as users recognize the need for arm life to match pole life.

Most Douglas fir arms used for communication and power-distribution lines are manufactured from timber selected for the purpose. Dense and close-grain lumber is used. *Publication* 14 of the West Coast Lumberman's Association sets forth grading and dressing requirements.

There is no grade of southern pine timber designated as crossarm stock, and crossarm users depend on the limitations set forth in their specifications to obtain a satisfactory quality of product. Pine arms are usually small boxed heart timbers.

Laminated arms are coming into use, but a buying specification is not available. Large transmission-line arms may be laminated structures or framed, treated round poles.

18-41. Crossarm Specifications. The most widely used specifications for power-distribution crossarms have been prepared by the Transmission and Distribution Committee of the Edison Electric Institute. For fir crossarms: Specification TD-90, which combines both dense and close-grain grades; Specification TD-92, Heavy-Duty Douglas Fir Crossarms; Specification TD-93, Heavy-Duty Douglas Fir Braces. For pine crossarms: Specification TD-91, Dense Southern Pine Crossarms Preservative Treated. Widely used specifications for communication crossarms are American Telephone and Telegraph Co. Specification AT-7298, Crossarms.

18-42. Strength of Crossarms. The most reliable source of information on the strength comes from tests made under conditions to simulate crossarms in service. Some tests have been made, and others are under consideration. Theoretical considerations, treating a crossarm as a beam, are valuable if those factors which control the actual strength are taken into account. Tests made several years ago on 84 six-pin, 3¼ by 4¼-in by 6-ft crossarms, with a uniformly distributed vertical load, gave average results shown in Table 18-9 (*U.S. Forest Service Circ. 204*, by T. R. C. Wilson). The maximum bending load shown in Table 4-91 is the total distributed vertical load. The maximum crushing strength is under compression parallel to the grain. Methods of tests are covered by ASTM specifications.

18-43. References on Wood Products

1. American Wood-Preservers' Association, *Manual of Recommended Practice.*

2. Appropriate standards of the ASTM.

3. Eggleston, R. C.: Evaluating the Relative Bending Strength of Crossarms; *Bell Systems Tech. J.,* January 1945.

4. Gurfinkel, G.: *Wood Engineering:* New Orleans, Southern Forest Products Association, 1973.

5. Nicholas, D. D. (ed.): *Wood Deterioration and its Prevention by Preservative Treatments;* Syracuse, New York, University Press, 1973.

6. MacLean, V. P.: Preservative Treatment of Wood by Pressure Methods; *U.S. Dept. Agr., Agr. Handbook* 40, 1960.

7. *Manual of Recommended Practice;* Washington, D.C., American Wood-Preservers' Association.

8. Ostman, H. F.: Crossarm Loading Studies; *Bull. EEI,* June 1945.
9. Publications of the U.S. Forest Products Laboratory, Madison, Wis.
10. Rasmussen, E. F.: Dry Kiln Operators Manual; *U.S. Dept Agr., Agr. Handbook* 188, Forest Products Laboratory, 1968.
11. *Timber Design and Construction Manual;* American Institute of Timber Construction, Current Edition.
12. Wood Handbook; *U.S. Dept. Agr., Agr. Handbook* 72, Forest Products Laboratory, 1974.
13. *Wood Structures;* New York, American Society of Civil Engineers, 1975.

INDEX

ABS polymers, 11-8
AC dielectric strength, 5-10
Acetals, 11-8
Acrylic resins, 16-5
Acrylics, 11-3, 11-8
ACSR (aluminum-cable, steel-reinforced) conductors, 2-25
ACSR conductor:
  electrical properties of, 2-69
  physical properties of, 2-60
ACSR self-damping conductors, 2-37
Aeolian vibration, conductors for, 2-37
Aging, 17-5
Alkaline silicate cements, 13-15
Alloys, 17-6
  cobalt-base, 17-15
  constant-permeability, 4-22
  ductile, 4-31
  fusible, 2-44, 2-76
  Heusler's, 4-25
  iron-cobalt, 4-24
  iron-silicon aluminum, 4-24
  nickel-base, 17-15
  quench-hardened, 4-28
  silicon-iron, 4-26
  temperature-sensitive, 4-25
  (*See also specific types of alloys, for example*: Aluminum alloys)
Alternating-current resistance of conductors, 2-27
Alumina substrates, 6-4
Aluminum, 17-25
  alloying, 2-13
  description of, 2-13
  temperature-resistance coefficient of, 2-8
  thermal conductivity of, 2-10
Aluminum alloys, 17-29
  density and weight of, 2-3
Aluminum cable, uses of, 2-57

Aluminum-clad cable:
  electrical properties of, 2-72
  properties of, 2-62
Aluminum conductor, physical properties of, 2-58
Aluminum wire:
  density and weight of, 2-2
  tables for, 2-57
American wire gage:
  definition of, 2-17
  use of, 2-16
Amorphous metal alloys, 4-27
Amorphous metals, 1-6
Ampere-turn, definition of, 4-1
Anisotropic material, 4-1
Annealing, 4-16, 17-10
Antiferromagnetic material, 4-2
Apparent power, definition of, 4-2
Asbestos, sources of, 13-1
Askarels, 12-3
Asphalt, 16-1

Batteries, long-lasting, development of, 1-7
Bearing metals, 17-24
Bend test, 17-3
Beryllia, thermal conductivity of, 6-5
Beryllium, description of, 2-14
Beryllium-copper alloys, 17-23
Beta rays, 8-3
Birmingham wire gage:
  establishment of, 2-17
  use of, 2-16
Bitumens, 16-1
Brass, use of, in structural materials, 17-22
Brick, 17-31
Bridge wire, 17-16
British standard gage, use of, 2-18

Brittleness, 17-4
Bronze, 17-22
  commercial grades of, 2-12
Brown & Sharpe gage, definition of, 2-17
Brushes:
  carbon, 3-1
  properties of, 3-4
Building stone, 17-30
Bus conductors, 2-31

Cable, preformed, use of, 2-21
Capacitive reactance, 2-30
Capacitive-reactance spacing factors, 2-73
Capacitive susceptance, equation for, 2-31
Capacitors:
  dielectrics in, 13-17
  ferroelectric, 13-16
  piezoelectric properties of, 13-18
Carbon:
  arc lamp, 3-5
  forms of, 3-1
  temperature coefficient of resistance for, 3-1
  typical properties of, 3-2
Carbon-brush applications, 3-1
Cast iron, 4-10, 17-7
  thermal properties of, 17-9
Cellulose, 12-1
Cellulosic, 11-3, 11-8
Cements, 13-15
Ceramic insulators, 13-8
Ceramic magnet material, 4-28
Ceramics, 1-1, 13-3
  properties of, 13-5
  thermal conductivity of, 6-4
Charging current, equation for, 2-31
Charpy impact test, 17-4
Chromium steels, 17-11
Circuit, definition of,
  magnetic, 4-4
Clay firebricks, 17-31
Coating powders, 15-1
Coating of metals, 17-20
Cobalt-base alloys, 17-15
Coefficient of expansion of metals, 2-9
Coercive force, definition of, 4-2
Cold-molded compounds, 11-9
Cold-molded materials, properties of, 11-16
Composite dielectrics, 5-7
Compounds, impregnating and filling, 16-1
Concrete, 17-32

Conductivity, percent, calculation of, 2-9
Conductor data tables, 2-44
Conductor insulation, 13-14
Conductor materials, 2-1
  conductivity of, 2-3
Conductors:
  bare, current-carrying capacity of, 2-31
  definition of, 2-1, 2-16
  effect of temperature changes on, 2-3
  elastic limit of, 2-24
  electrical, definitions of, 2-15
  electrical resistivity of ferrous, 2-7
  foil, 7-2
  general properties of, 2-1
  inductive reactance of, 2-29
  industry specifications for, table, 2-33
  insulated, current-carrying capacity of, 2-31
  internal inductance of, 2-26
  magnetic permeability of, 2-25
  prestressed, 2-25
  properties of, 2-15
  resistance-temperature relationship, 2-9
  size designation, 2-18
  solid, 2-32
  specific heat of, 2-10
  stranded:
    definition of, 2-16
    direction of lay, 2-20
    effect of, 2-19
    use of, 2-18
  temperature-expansion coefficients, table, 2-10
  types of, 2-2
  yield point of, 2-24
  Young's moduli for, 2-22
Contact metals, 2-76
Copper:
  alloying of, 2-13
  commercial, 17-21
  density and weight of, 2-2
  description of, 2-11
  temperature-resistance coefficient of, 2-7
Copper alloys, density and weight of, 2-2
Copper cable:
  classes of, 2-52
  dimensions of, 2-49
  electrical properties of, 2-64
  uses of, 2-48
Copper castings, 17-21
Copper-clad cables, 2-56
  electrical properties of, table, 2-66

INDEX

Copper wire:
  electrical properties of, 2-64
  physical properties of, 2-38
  tensile strength of, 2-45
  weight, breaking strength, and dc
    resistance of, 2-46
Copperweld products, 17-21
Cord, definition of, 2-16
Core loss, definition of, 4-2
Core steels, grade designations of, 4-16
Cores:
  annealing of, 4-16
  ferrite, 4-27
  laminated, materials for, 4-12
Corrosion:
  of materials, 1-7
  prevention of, 17-18
Corrosion resistance of structural
  materials, 17-5
Creosote, 18-7
Crossarms, 18-10, 18-16
Curie temperature, 4-2
Czochralski process of making silicon
  semiconductors, 1-5

Demagnetization curve, definition of, 4-2
Department of Commerce, materials
  research by, 1-9
Diallylphthalate (DAP), 11-9
Diamagnetic material, 4-2
Dielectric:
  breakdown strength of, 5-11
  circuit analogy of, 5-1
  composite, 5-7
  corona threshold voltage of, 5-8
  typical dc current behavior, 5-4
  variation of:
    with frequency, 5-6
    with temperature, 5-7
Dielectric breakdown in gases, 8-3
Dielectric constant, 5-2, 8-2
Dielectric permittivity, 5-3
Dielectric strength, 5-9
Domains, ferromagnetic, 4-2
Drying oils, 14-4
Ductile cast iron, 4-11
Ductility, 17-3

Eddy-current loss, 4-2
Elastic limit, 2-24, 17-2

Elastic strength, 17-2
Elasticity:
  modulus of, 2-22
  of materials, 2-21
Electric Power Research Institute (EPRI),
  1-1
  materials research, 1-5
Electric utility materials research, 1-5
Electrical conductivity and resistivity of
  conductors, table, 2-4
Electrical conductors, 2-15
Electrical insulating paper, 12-1
Electrical insulation, application of, 5-14
Electrical resistivity:
  of ferrous conductors, table, 2-7
  of materials, 2-3
Electrical steels, 4-12
  flat rolled, 4-22
  general properties of, 4-20
  grading of, 4-12
  grain-oriented, 4-14
Electrochemical trees, 5-12
Electronics, 1-4
Elongation, 17-3
Embedding compounds, thermal conduc-
  tivity of, 6-7
Epoxies, 11-9
Epoxy resins, 16-2
EPRI (Electrical Power Research Insti-
  tute), 1-1
  materials research, 1-5
Ester fluids, 9-6
Excitation of magnetic materials, 4-1
Exciting power, definition of, 4-2

Fabrics:
  insulating, 12-4
  varnish impregnated, 12-6
Fatigue strength, 17-4
Ferrimagnetic material, 4-2
Ferrite cores, 4-27
Ferritic stainless steel, 17-14
Ferroelectric capacitors, 13-16
Ferromagnetic material, 4-3
Ferrous metals, modulus of elasticity of,
  17-9
Fiberoptic communications, 1-6
Films:
  manufacturing of, 11-25
  properties of, 11-22
Flexible sheet insulation, 13-14

INDEX

Fluorocarbon liquids, 9-6
Fluorocarbons, 11-3, 11-8
Flux direction, 4-9
Foil conductors, 7-2
Fracture mechanics, 17-4
Free-machining steels, 17-6
Fuel cells, development of, 1-7
Fused silica, 13-3
Fuses, metals used for, 2-44
Fusible alloys, 2-44
    compositions and melting points of, 2-76
Fusing currents of wire, 2-73

Galvanic series of alloys, 17-19
Galvanized steel, properties of, 2-63
Gases:
    corona breakdown in, 8-6
    dielectric strength of, 8-4
    effect of frequency on dielectric strength of, 8-8
    effect of pressure on dielectric strength of, 8-8
    flashover in, 8-7
    ionization in, 8-2
    vacuum breakdown of, 8-10
Gauss, definition of, 4-3
German wire gage, use of, 2-18
Gilbert, definition of, 4-3
Glass:
    composition of, 13-1,
        table, 13-4
    dielectric strength of, 13-2
    electrical and mechanical properties of, 13-3
        table, 13-7
Glass bricks, 17-32
Gold, definition of, 2-78
Graphite, properties of, 3-3
Gypsum products, 17-32

Hard metals, 2-76
Hardness, 17-4
Hardwoods, 18-1
Heat, effect of, on materials, 1-7
Heusler's alloys, 4-25
High-energy radiation, 8-2
History of materials research, 1-1
Hooke's law, 2-22
Hysteresis loop, definition of, 4-3

IACS (International Annealed Copper
    Standard), 2-3
Impact resistance, 17-4
Impedance, definition of, 4-6
Impregnated paper, dielectric strength of,
    12-3
Inductance, internal, of conductors,
    2-26
Induction, definition of, 4-3
Inductive reactance of conductors, 2-29
Ingot iron, 17-6
Inorganic insulating systems, 13-9
Insulating fabrics, 12-4
Insulating materials:
    general properties of, 5-1
    inorganic, 13-1
    thermal conductivity of, 6-1
Insulating oils, 9-1
    treatment of, 9-5
Insulating varnishes (*see* Varnishes
    insulating)
Insulation:
    arc tracking of, 5-12
    conductivity of ions in, 5-5
    flexible sheet, 13-14
    magnet-wire, 7-1
    rigid sheet, 13-14
    surface resistivity of, 5-5
    thermal aging of, 5-12
    thermal conductivity conversion factors
        of, 6-3
    tracking resistance of, 5-13
    water penetration of, 5-12
    wire and cable, 7-2
International Annealed Copper Standard
    (IACS), 2-3
Ionization in gases, 8-2
Ionizing radiation, 5-12
Iron:
    corrosion of, 17-18
    density and weight of, 2-3
    description of, 2-15
Isotropic material, 4-4
Izod impact test, 17-4

Jet engine, development of, 113
Johnson elastic limit, definition of, 2-24

Kraft insulating paper, 12-2

INDEX

Laminated cores, materials for, 4-12
Laminated sheet, 11-12
  properties of, 11-17
Laminates, copper-clad, 11-25
Lead, 17-24
Liquids:
  insulating, properties of, 9-2
  synthetic insulating, 9-6

Magnesium, description of, 2-14
Magnesium alloy, 17-23
Magnet materials, properties and chemical
  composition of, 4-29
Magnet strip, 7-2
Magnet-wire insulation, 7-1
Magnetic circuit, 4-4
Magnetic core materials, properties of, 4-26
Magnetic materials:
  commercial, 4-10
  comparative power losses of, 4-25
  definition of, 4-1
  special purpose, 4-19, 4-23
Magnetic properties, 4-8
Magnetic steels, 4-17
Magnetism, types of, 4-9
Magnetizing force, definition of, 4-5
Magnetodynamic, definition of, 4-5
Magnetostatic, definition of, 4-5
Magnets:
  powder compacts for, 4-31
  sintered alnico, 4-31
Manganese steels, 17-11
Martensitic stainless steel, 17-14
Massachusetts Institute of Technology,
  materials research, 1-3
Materials:
  chronological development of, 1-2
  for light and electronics, 1-8
  production of, 1-8
  strength of, 1-7
  used for conductors, 2-1, 2-3
  (*See also specific types of materials, for*
    example: Anisotrophic materials)
Materials research, 1-1
  for the future, 1-7
  history of, 1-1
  at MIT, 1-3
Maxwell, definition of, 4-5
Melamine, 11-9
Metal coatings, 17-20
Metal phosphate cements, 13-15

Metallurgy, powder, 17-5
Metals:
  coatings of, 17-20
  coefficient of expansion of, 2-9
  commercial grades of, 2-76
  creep strength of, 17-15
  fusible, 2-44
  hard, 2-76
  high-melting-point, 2-78
  highly conductive, 2-76
  influence of chemical composition on,
    2-11
  influence of mechanical treatment on,
    2-11
  melting points of, 2-79
  noncorroding, 2-76
  properties of, 2-2
  resistance of, 2-74
  (*See also specific types of, for example*:
    Amorphous metals)
Mica, 10-1, 13-3
  dielectric properties of, 10-2
  physical properties of, 10-1
  sources of, 10-1
  synthetic, 10-3
Mica paper, 10-3
Microelectronic materials, development of,
  1-3
Mil, definition of, 2-17
Millimeter wire gage, use of, 2-18
Mineral oils, 9-2
  dielectric properties of, 9-3
Moduli for conductors, 2-22
Modulus of elasticity, 2-22, 17-2
Modulus of rupture, 17-3
Molded plastics, 11-1
Molybdenum, definition of, 2-78
Monel metal, 4-24, 17-23
Mortar, 17-32
Muscovite mica, 10-1

Nickel, 17-23
Nickel steels, 17-11
Nonferrous metals, properties of, 17-26
Nylons, 11-3, 11-8

Oersted, definition of, 4-6
Oils:
  drying, 14-4
  insulating, 9-1, 9-5

INDEX

Oils (*Cont.*):
mineral insulating, 9-2, 9-3
Old English wire gage, use of, 2-18

Paints for metal protection, 17-20
Palladium, definition of, 2-78
Paper, dielectric loss factor of, 12-2
Paper-based phenolic, 11-12
Paramagnetic material, 4-6
Pentachlorophenol, 18-7
Permanent-magnet materials, 4-28
Permeability, 4-6
Phenolic, 11-9
paper-based, 11-12
Photovoltaic cells, 1-8
Pitch, 16-1
Pitch diameter of stranded conductors, 2-19
Plastic coatings, thermal conductivity of, 6-6
Plasticity, 17-3
Plastics:
definition of, 11-1
filament-wound, 11-21
flame resistant, 11-21
fluorocarbon-based, 11-25
thermal conductivity of, 6-3
Platinum:
alloys of, 2-78
definition of, 2-78
Pole stubbing, 18-15
Poles, wood (*see* Wood poles)
Polyarylester, 11-3, 11-8
Polybutadiene, 16-5
Polybutadiene impregnating resins, properties of, 16-6
Polybutylene, 11-8
Polycarbonates, 11-8
Polyester organic film, 5-10
Polyester resins, 16-2
Polyesters, 11-8
Polyimide films, 7-7
Polyimides, 11-8
Polymer concretes, 1-7
Polymer fibers, use of, in insulations, 12-5
Polymers, 1-4
Polypropylene plastic, 7-7
Polystyrene, 11-8
Polysulfones, 11-9
Porcelain:
casting process for, 13-8

Porcelain (*Cont.*):
composition of, 13-6
dry-process, 13-8
glazes on, 13-8
wet-process, 13-6
Portland cement, 17-32
Potting resins, 16-7
Powder metallurgy, 17-5
Powdered-iron cores, 4-27
Powders, coating, 15-1
Preservatives, wood, 18-4
Properties of metals, 2-2
Proportional limit, 17-2

Quantum physics in materials research, 1-1
Quench-hardened alloys, 4-28

Reactance, inductive, of conductors, 2-29
Reactive power, 4-8
Real power, definition of, 4-1
Refractories, properties of, 13-10
Refractory brick, 17-31
Reinforced plastics, properties of, 11-10
Remanence, 4-8
Research of materials by electric utilities, 1-5
Resilience, 17-4
Resins, 14-2, 16-2
acrylic, 16-5
Resistance:
ac, of conductors, 2-27
definition of, 2-25
Resistance metals and alloys, 2-74
Resistivity of electrical materials, 2-3
Resistivity-temperature constant, 2-9
Retentivity, 4-8
Rigid sheet insulation, 13-14
Roebling gage, use of, 2-17
Ruby-crystal laser, 1-1

Selenium, definition of, 2-78
Semiconductors, making of, 1-5
Shearing strength, 17-3
Silica brick, 17-31
Silicon, description of, 2-14
Silicon steels, 17-11
Silicone liquids, 9-6
Silicones, 11-9

INDEX

Silver, definition of, 2-78
Skin effect:
  of conductors, 2-2
  definition of, 2-26
Skin-effect ratios, 2-27
Sodium, description of, 2-14
Softwood, 18-1
Solar cells, materials for, 1-8
Solid conductors, 2-32
Solvents, 14-4
Spacing factors, capacitive-reactance, 2-73
Stainless steels, 17-11
  properties of, 17-14
Standard wire gage, use of, 2-17
Steatite ceramics, 13-8
Steel:
  AISI-SAC designation of, 17-8
  copper-clad, density and weight of, 2-2
  corrosion of, 17-18
  description of, 2-15
  electrical (*see* Electrical steel)
  heat treatment of, 17-9
  mechanical properties of, 17-12
  tempering of, 17-10
  (*See also specific types of steel, for
     example*: Chromium steels
Steel rails, 2-15
Steel rope, 17-16
Steel wire, galvanized, density and weight
  of, 2-3
Steel wire gage, use of, 2-17
Sterling silver, definition of, 2-78
Strain, definition of, 2-21, 17-1
Stranded conductor:
  definition of, 2-16
  direction of lay, 2-20
  effect of, 2-19
  use of, 2-18
Strength of materials, 1-7
Stress, definition of, 2-21, 17-1
Stress-strain curves, 2-22
Stress-strain diagram, 17-1
Structural iron, 17-6
Structural materials, 17-1
  chemical analysis of, 17-5
Structural steel, 17-6
Stubs' wire gage, use of, 2-17
Synthetic fiber insulating paper, 12-3
Synthetic mica, 10-3

Tape, nickel-alloy, 4-26

Technology transfer, 1-3
Temperature changes, effect of, on
  conductors, 2-3
Tempering of steel, 17-10
Tesla, definition of, 4-8
Thermal aging of insulation, 5-12
Thermal conductivity of insulating
  materials, 6-1
Thermal conductivity measurements, 6-8
Thermoplastic molding compounds, 11-1
Thermoplastic standards, 11-2
Thermoplastics:
  properties of, 11-3
  reinforced, 11-9
Thermosetting compounds, 11-9
Thermosetting materials, properties of,
  11-13
Thinners, liquid, 14-4
Tin, 17-24
Tin coating, 17-19
Titania bodies, 13-8
Titanium alloy, 17-25
Toughness, 17-4
Transformer cores, 1-6
Tungsten, definition of, 2-78
Turbine blade development, 1-5

Ureas, 11-9
Urethane, 16-5
Urethane impregnating resins, properties
  of, 16-6

Vanadium steels, 17-11
Var, definition of, 4-8
Varnishes, insulating, 14-1
  manufacture of, 14-2
  methods of applying, 14-5
  methods of testing, 14-6
  oleoresinous, 14-1
  resins used in, 14-2
Volt-ampere, definition of, 4-8

Washburn & Moen gage, use of, 2-17
Watt, definition of, 4-8
Waveform, magnetic, 4-9
Waxes, 16-1
Weber, definition of, 4-8
Wire diameters, measurement of, 2-18
Wire rope, 17-16

INDEX

Wire sizes, definition of, 2-16
Wires:
contact (trolley), 2-31
  enamel-insulated, 7-3
  fibrous-covered, 7-5
Wood:
  American Lumber Standards for, 18-8
  annual rings, in, 18-2
  commercial names for, 18-9
  composition of, 18-2
  decay of, 18-4
  description of, 18-1
  dielectric strength of, 12-4
  effect of moisture on, 18-4
  electrical properties of, 18-3
  mechanical properties of, 18-5
  methods of treating, 18-7
  moisture in, 18-2
  specific gravity of, 18-2, 18-4
  thermal properties of, 18-3

Wood crossarms, 18-10, 18-16
Wood poles, 18-10
  classification of, 18-15
  conductivity of, 18-15
  dimension of, 18-12
  inspection of, 18-15
  standards for, 18-10
Wood seasoning, 18-3
Wrought carbon steel, 4-11
Wrought iron, 4-10

Yield point, 2-24, 17-2
Yield strength, 17-2
Young's modulus, 2-22

Zinc, 17-24
Zinc coating, 17-18

ABOUT THE AUTHOR

H. Wayne Beaty graduated from the University of Houston with a Bachelor of Science degree in Electrical Engineering. He is a member of the Institute of Electrical and Electronics Engineers. During his career he was a Senior Editor with McGraw-Hill's *Electric World* magazine, Manager of Member Services for the Electric Power Research Institute, and Vice President of Loadmaster Systems, Inc. Mr. Beaty is now a consultant in the electrical engineering field. He resides in Fairfax, Virginia.